Proceedings in Life Sciences

Evolution of Insect Migration and Diapause

Edited by Hugh Dingle

With Contributions by

H. Dingle M.A. Hoy C.A. Istock
J. Lumme S. Masaki R.C. Rainey
M.A. Rankin C. Solbreck T.R.E. Southwood
C.A. Tauber M.J. Tauber K. Vepsäläinen
G.P. Waldbauer

With 103 Figures

Springer-Verlag
New York Heidelberg Berlin

Hugh Dingle
University of Iowa
Department of Zoology
Iowa City, Iowa 52242
USA

Library of Congress Cataloging in Publication Data. Main entry under title: Evolution of insect migration and diapause. (Proceedings in life sciences) "An outgrowth of a symposium entitled 'Evolution of escape in space and time' held at the XV International Congress of Entomology in Washington, D.C. ... August, 1976." Includes index. 1. Insects—Evolution—Congresses. 2. Insects—Migration—Congresses. 3. Diapause—Congresses. I. Dingle, Hugh, 1936– II. International Congress of Entomology, 15th, Washington, D.C., 1976.
QL468.7.E96 596.7'05'2 77-17857

Printed in the United States of America.

9 8 7 6 5 4 3 2 1

ISBN 0-387-90294-5 Springer-Verlag New York Heidelberg Berlin
ISBN 3-540-90294-5 Springer-Verlag Berlin Heidelberg New York

This volume is dedicated to
PROFESSOR JOHN S. KENNEDY, F.R.S.
upon his retirement from the
Chair of Animal Behavior
Imperial College
London

Preface

This volume is an outgrowth of a Symposium entitled "Evolution of Escape in Space and Time" held at the XV International Congress of Entomology in Washington, D. C., USA in August, 1976. The choice of topic was prompted by recent advances in evolutionary ecology and the apparent suitability of insect migration and diapause as appropriate material for evolutionary studies. In the event, that choice seems amply justified as I hope a perusal of these papers will show. These Symposium papers hardly cover the topic of the evolution of escape mechanisms exhaustively, and I am sure everyone will have his favorite lacuna. Some of the more obvious ones are indicated by Professor Southwood in his Concluding Remarks at the end of the book. The purpose of the Symposium, however, was not complete coverage, but rather to indicate the potential inherent in insect migration and diapause for the study of evolutionary problems. In that I think we have succeeded reasonably well.

These papers are expanded and in some cases somewhat altered versions of the papers delivered in Washington. This has allowed greater coverage of the topics in question. I suggested a format of a general overview of a topic emphasizing the author's own research contributions. In general the papers follow this outline although emphases vary. Two of the authors, Dr. Rainey and Dr. Lumme, were unable to attend the Symposium. Dr. Rainey's paper was read by Mr. Frank Walsh, but Dr. Lumme's was not on the Washington program. I am pleased, however, that he was able to contribute a manuscript for this book.

I have divided the papers into three sections. The first includes two papers dealing with migration, the one concentrating on physiology and the other ecology. The second section concentrates on diapause and its contribution to phenology and other aspects of life histories. A number of aspects, genetic, physiological, and ecological are discussed in a variety of species. The final section includes papers which analyze both diapause and migration and their interactions with respect to the evolution of life histories. Professor Southwood's Concluding Remarks are similar to those he gave at the Symposium. One of the delights of an editorship of this sort is the opportunity to expand one's horizons. Certainly the breadth and depth of my understanding of "escape mechanisms" has increased enormously, and I sincerely hope readers will profit from these papers at least in part as much as I have.

No volume such as this comes to fruition without the help of many people. I should like to thank, first, Dr. Kenneth Hagen of the University of California, Berkeley who was my co-organizer. Although his name does not appear on this book, his presence is felt throughout. Drs. J. P. Hegmann, C. A. Istock, S. Masaki, and M. J. Tauber graciously agreed to review manuscripts. Barbara O'Donnell, Robin Spicher, and Robin Wagner helped with typing and retyping in Iowa City. The book was shepherded through publication by Dr. Mark Licker of Springer-Verlag. Finally, I should like to thank the contributing authors who made the editor's task such a pleasurable one.

Iowa City
June, 1977

Hugh Dingle

Contents

List of Contributors

HUGH DINGLE, Department of Zoology, University of Iowa, Iowa City, Iowa 52242, USA

MARJORIE A. HOY, Department of Entomological Sciences, University of California, Berkeley, California 94720, USA

CONRAD A. ISTOCK, Center for Evolution and Paleobiology and Department of Biology, University of Rochester, Rochester, New York 14627, USA

JAAKKO LUMME, Department of Genetics, University of Oulu, SF-90100 Oulu 10, Finland

SINZO MASAKI, Laboratory of Entomology, Faculty of Agriculture, Hirosaki University, Hirosaki 036, Japan

R. C. RAINEY, Centre for Overseas Pest Research, College House, Wrights Lane, London W8 5SJ, UK

MARY ANN RANKIN, Department of Zoology, University of Texas, Austin, Texas 78712, USA

CHRISTER SOLBRECK, Department of Zoology, University of Stockholm, Box 6801, 113 86 Stockholm, Sweden

T. R. E. SOUTHWOOD, Imperial College Field Station, Silwood Park, Ascot, Berks, SL5 7PY, UK

CATHERINE A. TAUBER, Department of Entomology. Cornell University, Ithaca, New York 14853, USA

MAURICE J. TAUBER, Department of Entomology, Cornell University, Ithaca, New York 14853, USA

KARI VEPSÄLÄINEN, Department of Genetics, University of Helsinki, P. Rautatiekatu 13, SF-00100 Helsinki 10, Finland

GILBERT P. WALDBAUER, Department of Entomology, University of Illinois, Urbana, Illinois 61801, USA

Introduction and Dedication

My interest in the study of migration was kindled by J. S. Kennedy's seminal paper which appeared in *Nature* in 1961 (Kennedy, 1961). That paper was aptly titled "A Turning Point in the Study of Insect Migration," and it made the claim that recent advances permitted the characterization of insect migration ecologically, behaviorally, and physiologically as C. B. Williams, one of the pioneers in the study of migration, had anticipated (Williams, 1958). Drawing on the work of Rainey (1951), Johnson (1960), and Southwood (1962), Kennedy advanced what was then a somewhat radical thesis, that migration was the result of the evolution of specialized behavior involving straightened out movement and the suppression of what he called "vegetative" activities such as feeding and reproduction. The function of this specialized behavior is adaptive change of location to achieve coverage and exploitation of available habitats over a wide area.

It was my privilege, as a result of reading the 1961 paper, to spend a postdoctoral year with John Kennedy at the Entomological Field Station of the Agricultural Research Council's Unit of Insect Physiology in Cambridge, and to initiate there my own studies of migration. Those were exciting times in Kennedy's laboratory, for he was in the midst of the long series of exacting experiments with aphids, then in collaboration with C. O. Booth, demonstrating the central nervous interaction between vegetative settling responses and locomotory migratory ones (see below). These experiments have, of course, demonstrated that indeed migration is specialized behavior. But they have also demonstrated something more, and that is that migration cannot be considered separately from the life history of which it is an integral part.

Indeed that point had been made in the 1961 paper with its discussion of the relations between migration and diapause. On the face of it, migration looks like an alternative to diapause taking as it does the insect to a new habitat while diapause allows it to remain in the old. But as Kennedy pointed out, both result in suspension of growth, development, or reproduction and therefore bear many physiological similarities. In fact a commonly observed phenomenon is the migration to and from diapause sites. In the intervening years, still more intimate relationships including the promotion of migration by diapause (Dingle, this volume) and a common hormonal basis (Caldwell and Rankin, 1972; Rankin, 1974 and this volume) have been demonstrated.

There thus seems to be ample justification for a symposium which deals with both migration and diapause, their interrelationships, and their roles as evolved adaptations (Johnson, 1969) in insect life histories.

The 15 years between 1961 and 1976, when this symposium on "Evolution of Escape in Space and Time" was held at the XV International Congress of Entomology, have also firmly established the behavioral, ecological, and physiological basis of migration and the similarity of the syndrome across insect groups. This has been in no small part due to Kennedy's own elegant work on the migratory phase of the black bean aphid, *Aphis fabae* Scop. (Kennedy, 1965, 1966; Kennedy and Booth, 1963a, b, 1964; Kennedy and Ludlow, 1974). Not that the issue is permanently settled, for in 1975 Kennedy was still moved to open his address to a Cornell symposium on "Insects, Science, and Society" with the statement: "This essay is focused upon what has been perennially the most controversial aspect of insect dispersal, the involvement of specialized behavior" (Kennedy, 1975). But even though old ideas die hard, resistance is crumbling as the weight of the evidence becomes overwhelming. The notion of migration and diapause as integral parts of life histories also seems secure (Dingle, 1974; Vepsäläinen, 1974; Kennedy, 1975; Southwood, 1975).

Perhaps we have reached a new "turning point" where we can now concentrate on differences in evolved migration and diapause strategies. Certainly it has become apparent that there is a great deal of natural variation in these strategies. For example, as much observation and experimentation has shown, the range of flight durations within species can be as great as that between species (Johnson, 1976). There is a similar range of variation with respect to diapause (e.g. Danilevskii, 1965). This variation must be important in the evolution of adaptive strategies and has important implications for evolutionary biology, physiology, and applied entomology as the papers in this volume indicate. The increasing presence of modern genetic theory in studies of migration and diapause, as is also evident in this volume, may presage a new era in the analysis of the evolution of these responses.

It is no accident that evolutionary approaches are the focus of this symposium, for insect migration and diapause would seem to provide rich material for the evolutionary biologist. There are many theories to be tested for which these two behaviors seem suitable. Some of these theories, such as "Fisher's fundamental theorem" for the evolution of fitness (Fisher, 1958) have been around for some time and are undergoing a revival. The relation of migration and diapause to fitness is apparent, and their experimental manipulation can lead to new insights (Istock, this volume). Other theoretical work is much newer, such as the predictions concerning life history strategies arising from island biogeography (MacArthur and Wilson, 1967) or the development of "evolutionarily stable strategies" (Parker and Stuart, 1976). The notion that populations can fluctuate in space as well as time (Taylor and Taylor, 1977) lends new importance to the analysis of migration and diapause as integral parts of life histories. So does the emphatic need for detailed empirical studies of life history evolution (Stearns, 1976).

It is with these exciting prospects in mind that we honor John Kennedy upon his retirement in October, 1977 from the Professorship of Animal Behavior at Imperial College, London. The debt we owe is obvious, for he has done much to define our own life history strategies. As does the change from flight to settling in *Aphis fabae*, "retirement" represents a change in life history, a change in the threshold level of different activities. Whether John will now settle down is not for me to say, but I can make one prediction. As with the settling aphid, we are about to see a new phase of productivity with the exploitation of the resources of a new habitat, and I suspect another turning point to set the course for the next several years.

Iowa City
June, 1977

Hugh Dingle

References

Caldwell, R. L., Rankin, M. A.: Effects of a juvenile hormone mimic on flight in the milkweed bug, *Oncopeltus fasciatus*. Gen. Comp. Endocrinol. *19*, 601-605 (1972).

Danilevskii, A. S.: Photoperiodism and Seasonal Development of Insects. Edinburgh: Oliver and Boyd (1965).

Dingle, H.: The experimental analysis of migration and life-history strategies in insects. *In* Experimental Analysis of Insect Behaviour (ed. L. Barton Browne). New York: Springer-Verlag (1974).

Fisher, R. A.: The Genetical Theory of Natural Selection. New York: Dover (1958).

Johnson, C. G.: A basis for a general system of insect migration and dispersal by flight. Nature, Lond. *186*, 348-350 (1960).

Johnson, C. G.: Migration and Dispersal of Insects by Flight. London: Methuen (1969).

Johnson, C. G.: Lability of the flight system: a context for functional adaptation. *In* Insect Flight (ed. R. C. Rainey). R. E. S. Symposium 7. Oxford: Blackwell (1976).

Kennedy, J. S.: A turning point in the study of insect migration. Nature, Lond. *189*, 785-791 (1961).

Kennedy, J. S.: Co-ordination of successive activities in an aphid. Reciprocal effects of settling on flight. J. Exp. Biol. *43*, 489-509 (1965).

Kennedy, J. S.: The balance between antagonistic induction and depression of flight activity in *Aphis fabae* Scopoli. J. Exp. Biol. *45*, 215-228 (1966).

Kennedy, J. S.: Insect dispersal. *In* Insects, Science, and Society (ed. D. Pimentel). New York: Academic Press (1975).

Kennedy, J. S., Booth, C. O.: Free flight of aphids in the laboratory. J. Exp. Biol. *40*, 67-85 (1963a).

Kennedy, J. S., Booth, C. O.: Co-ordination of successive activities in an aphid. The effect of flight on the settling responses. J. Exp. Biol. *40*, 351-369 (1963b).

Kennedy, J. S., Booth, C. O.: Co-ordination of successive activities in an aphid. Depression of settling after flight. J. Exp. Biol. *41*, 805-824 (1964).

Kennedy, J. S., Ludlow, A. R.: Co-ordination of two kinds of flight activity in an aphid. J. Exp. Biol. *61*, 173-196 (1974).

MacArthur, R. H., Wilson, E. O.: The Theory of Island Biogeography. Princeton: Princeton University Press (1967).

Parker, G. A., Stuart, R. A.: Animal behavior as a strategy optimizer: evolution of resource assessment strategies and optimal emigration thresholds. Am. Natur. *110*, 1055-1076 (1976).

Rainey, R. C.: Weather and the movement of locust swarms: a new hypothesis. Nature, Lond. *168*, 1057-1060 (1951).

Rankin, M. A.: The hormonal control of flight in the milkweed bug, *Oncopeltus fasciatus*. *In* Experimental Analysis of Insect Behaviour (ed. L. Barton Browne). New York: Springer-Verlag (1974).

Southwood, T. R. E.: Migration of terrestrial arthropods in relation to habitat. Biol. Rev. *37*, 171-214 (1962).

Southwood, T. R. E.: The dynamics of insect populations. *In* Insects, Science, and Society. New York: Academic Press (1975).

Stearns, S. C.: Life-history tactics: a review of the ideas. Quart. Rev. Biol. *51*, 3-47 (1976).

Taylor, L. R., Taylor, R. A. J.: Aggregation, migration, and population mechanics. Nature, Lond. *265*, 415-421 (1977).

Vepsäläinen, K.: The life cycles and wing lengths of Finnish *Gerris* Fabr. species (Heteroptera, Gerridae). Acta Zool. Fenn. *141*, 1-73 (1974).

Williams, C. B.: Insect Migration. Collins: London (1958).

1 Migration

Migration in insects serves not only for escape from old habitats but also for reproduction and colonization in new ones. As a result there is an intimate relation between physiology and ecology in the syndrome. The two papers in this section take, respectively, physiological and ecological approaches to migration. M.A. Rankin reviews studies concerning the role of hormones, especially juvenile hormone, in migratory physiology, while R.C. Rainey discusses the use of wind patterns in the orientation and displacement of migrating insects.

Various workers had suggested that migration might result from a deficiency of juvenile hormone because this hormone was known to stimulate reproduction which in turn terminated long-distance flight. Rankin discusses evidence indicating that juvenile hormone in fact has a positive influence on migration. She first reviews studies on different insects that suggest a possible influence of changing hormonal titers. She then summarizes her own work on juvenile hormone in the milkweed bug (*Oncopeltus*) including assays which suggest that intermediate titers of JH promote migration while high titers stimulate reproduction with resultant cessation of flight. She concludes with a descriptive model of the way the hormone acts in this insect. In other insects such as the cotton stainer bugs (*Dysdercus*), the hormone may act somewhat differently on the flight system suggesting variation between species in hormonal mechanisms.

Rainey summarizes the influence of interactions between migration and wind patterns on the ecology of migrants and suggests that flight may have evolved in insects as a response to wind-aided transport. His own classic studies with locusts have shown that by moving downwind the swarms eventually arrive at breeding areas resulting from rainfall caused by convergent winds. Recent radar studies also indicate that a number of other insect species undertake dense migrations downwind although the ecological significance is still uncertain. Rainey concludes that long-distance migrations cannot be fully understood until meteorological factors are considered as thoroughly as oceanographers study the physical environment of the ocean.

Hormonal Control of Insect Migratory Behavior

M. A. Rankin

Most insect migrants seem to share in common certain behavioral, physiological, and ecological characteristics (Dingle, 1972). Behaviorally, migrants are generally extremely responsive to stimuli which induce flight and less responsive or unresponsive to stimuli associated with reproduction and feeding (Kennedy, 1961). In many species migration occurs in response to short photoperiods and is an alternative to immediate reproduction. Often associated with and induced by conditions which produce diapause, it offers an escape from unfavorable conditions in space as diapause provides an escape in time. Migrants, like diapausing insects, are often found to have hypertrophied fat bodies and immature ovaries (Johnson, 1969). Since migration is associated with a distinct set of physiological characteristics similar to those which accompany adult diapause, the possibility of an hormonal component to its control was often suggested by earlier workers. In his work on insect migration, Johnson (1969) put forth a model in which adult migration was triggered when ecdysone was absent and the juvenile hormone (JH) titer was low. The rise in JH titer then caused both the cessation of migratory behavior and the onset of oogenesis.

There is, in fact, evidence from a few insect species that such control does exist. In *Dysdercus intermedius* where starvation stimulates flight while feeding results in ovarian development and wing muscle histolysis (Dingle, 1972; Dingle and Arora, 1973). Edwards (1970) has concluded that flight muscle degeneration and vitellogenesis are induced by the corpus allatum (CA) which is apparently activated by feeding. Edwards' conclusions are based on implantations of retrocerebral complexes rather than isolated CA. However, in the closely related *Dydercus fulvoniger* Davis (1975) has shown that active CA implanted into starved (flying) females stimulate flight muscle histolysis. Allatectomized females when fed and mated undergo neither flight muscle histolysis nor vitellogenesis. Similarly, topical application of synthetic JH initiates flight muscle histolysis in starved females and in males. Denervation of the flight muscles results in rapid hystolysis which indicates that the effect of JH on flight muscle degeneration may be expressed through the flight motor neurons.

In a similar study with *Ips confusus*, Borden and Slater (1968) have shown that topically applied synthetic JH induced degeneration of wing muscles within two days. Also the normal hystolysis of the wing musculature of the domestic cricket, *Acheta domestica*, was prevented by allatectomy (Chudakova and Bocharova-Messner, 1968). However, in many other insects which have been investigated, the data suggest that the reverse situation probably exists, that is, that JH stimulates migratory activity possibly through a direct effect on the level of excitability of the central nervous system or flight muscle maturation.

Control of Migratory Flight in Melolontha melolontha

In the cockchafer, *Melolontha melolontha*, the state of activity of the CA was found to affect the orientation of flight (Stengel, 1974;

Stengel and Schubert, 1970, 1972 a,b,c,d). The female adult beetle passes through two or three ovarian cycles, each of which is characterized by oriented migrations which lead her toward feeding areas and then back to egg-laying sites. When she emerges from the soil, the female makes an oriented flight toward a feeding area (usually the edge of a forest or thicket). At this time the ovaries are undeveloped. About two weeks later after mating and feeding and when the first eggs are fully developed, the female undergoes an oviposition flight in the reverse direction, back to the field in which she then lays her eggs. Afterwards she returns to the forest in a postoviposition flight. This sequence may be repeated two or three times (Stengel and Schubert, 1970). Orientation during the initial prefeeding flight and during the postoviposition flights is toward a dark silhouette of trees on the horizon. The oviposition flight occurs in the opposite direction to these flights, independent of features of the countryside (Stengel and Schubert, 1970). Males, like females migrate to the feeding site after emergence from the soil. The male, however, never shows the reversal in orientation of flight which is typical of the female.

The effect of the CA on *Melolontha* migratory behavior was investigated in a series of organ-transplant experiments (Stengel and Schubert, 1970, 1972 a,b,c). The CA of prefeeding females were implanted into either preoviposition or prefeeding females. Similarly, in a second series of experiments the CA of preoviposition females were implanted into either preoviposition or prefeeding females. Sham-operated controls were also done for each of the four groups. Prefeeding females which received CA from other prefeeding females did not change their migratory behavior at all. However, 70% of the prefeeding females receiving preoviposition CA showed a reversal of the direction of flight to that of control and sham-operated preoviposition females (Stengel and Schubert, 1970). In a subsequent experiment 73% of 216 males receiving CA from preoviposition females flew in a direction identical to that of preoviposition females even though the male never shows this reversal of flight direction in nature (Stengel and Schubert, 1972a). CA from prefeeding females or from males were ineffective in altering flight orientation of any host. Stengel and Schubert (1972 b,c) have further shown that neurosecretory cells of the pars intercerebralis (PI) have actions identical with those of the CA. These results may indicate that the brain neurosecretory cells are necessary to stimulate the activity of the CA in the preoviposition female. However, more control experiments involving flight tests after allatectomy or neurosecretory cell cautery as well as JH treatment and JH titer analyses are necessary to determine exactly what the relative influence of the brain and the CA may be in the hormonal control of flight in this species.

Somewhat similar results have been reported by Lebrun (1969) who has shown that the readiness of *Calotermes flavicolis* to embark on "nuptial swarming flights" depends upon high titers of JH.

Control of Migration and Diapause in Leptinotarsa decemlineata

A somewhat different control of migration has been suggested in *Leptinotarsa decemlineata*, the Colorado potato beetle. Short-day photoperiods induce reproductive diapause and migration of adult beetles from the solenaceous host plant into the forest soil. Diapause and the associated soil-positive behavior could be induced in long-day beetles by allatectomy (de Wilde and de Boer, 1961). Active Ca implants restored normal soil-negative behavior to these allatectomized long-day beetles. However, CA implants (as many as six active glands) into short-day beetles did not counteract the diapause syndrome, and soil-positive behavior persisted (de Wilde and de Boer, 1969). Also implants of a long-day brain-CA complex could reverse the diapause inducing effect of short days. The activity of the CA is apparently maintained by a neurosecretion from the brain which is released only under long-day conditions (de Wilde and de Boer, 1969). After implantation into a short-day host, the active glands quickly become inactivated because of the absence of this neurosecretion, and diapause is maintained. JH titer analyses in *Leptinotarsa* (de Wilde et al., 1968) indicate that migration to and from the diapause site is probably stimulated by intermediate titers of JH below the threshold for ovarian development but sufficient to stimulate migratory flight.

Ovariectomy apparently has little effect on diapause behavior. Allatectomized and ovariectomized females respond to implantation of active CA by switching from soil-positive to soil-negative behavior just as allatectomized females do. The action of JH is not mediated through its effects on the ovaries, although it is possible that the post-migration flight may be terminated by some aspect of ovarian development as it appears to be in *Oncopeltus fasciatus* (Rankin, 1974).

It should be noted with respect to these studies that they have all been investigating the hormonal control of diapause and diapause-related behavior rather than migratory flight behavior itself. They are probably highly relevant to the hormonal control of migration, however, since there is a close association in time of migration with diapause in this species.

As is true of many species of insect migrants, *Leptinotarsa* displays a flight polymorphism within the population. Some individuals fly briefly or not at all, remaining fairly close to the point of emergence, while others make very extensive flights just prior to and/or just after diapause (Johnson, 1969). The hormonal control of flight behavior itself and the physiological basis of the flight polymorphism should be investigated in this species before definite conclusions can be drawn concerning the hormonal control of migration in *Leptinotarsa*.

Hormonal Influences on Locust Migratory Behavior

Most studies of hormonal influences on migratory behavior in locusts have been done on the species *Locusta migratoria migratorioides* and *Schistocerca gregaria*. In addition to effects of JH on flight and orientation, other hormones have been implicated in the control of locust flight such as the adipokinetic hormone from the locust corpora cardiaca and possibly ecdysone.

A definite behavioral pattern results when locust hoppers are reared in isolation. The nymphs show only a moderate level of activity; adults are generally solitary, unresponsive to migratory stimuli and usually do not perform sustained long flights. Crowding of young locusts, on the other hand, results in highly active, gregarious hoppers which display marching behavior and which show rapid oriented locomotion in response to migratory stimuli. The adults are gregarious and readily make long sustained flights (Johnson, 1969).

Carlisle and Ellis (1959) observed that solitary hoppers have larger prothoracic glands than do gregarious hoppers. Furthermore, in the adult, the prothoracic glands persist in the solitary form but rapidly disappear in gregarious locusts. When prothoracic gland homogenates were injected into gregarious hoppers, marching was decreased (Carlisle and Ellis, 1963). It is therefore possible that at least one physiological difference between solitary and gregarious phase locusts might be the level of activity of the prothoracic gland. Michel (1972) tested this hypothesis by implanting prothoracic glands from solitary adults into gregarious adults and found that such treatement decreased the duration of tethered flight of the host by about 50% for about six days. However, the differences in flight behavior were not striking and no sham implants of other tissues were done. Because a decrease in activity could be caused by many factors including injury due to injection or implantation of foreign tissues, these results, though interesting are somewhat inconclusive.

The question of prothoracic gland control of locomotor activity in locusts was addressed earlier by Haskel and Moorehouse (1963). As a rule, molting larvae display greatly decreased locomotor activity. Haskel and Moorehouse (1963) examined this periodic cessation of activity in the desert locust, *Schistocerca gregaria* using a standard nerve cord preparation obtained from male locusts treated with applications of hemolymph taken from nymphs during either the molt or intermolt periods. Changes in levels of interneuron activity were determined by recording from the commissures between the first and second thoracic ganglia; motor neuron activity was monitored in the nerve which innervates the extensor tibialis muscle of the metathoracic leg. When hemolymph from intermolt nymphs was used to bathe the nerve cord, little change was noted in levels of either interneuron or motor neuron activity. By contrast, hemolymph obtained from larvae 12 hours prior to ecdysis caused a marked increase in interneuron firing, but substantially decreased motor output. This effect could be mimicked by adding to the intermolt blood *Bombyx mori* extracts which had high concentrations of α and β ecdysones. Evidently, the hormonal agent which acts on the epidermis to provoke the molt also acts on the CNS to suppress locomotor activity prior to ecdysis.

The relative importance of ecdysone in the regulation of the activity of locusts still remains to be demonstrated conclusively. What is needed is a demonstration that gregarious behavior can be produced in solitary locusts by extirpation of the prothoracic glands and that solitary activity can be restored to such locusts by administration of ecdysone or prothoracic gland implants.

Several other hormone sources have been implicated in the control of locust flight. The corpora cardiaca have been shown to be necessary for migratory flight in *Locusta* and *Schistocerca* (Goldsworthy, Johnson, and Mordue, 1972; Mayer and Candy, 1969). These glands have been shown to release an adipokinetic hormone (Mayer and Candy, 1969) which is necessary to mobilize fat body lipid reserves for flights longer than about 15 minutes. Locusts in which adipokinetic hormone release from the glandular lobes is prevented by severance of both sets of nerves from the brain to the corpora cardiaca (NCCI and NCCII) fly poorly compared with operated control locusts; flight activity can be improved by the injection of CC extracts (Goldsworthy, Johnson, and Mordue, 1972). Goldsworthy, Coupland, and Mordue (1973) have found that cautery of the median neurosecretory cells of the brain has no effect on the flight behavior of locusts, but isolation or removal of the glandular lobes of the CC results in complete lack of lipid mobilization. Thus they conclude that the adipokinetic hormone is produced by and released from the glandular lobes of the CC. Since severing either the NCCI or the NCCII alone does not diminish flight performance, it would appear that release of the adipokinetic hormone from the glandular lobes is governed by a double innervation from the brain via both of these nerves. The hormone appears to be released almost immediately as flight begins.

In the normal locust, the flight pattern is thought to reflect changes in substrate utilization (Weiss-Fogh, 1952). During the initial 5 to 10 minutes the flight speed is high but declines markedly for about 25 minutes until a "cruising speed" is attained. Weis-Fogh (1952) has suggested that during the initial high-speed flight carbohydrate is the predominant fuel and that the subsequent decrease in speed corresponds to a gradual switchover to the utilization of lipid -- a process which is dependent upon the release of the adipokinetic hormone from the CC. It would seem that flight speed is dependent to some extent on the nature of the substrate utilized by flight muscles. Locusts burning carbohydrate fly more quickly than those utilizing lipid. Intact locusts injected with CC extract fly more slowly than noninjected locusts during the early stages of flight. This has led Goldsworthy et al. (1973) to postulate that CC factors, possibly the adipokinetic hormone itself, may suppress carbohydrate utilization by the flight muscle or favor lipid oxidation to the extent that carbohydrate is at a disadvantage as a substrate. Studies on the nature of lipid mobilization during flight (Beenakkers, 1969; Mayer and Candy, 1969) have shown that it is the level of diglyceride in the hemolymph that is elevated at the expense of fat body triglyceride. Studies of lipid utilization by the flight muscle *in vivo* and *in vitro* (Robinson and Goldsworthy, 1974) have shown that the adipokinetic hormone has two major sites of action: fat body lipid is mobilized as diglyceride and the flight muscle is stimulated to utilize lipid in preference to carbohydrate. Extracts of glandular lobes or a partially purified extract containing adipokinetic hormone reduce the respiratory rate of intact dorsal longitudinal flight muscle *in vitro* when trehalose is the only substrate provided. However, these hormonal extracts stimulate the oxidation of dipalmitin (Robinson and Goldsworthy, 1974). It is thought that the adipokinetic hormone may act on lipid oxidation by stimulating the entry of lipoprotein or diglyceride into the muscle or of acyl groups into the mitochondria (Robinson and Goldsworthy, 1974).

There seems to be some disagreement as to the actual source of the adipokinetic hormone in locusts. Michel (1972, 1973 a,b) has done a series of extirpation-implantation experiments involving the implantation of CC (either glandular or neurohemal lobe) from locusts reared either in isolation or in dense groups (poor and good fliers, respectively). Under these conditions Michel found that only the neurohemal lobe of the CC from good (gregarious) fliers could stimulate flight in poor fliers. Glandular lobes were ineffective as were neurohemal lobes in which the connections to the brain had been severed 15 days before. Similarly, destruction of the pars intercerebralis by electrocoagulation (Michel and Bernard, 1973) drastically reduced the tendency to fly in the desert locust *Schistocerca gregaria*. (These results are in direct contrast to those of Goldsworthy and co-workers mentioned above.) However, transplantations of the PI or the entire brain from good fliers to poor ones are ineffective in restoring flight to operated locusts. Michel and Bernard (1973) conclude that the PI substances are inactive in the PI where they are formed and are activated in the neurohemal part of the CC which can also store them. The discrepancies in the experimental results of Goldsworthy and co-workers and Michel are difficult to resolve unless they are based upon some misinterpretation of the anatomy of the pars intercerebralis and/or the CC.

In addition to ecdysone and the adipokinetic hormone, JH has also been implicated in the control of locust flight. Cassier (1963, 1964) found that implantation of CA from mature male *Locusta migratoria* into a male of the same age led to an increase in the speed of walking associated with a stronger phototactic response. According to Odhiambo (1966) removal of the CA in male *Schistocerca gregaria* resulted in a decrease in locomotor activity which could be reversed by implanting active CA into allatectomized males. Odhiambo went so far as to propose a direct effect of the CA hormone on the level of excitability of the CNS. However, Strong (1968 a,b) reported that *Locusta*, unlike *Schistocerca*, did not show any obvious reduction in locomotor activity after removal of the CA. Strong (1968a) suggested that the reported effects of allatectomy on locomotor activity were simply the result of the reduction in sexual behavior. Since caged locusts spend most of their time in sexual activity, he concluded that operations which inhibited this behavior would also appear to reduce locomotor behavior. However, the converse could also be true, that in allatectomized locusts sexual behavior may be fully developed but simply not expressed because of the sluggishness and inactivity caused by allatectomy. Wajc and Pener (1971) disproved Strong's hypothesis by testing the performance of allatectomized *Locusta* males on a flight roundabout. The allatectomized locusts consistently flew less intensely than did sham-operated controls. More recent work in flight-testing allatectomized locusts has produced somewhat more variable results.

Wajc and Pener (1971), Goldsworthy et al. (1972a), and Michel (1972a) have shown that the allatectomy of immature adult locusts leads to a marked decline in flight performance on roundabouts. Flight performance is, however, not affected in locusts allatectomized when mature (Goldsworthy et al, 1972a). Wajc (1973) also found that locusts allatectomized when immature are capable of flight performance similar to that of controls when flown on a flight balance in a wind tunnel under certain conditions. Lee and Goldsworthy (1975, 1976) have recently made a detailed study of the influence of the CA on locust flight in which they find that in adult male *Locusta* flight performance is a function of age, reaching its peak about 18 days after emergence. Allatectomy of immature locusts has the effect of retarding the development of normal flight capability so that for a short period after the operation allatectomized locusts fly poorly compared with operated controls. However, their flight performance is subsequently superior to that of operated controls of the same age This difference is due to the fact that there is a rapid decline in flight performance with age in the former which is absent in the latter. Periodic topical applications of JH to mature allatectomized locusts decreases flight performance somewhat although this effect decreases with age. The period of optimum flight performance is prolonged in locusts deprived of their CA when mature but the effect is not so pronounced as that seen in immature allatectomized locusts.

Lee and Goldsworthy (1975, 1976) suggest that the initial decrease in flight performance after allatectomy is due to an indirect effect of JH on this behavior via some process associated with flight muscle development. This effect would be transient, however, since allatectomized locusts eventually develop a flight performance comparable to the best attained by younger control locusts. It may be that some aging process involved in the metabolism of flight muscles is affected. Allatectomized locusts live longer than control, intact locusts and may in fact be physiologically younger than intact locusts of the same chronological age (Mordue and Goldsworthy, 1973).

Ovariectomy leads to a drastic suppression of flight performance in female locusts (Lee and Goldsworthy, 1976). This effect is probably due to high concentration of hemolymph diglyceride which may inhibit trehalose utilization in the flight muscles. These workers also suggest, on the basis of cuticle changes, that ovariectomized females age much faster than controls. Poor flight performance of ovariectomized locusts may be a result of premature aging. An injection of CC extract containing the adipokinetic hormone improves the flight performance of ovariectomized females. It is possible that the adipokinetic hormone introduced in the CC extract facilitates the utilization of the lipids by flight muscles.

Locusts which have been both ovariectomized and allatectomized display a flight performance which is similar to that of allatectomized locusts. Removal of the CA may counteract the aging effect of ovariectomy (Lee and Goldsworthy, 1976).

Migratory Behavior of the Monarch Butterfly, *Danaus plexippus*

Populations of monarch butterflies in North America undergo a dramatic migration in the spring and fall of each year. The monarch flies up to 2000 miles from Canada and the northern United States to overwintering sites in southern California, Florida, and Mexico and migrates back again to the north the following spring. The southward flight from the northern breeding areas in the fall occurs in large aggregations at a fairly leisurely pace. The butterflies move south with the weather fronts, flying primarily in the middle of the day, accumulating in overnight roosts at night and during bad weather, and replenishing fuel reserves by feeding on nectar. Populations of migrating monarchs will often remain in an area 2 to 3 weeks or more feeding and occasionally ovipositing if weather conditions are favorable and food supply is abundant (if milkweed is present, under favorable conditions, an occasional female will oviposit -- Rankin, unpublished observations). It is thought that most fall migrants are newly emerged individuals from the previous summer generation, most of whom are in an adult reproductive diapause induced by short fall photoperiods and cool temperatures (Herman, 1973; Urquhart, 1960). Brower (1977) found that 85% of 111 overwintering females taken from a huge, newly discovered Mexican hibernation site were virgin, 11% had mated once, and 2% had mated two or more times. They therefore probably began migrating very soon after emergence prior to reproduction and remained in reproductive diapause throughout the cool winter in the Mexican mountains. The winter diapause seems to be highly facultative however, because overwintering populations in southern Florida appear not to experience a cessation of reproductive activity (Brower, 1961, 1962; Urquhart and Urquhart, 1976) probably due to the warmer winter temperatures in that area.

Less is known about the monarch's northern flight. Some of the same individuals are thought to return to the northern breeding grounds the following season. The northward flight is said to be more rapid, wing-beat amplitude being about 190° as compared to about 30° in the southward cruising flight. Butterflies apparently fly alone by night as well as by day, resting singly rather than in aggregations. They apparently stop and feed less often and for shorter periods of time, using as fuel the fat stored during the winter hibernation (Urquhart, 1960; Johnson, 1969). Females apparently mate just prior to or during migration with the ovaries still in an undeveloped condition. They therefore migrate from overwintering sites initially unencumbered with large numbers of eggs, but with an increasing rate of egg maturation as both seasonal temperature and day length increase (Brower, 1977). Presumably this maturation would be gradual since the animals would be flying north into cooler temperatures. Some oviposition has been noted along the route of the spring migration. It is not known whether the collective migration is done only by nongravid insects, the gravid ones merely terminating their migration at the point of oviposition, or whether both gravid and nongravid individuals can make long flights. Since the Florida populations do not undergo a winter diapause but do contribute to the spring migration, it would seem that the latter possibility is more likely. It could, however, be only newly emerged Florida butterflies that migrate north in the spring. The question of monarch migration after or during the reproductive period has long been debated and is important to understanding the physiology of migration and to explaining the apparent behavioral differences between the spring and fall movements.

The summer breeding population undergoes several nonmigratory generations in the northern breeding areas. In the fall, decreasing day lengths and lowered temperatures appear to affect the neuroendocrine system in such a way that the activity of the CA and possibly the brain neurosecretory cells is diminished (Herman, 1973; Barker and Herman, 1976). It is clear that monarchs of the fall generation are physiologically and behaviorally distinct from those of the summer generations in ways that make their southward migration possible. These differences include the repression of gonadal development, a marked decrease in courtship and mating activities, a loss of oviposition response to milkweeds, the development of social nectaring assemblages, and the development of late afternoon aggregation behavior that results in temporary nocturnal cluster formation (Brower, 1977). These behaviors develop along with an increased tendency to fly southward, and by mid-September the mass migration is under way.

The mechanism of the stimulation of flight behavior has not been worked out, although Herman (1973) has shown that winter animals from the California populations are reproductively inactive owing at least in part to inactive CA. He speculates that the cycles of north-south migration and the cycles of reproductive activity that occur in monarchs may be due to cycles of allatal activity and inactivity. By his criterion of ovarian maturity, the CA are active in both northward migrants and overwintering animals in some areas. Collectively, these observations suggest that the CA may play a major role in regulating both migratory behavior and reproduction in this species. One difficulty with Herman and Barker's speculations is that ovarian development was the only criterion used to determine allatal activity. If lower titers of JH stimulate flight activity while higher titers or longer exposure to the hormone are necessary for ovarian development, it is possible that changing titers of JH during the migratory and reproductive periods govern which activity will predominate as is the case with the milkweed bug, *Oncopeltus fasciatus* (Rankin and Riddifor, 1977b). This possibility is currently being investigated in our laboratory.

Hormonal Control of Flight in Oncopeltus fasciatus

The milkweed bug, *Oncopeltus fasciatus* is a fairly typical insect migrant. Its summer range extends from central Canada to Central America but it does not overwinter in the central United States north of the 40th parallel (Dingle, 1965). Dingle (1966, 1968) has shown, on the basis of laboratory flight-testing and some highly suggestive field evidence, that at least 30% of the resident southern population migrates north in the spring and colonizes the very abundant habitat of milkweed fields in the north. It appears in Iowa milkweed patches in late June or early July and the population rapidly expands, reaching a peak in early September. By October, *O. fasciatus* has nearly disappeared from this area, apparently due to southward flights of much of the population. In the laboratory the duration of tethered flights in long-day photoperiods (16L:8D, 23°C) is greatest 8 to 12 days after adult eclosion (Dingle, 1966, 1968). Since cuticle hardening occurs from days 0 to 5 or 6, and oviposition in the female begins about day 15, long flight occurs in the post-teneral, pre-oviposition period in female *Oncopeltus*, as does migratory flight in many insects, but may continue considerably longer in males (Dingle, 1965; Caldwell, 1974; Rankin, unpublished observations). Under most environmental conditions, copulation takes place prior to the onset of flight and continues throughout the presumptive migratory period. Mating and flight, however, generally occur at different times of the day, the former being most frequent in the evening while flight occurs most frequently in the afternoon. Individuals may therefore copulate late in the day without this activity interfering with daily afternoon migratory flights (Caldwell and Rankin, 1974). Females are capable of mating repeatedly and of utilizing sperm from several matings (Gordon and Gordon, 1971). Thus mating prior to or during the migratory period would greatly increase the genetic variance and plasticity of the colonizing population and reduce any "founder effects."

Environmental photoperiod, temperature, and food availability appear to influence the amount of migratory behavior displayed by an *Oncopeltus* population. When reared in short photoperiods (12L:12D, 23°C), *Oncopeltus* enters an adult reproductive diapause (Dingle, 1974). The delay in reproduction which results provides a longer period of time prior to egg maturation during which females may be stimulated to make long flights and also results in a greater percentage of individuals of both sexes in laboratory populations making a long flight (Caldwell, 1974; Rankin, 1974). In the field, short fall photoperiods would presumably have a stimulatory effect on migratory behavior resulting in the southward migration of much of the northern population. In addition to photoperiod, temperature may also affect migration. Above about 25°C female migratory flight is at least partially inhibited at most natural photoperiods due to the stimulatory effect of high temperatures on ovarian development. The combination of high temperature and long photoperiod brings on reproduction very rapidly and virtually eliminates the migratory phase in female *Oncopeltus* (Caldwell, 1974).

Food quality or availability can also greatly affect flight and reproduction. *Oncopeltus* generally undergoes an intense period of almost constant feeding during the teneral period; in post-teneral adults feeding becomes periodic, occurring primarily during the evening, often in association with mating and with the same advantage of not interfering with migratory activity (Caldwell and Rankin, 1974). When fed on suboptimal food (green pods or flowers), female *Oncopeltus* delay reproduction and are stimulated to make long flights (Rankin and Riddiford, 1977a). These results along with some earlier work by Caldwell (1969, 1974) indicate that even in a long day (16L:8D) and relatively high temperature (27°C) regime (normally a reproduction inducing regime), *Oncopeltus* delays reproduction and exhibits migratory flight behavior in response to suboptimal food (green pods) or starvation (Rankin and Riddiford, 1977a).

Since the first northern colonizers typically arrive in Iowa in late June or early July when temperatures are high and photoperiods are approaching the longest of the year (Dingle, 1966), it is likely that food quality (flowers and green pods at this time) is important in stimulating northward flights in the spring. By this same mechanism flight can also be stimulated among the reproductively mature residents by a food shortage, greatly increasing the probability of finding new food sources (Rankin and Riddiford, 1977a).

In order to determine the hormonal basis of migratory behavior in *Oncopeltus fasciatus*, a mixture of JH analogs (Law et al., 1966) was topically applied to males and females five days after the adult molt and flight tests were administered at three-day intervals. Animals were maintained under a 12L:12D, 24°C (diapause-inducing) regime and were checked daily for reproductive activity.

Topical application of the JH analogue significantly increased flight behavior of males but not of females in reproductive diapause (Caldwell and Rankin, 1972). JH application to intact females decreased flight activity apparently due to the induction of precocious ovarian development since ovariectomized females responded to JH treatment as did the males (Fig. 1). Similar responses were elicited by the implantation of active CA into intact males and females and into ovariectomized females (Fig. 2), indicating that JH from the CA was important in the control of migratory behavior in this species (Rankin, 1974).

To examine the effects of the corpus cardiacum on flight, three pairs of CC were implanted into eight-day-old *Oncopeltus* and these animals

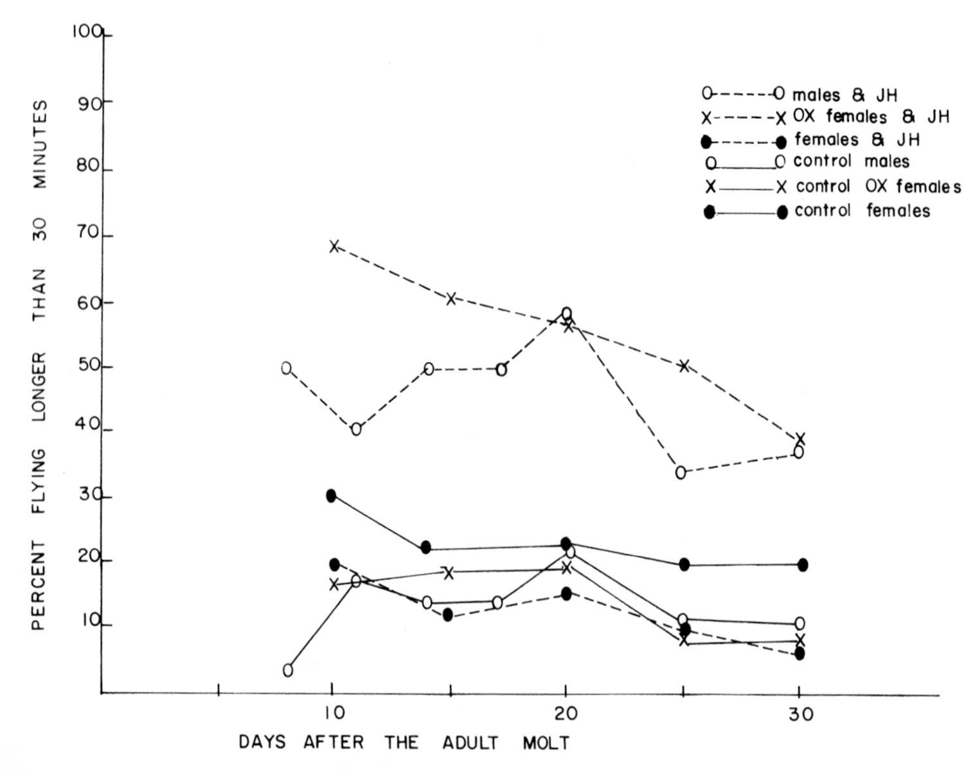

Fig. 1. The effect of JH mimic (Calbiochem) on flight of males (O----O), ovariectomized females (X----X), and intact females (●----●). Hormone was topically applied to abdominal sternites on day 5 after the adult molt. Environmental regime was 12L:12D, 23°C. Controls consisted of untreated males (O——O), untreated ovariectomized females (X——X), and untreated intact females (●——●) (from Caldwell and Rankin, 1972).

were flight-tested on day 10. Figure 3 shows that in neither males nor females was there any significant enhancement of flight activity over the unoperated or sham-implanted control animals. In a second experiment CC extracts were prepared by homogenizing five pairs of freshly excised glands in 0.2 ml sterile insect ringers (Pringle's) solution. Ten microliters of this extract was injected between the abdominal sternites or between the thorax and the abdomen. Injection of CC extracts into intact animals had no effect on flight activity. By contrast, three CA implanted on day 8 after the adult molt stimulated about 60% of the treated individuals to fly longer than 30 minutes when tested on day 10 (Fig. 3). Similarly, topical application of the JH mimic, ZR512 (5 μg) to animals on day 8 stimulated flight activity two days later (Fig. 3) (Rankin, 1974)

Effects of Starvation

Results of flight-testing allatectomized *Oncopeltus* were inconclusive since because of an intimate association of the CA with the dorsal aorta and esophagus in this species survival of allatectomized animals was extremely poor. Because Johansson (1958) had shown starvation to inhibit egg maturation through a neuornal inhibition

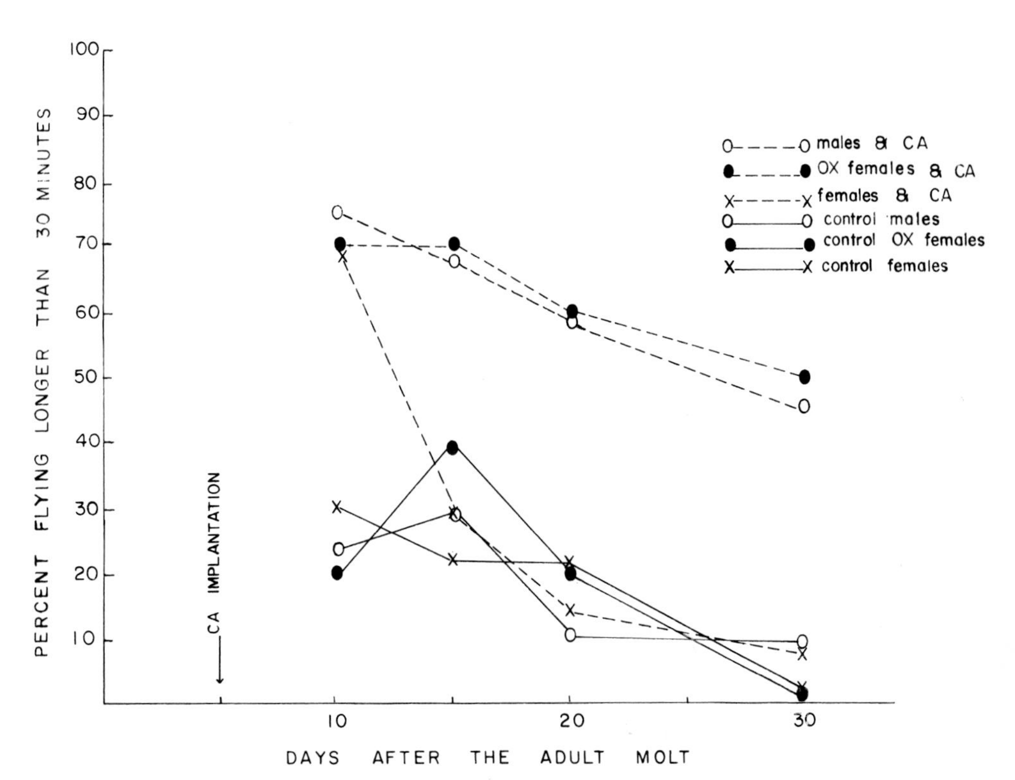

Fig. 2. The effect of implantation of corpora allata (CA) on males (0----0), ovariectomized (OX) females (●----●), and intact females (X----X). Three CA were implanted into the donor abdomens on day 5. Environmental regime was 12L:12D, 23°C. Controls were sham operated but received no implanted glands (from Rankin, 1974).

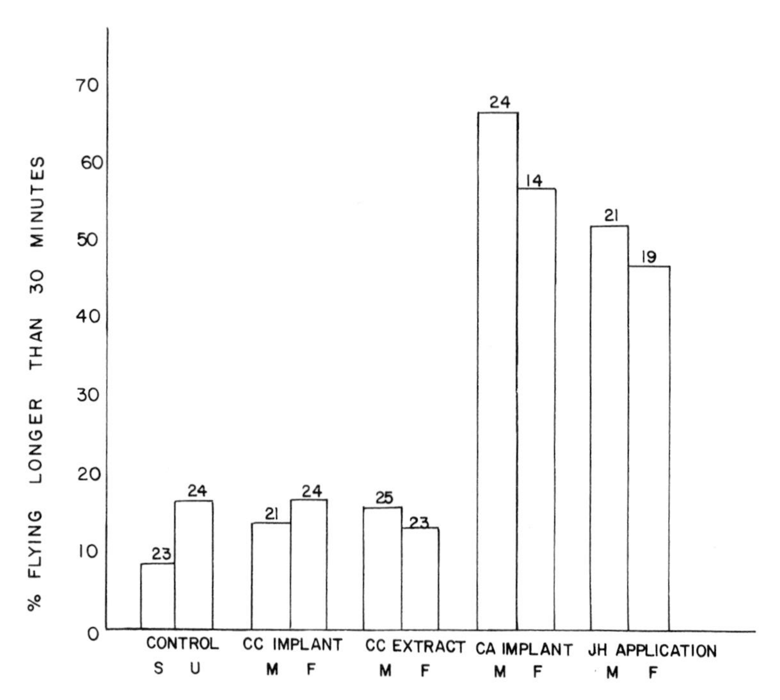

Fig. 3. The effect of implanted corpora cardiaca (CC) and corpora allata (CA), injection of CC extract, and JH (ZR512) application on flight in *O. fasciatus* (from Rankin and Riddiford, 1977a).

of the CA, starvation was employed as a "pseudoallatectomy" procedure. Since starvation had been shown to inhibit allatal activity (Johansson, 1958), it was thought that it would lead to sharply decreased flight activity. When *Oncopeltus* were given only water after day 5 and flight-tested on days 10, 20, and 30, it was found that at day 10 flight activity increased followed by a sharp decrease. By day 20 after starvation flight activity had declined below that of fed controls and as expected, flight activity was suppressed by day 30 in all individuals (Rankin, 1972, 1974; Rankin and Riddiford, 1977a).

Because of this increase then decrease in flight activity after the beginning of starvation, it was of interest to determine the JH titer in these starved animals. *Oncopeltus* "superfliers" were starved and their hemolymph assayed for JH at varying times thereafter.

Hemolymph was obtained from *Oncopeltus* by cutting off the tips of the antennae and thoracic appendages and collecting the exudate from the cut tips in a calibrated capillary tube. JH was extracted from the hemolymph samples according to the method of Fain and Riddiford (1975) and was assayed using the *Manduca* pigmentation bioassay (Truman, Riddiford, and Safranek, 1973). To standardize the assay of unknown extracts, the mean score obtained from 25 assay animals each having received 2 μl of hemolymph extract in cyclohexane was compared to results obtained when known concentrations of JH III were extracted, redissolved, and applied in a similar way (Fig. 5). JH III was selected as a reference hormone even though the *Manduca* pigmentation bioassay has been shown to be more sensitive to JH I (Truman et al., 1973) because recent work by several laboratories has shown JH III to be the more prevalent naturally occurring hormone in hemimetabolous insects (Dahm et al., 1976; Trautmann et al., 1976; Judy et al., 1973; Muller et al., 1974). It is important to point out in this regard that while the relative JH titers after starvation can be determined with confidence, the actual JH concentration in *Oncopeltus* hemolymph can only be estimated from these data and thus must be considered an approximation.

Determination of the relative JH concentration in *Oncopeltus* hemolymph for 30 days following starvation (Fig. 4) revealed a dramatic decline in detectable JH after two to three days of starvation. The inactivation appears to be gradual as JH was still detectable after four to six days. Of interest was the correlation between the decline in JH to an intermediate level (Fig. 5) and the increase in flight activity two days after starvation was begun. Later when the JH level was nearly undetectable, no flight activity was observed. Yet, as seen in Figure 5, topical application of 10 μg of ZR512, a JH mimic, nine days after the onset of starvation restored flight activity to its original level. Two of the eight females treated with JH mimic subsequently laid eggs after the flight activity had again declined.

Rankin (1974) showed that implants of three CA per individual on day 11 to starved *Oncopeltus* followed by flight tests on days 15, 20, and 30 induced a significant increase in the percentage of individuals in

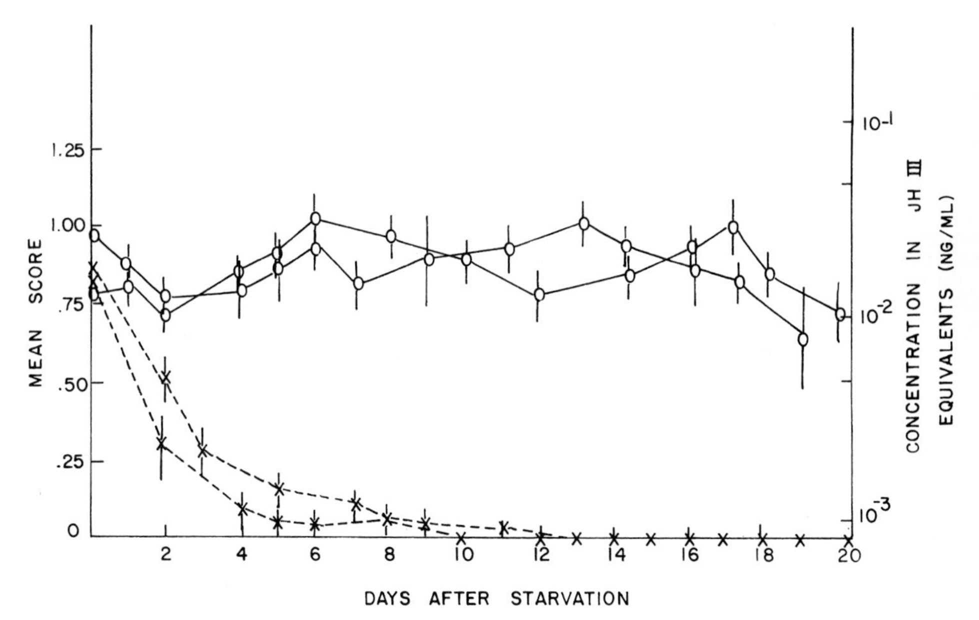

Fig. 4. Effect of starvation on hemolymph titers of juvenile hormone. The population was divided into two groups, controls (solid lines) continued to receive food, experimentals (broken lines) were deprived of food. Hemolymph from 25 animals was pooled for each point in both groups. JH titer was assayed using the *Manduca* pigmentation bioassay. Environmental regimen was 12L:12D, 24° C in this experiment. Approximate hemolymph JH concentrations were estimated using the dose-response curve obtained with known doses of JH III. Bars indicate ± one standard deviation among the assay larvae. The experiment was repeated with essentially identical results in a second series of determinations (from Rankin and Riddiford, 1977a).

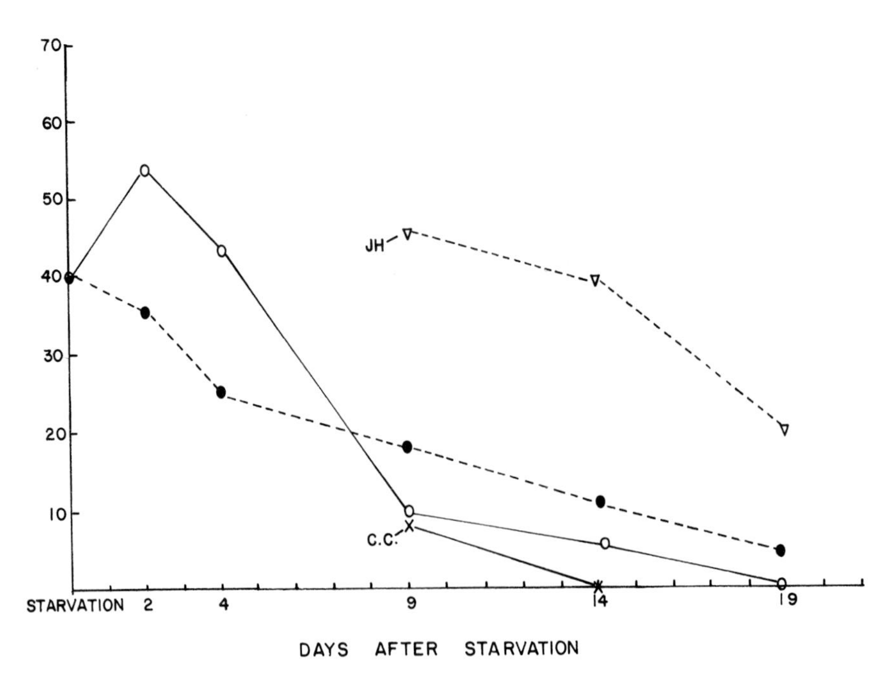

Fig. 5. The effect of starvation, CC implantation, and JH application on flight. A group of 45 animals was starved from day 10 after the adult molt and flight-tested along with controls at intervals after the initiation of starvation. Flight initially increased and then decreased to a very low level in the starved population. On day 9 the starved group was divided into three parts. Eleven animals received implanted corpora cardiaca (X——X), 15 received applications of 10 μg JH (ZR512) (∇----∇) applied topically in acetone to the abdominal sternites, 10 received only acetone applied in a similar manner, and 10 were untreated. Untreated and acetone treated controls were nearly identical and are thus graphed together as starved controls (0——0). ●----● indicates fed controls. The environmental regime in this experiment was 12L:12D, 24°C (from Rankin and Riddiford, 1977a).

the allatal-implanted group making long flights at each flight test over the control group (Fig. 6). Reproductive activity was also stimulated by the implanted glands; seven of 18 starved implanted females laid eggs at least once and of these, three laid several large clutches, while none of the control females had oviposited. Flight tests on days 20 and 30 indicated that females having laid eggs were much less willing to make long flights. In fact, none of these females made a flight greater than 10 minutes though females that had not oviposited made flights greater than 30 minutes on that day. Seventy percent of the implanted males tested on day 30 made long flights while only 17% of the unimplanted controls did so (both sexes) (Rankin, 1974). By contrast, implants of three pairs of CC did not enhance flight activity of starved animals (Fig. 5) (Rankin and Riddiford, 1977a). It would appear, therefore, that in *Oncopeltus* the CC plays no role in flight behavior, while JH appears to be necessary and sufficient for increased flight activity.

The increase in flight observed after starvation seems to be correlated with a decrease in JH titers below that necessary to stimulate ovarian development but sufficient to induce migratory flight. A slow turning-off of the CA would be highly adaptive in inducing a period

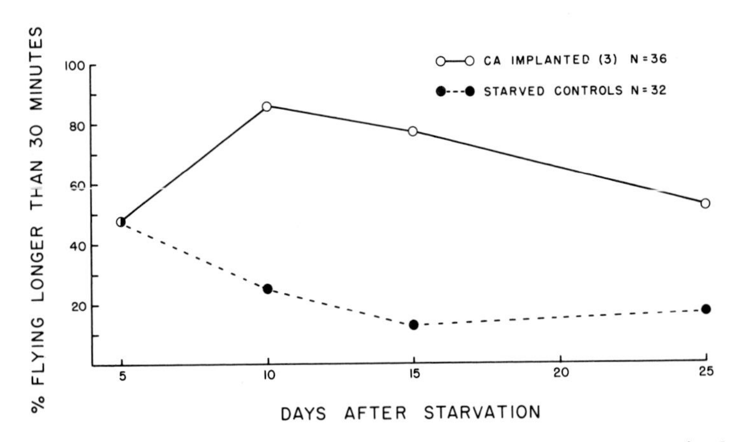

Fig. 6. Effect of CA implantation on flight of starved *Oncopeltus*. Animals were starved from day 5 of adult life and received CA implants on day 11 of adult life (six days after starvation began). Controls were sham operated but received no implanted tissue. Three CA were implanted per host. Environmental regime was 12L:12D, 24° C (from Rankin, 1974).

of flight activity following starvation. The subsequent drop in JH titers to very low levels insufficient to stimulate flight would also limit the investment in flight to a few days after food shortage was encountered and assure conservation of ultimate reserves after the initial investment in flight.

An adipokinetic hormone such as that demonstrated in locusts (Goldsworthy et al., 1972) does not appear to be involved, at least in the stimulation or timing of migratory behavior in *Oncopeltus*, since neither CC implants nor extracts had any effect on flight activity (Figs. 3 and 5). However, the possibility that a constitutively produced hormone is involved in the continual regulation of fat mobilization has not been eliminated since flight tests have not been administered to cardiacectomized animals.

JH Titer Determinations

These data (Rankin and Riddiford, 1977a) and those of Rankin (1974) indicate that JH is the primary hormone responsible for the stimulation of both migration and reproduction in *Oncopeltus*. Since JH induces both migration and ovarian development and the latter inhibits migratory flight behavior, these two activities seem to be mutually exclusive. It thus became important to determine the normal hormonal control of these two events in the life of this migratory insect. To this end, the JH titer throughout pre- and early reproductive life was determined and correlated with flight and the onset of reproductive behavior in *Oncopeltus* (Rankin and Riddiford, 1977b).

Hemolymph (obtained as above) from 25 animals (approximately 125 μl) was pooled for each hormone titer determination. In order to minimize within-sample variance, pooled animals were full siblings from the same or consecutive clutches of eggs. After extraction and purification [according to the method of Fain and Riddiford, 1975)] the JH was assayed using the *Manduca* pigmentation bioassay (Truman et al., 1973). Again the assay was standardized to known concentrations of JH III.

Determination of JH titer changes from 0 to 28 days after the adult molt in animals maintained under a short-day (12L:12D, 24°C) regime showed that in males there is a brief rise in JH titer shortly after adult emergence at about day 4-6 which is followed by a decline in JH titer to very low levels until day 12. Then a subsequent gradual increase to higher titers occurs by day 18 to 20 (Fig. 7). In females the JH titer begins to rise after eight days, then plateaus about day 16. Males seemed to reach and maintain a somewhat higher JH titer than females by day 20.

In contrast, the JH titer in animals maintained under a long-day (17:7D, 24°C) regime (Fig. 8) increased rapidly from an initial low level beginning about day 5. By day 8 under the long-day, JH titers of the females had surpassed those seen by day 20 in either sex reared under short-day lengths and began to level off. Of interest is the apparent decline in JH titer among reproductively mature females after about day 16 similar to the decline seen in the short-day diapausing females during the same period. JH titers in males under long days increased more slowly and seemed to plateau about day 18 (Rankin and Riddiford, 1977b).

In order to correlate flight and oviposition behavior with JH titer changes throughout early adulthood, siblings of the *Oncopeltus* set aside for JH titer determination were set up in small containers as single pairs and monitored for flight and oviposition behavior. Flight tests were administered every two days from day 5 until the termination of the experiment.

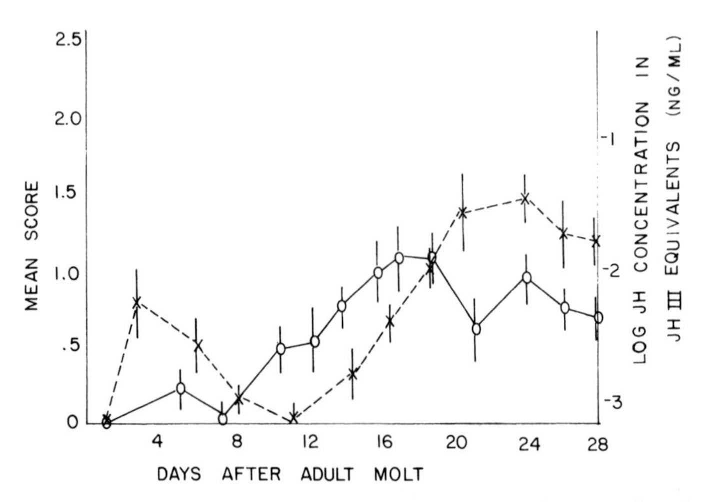

Fig. 7. JH titer determinations on hemolymph taken from animals maintained under short-day (12L:12D, 24°C) conditions. Hemolymph from 25 animals was pooled for each determination. Titer determinations were done on both males (X----X) and females (0——0). Bars indicate ± one standard deviation among assay animals (Rankin and Riddiford, 1977b).

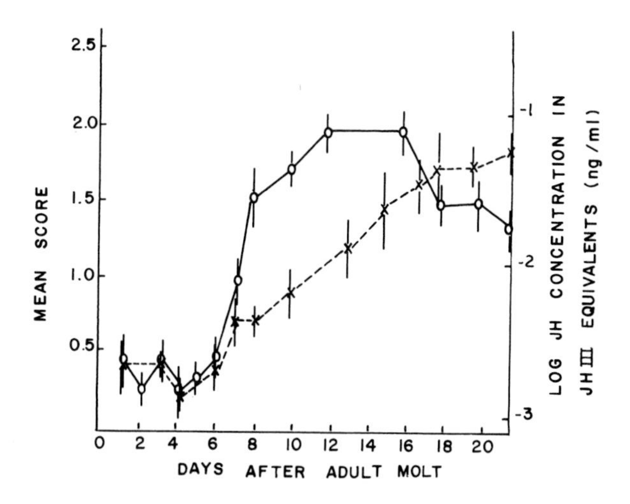

Fig. 8. JH titer determinations on hemolymph taken from animals maintained under long-day (17L:7D, 24°C) conditions. Hemolymph from 25 animals was pooled for each determination. Titer determinations were done on both males (X----X) and females (O——O). Bars indicate ± one standard deviation among assay animals (Rankin and Riddiford, 1977b).

Flights longer than 30 minutes did not occur in the short-day population until day 10 (Fig. 9); subsequent flight tests revealed a steady increase in the proportion of individuals in the population which would make a long flight - until days 14 to 16, at which time 49% of the population made flights longer than 30 minutes. By days 18 and 20 the proportion making long flights had declined to 18% and by day 25 no further long flights were made. As flight activity was declining, Fig. 9 also shows that oviposition activity was beginning in the population. By day 24, 30% of the females had oviposited and by day 26, when flight activity had virtually ceased, 55% of the females had oviposited at least once. This confirms earlier work by Dingle (1965) and Rankin (1974) which indicated that ovarian development inhibits further long flights in female *Oncopeltus*. The most interesting relationship to be noted here, however, is the fact that as Figs. 7 and 9 show, migratory flight is most frequent during the period of increasing JH titers from days 10 to 18 in the sibling population. The JH titer then plateaus at which time flight is ceasing and oviposition beginning. Therefore, it appears that here as in starved animals (Rankin and Riddiford, 1977a) flight occurs at an intermediate JH titer. When the JH titer becomes sufficiently high, it initiates oogenesis. When flight behavior and oviposition activity are examined in a long-day population (Fig. 10), quite a different activity pattern emerges. Very little flight activity occurred in the long-day population at any time. The maximum occurred on days 8 and 10 of adult life when 11% of the population made flights of over 30 minutes. By day 14, 55% of the female population had oviposited and by day 20, 87%. Once again the correlation among JH titer and ovarian development is of interest. As Fig. 8 shows, JH titer rises very rapidly in the long-day population, reaching a maximum by day 12. Very shortly thereafter in the sibling long-day population the number of females ovipositing increases dramatically. The very brief period of minimal migratory flight activity from about day 8 to 10 again coincides with an intermediate JH titer, one quite similar to that found during peak flight activity in short-day-reared adults (Figs. 8 and 10).

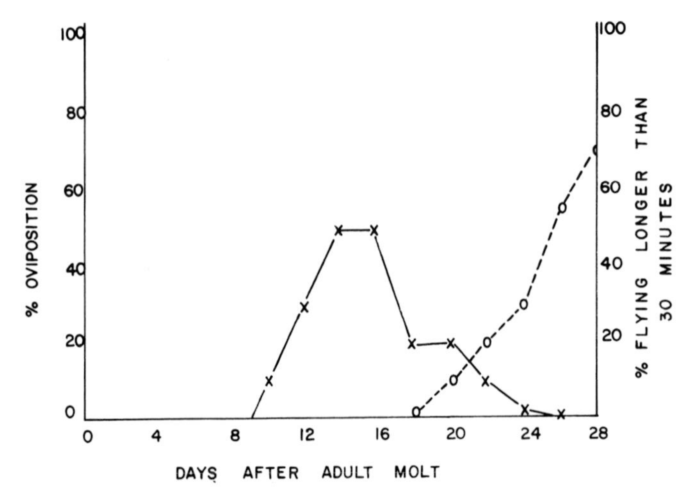

Fig. 9. Percentage of individuals reared under short-day conditions (12L:12D, 24°C) making flights longer than 30 minutes (n = 95) (X——X) and percentage of females ovipositing (n = 40) (0----0) (from Rankin and Riddiford, 1977b).

These results tend to indicate that in *Oncopeltus* intermediate titers of JH stimulate migratory flight and higher titers stimulate ovarian development culminating in oviposition. We have not eliminated the possibility that it is either the necessary time of exposure to JH that differs between the flight system and the ovary or that both behaviors can be initiated by the same minimum level of JH but during the normal course of events oogenesis requires a sustained exposure to a high level of JH for completion. Increased flight behavior is likely a faster response to a certain level of JH than is oviposition since JH is necessary for the initiation of vitellogenesis in *Oncopeltus* and this process requires approximately four days (Johansson, 1958). However, the fact that after starvation oogenesis has ceased while flight is still continuing (Rankin, 1974; Rankin and Riddiford, 1977a) argues against this hypothesis. Similarly, the fact that the peak in flight activity in Fig. 9 precedes the peak in oviposition

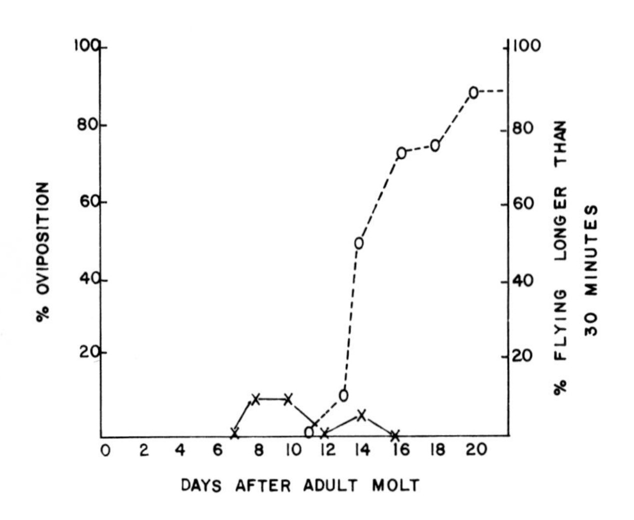

Fig. 10. Percentage of individuals reared under long-day conditions (17:7D, 24°C) making flights longer than 30 minutes (n = 95) (X——X) or percentage of females ovipositing (n = 40) (0----0) (from Rankin and Riddiford, 1977b).

by approximately 12 days would seem to indicate that response thresholds for the two systems are different. Experiments are now in progress to clearly distinguish between the above two possibilities.

Although the animals in these experiments were not flight-tested after day 30, Dingle (1965), Caldwell (1974), and Rankin (unpublished) have shown that males typically have a second period of flight activity or continue to fly after female flight has declined and oviposition has begun. Thus it is probable that some physiological change associated with the onset of egg maturation is responsible for the cessation of flight in the female although some inhibitory effect of high JH titer on flight activity has yet to be entirely excluded.

The initial small rise in hemolymph JH immediately after adult ecdysis is similar to that observed in adult *Leptinotarsa* (de Wilde et al., 1968) and in *Schistocerca gregaria* and *Locusta migratoria* (Johnson and Hill, 1973). It may represent the initial reactivation of the gland after the final ecdysis. The subsequent decline in JH titer observed in the short-day population is very similar to that observed in diapausing *Leptinotarsa*. Indeed, de Wilde et al. (1968) propose a declining titer of JH as one factor in the induction of diapause behavior in that species. Similarly, in both *Leptinotarsa* and *Oncopeltus* the daylengths eliminated the subsequent decline in JH titer.

Selection for Late Onset of Flight

In a series of selection experiments (Rankin and Riddiford, 1977c) *Oncopeltus* were selected for late age of first flight for four generations. Mating, oviposition, and flight activity were monitored as in previous experiments (Rankin, 1974; Rankin and Riddiford, 1977a, b) and JH titer determinations were performed as outlined above (Fain and Riddiford, 1976) on a portion of the fourth-generation population.

Selection for late onset of flight resulted in a correlated delay in onset of reproduction as compared with the unselected population (Fig. 11). After four generations of selection the age of peak flight activity was day 32 as compared with day 14 in the unselected population. Similarly, the mean age of first oviposition was day 46 in the selected strain and day 24 in the nonselected group. When JH titer analyses were done on the selected population, it was clear that a similar delay in JH titer rise after the adult molt was correlated with the observed delay in onset of reproduction and flight (Fig. 12). Although correlation does not necessarily imply cause, these results do add further support to the premise that rising titers of JH after adult emergence stimulate first migratory flight and then, at higher titers, ovarian development (Rankin and Riddiford, 1977c).

JH Specific Esterase in Oncopeltus

A decrease in JH titer or a continued low titer after adult emergence could be due to a diminished activity of the CA. It could also be due to an increased metabolism of the hormone at these times.

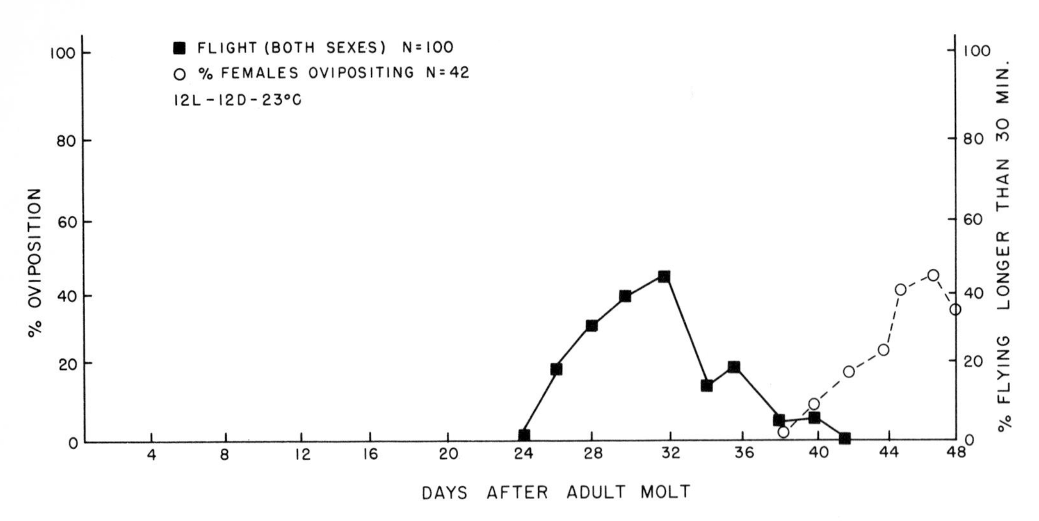

Fig. 11. Age at which flight (■----■) and oviposition (O----O) occur among individuals selected for four generations for late first flight. Environmental regime for all generations was 12L:12D, 24° C. (from Rankin and Riddiford, 1977c).

JH has been shown to be transported and protected in the hemolymph by a specific carrier protein in several insect species (Ferkovich et al., 1975; Kramer et al., 1976). Only specific JH esterases which may appear or increase in activity at particular times during development are able to metabolize JH when it is bound to the carrier protein in the hemolymph. In several insect species increased activity of JH specific esterases in the hemolymph has been

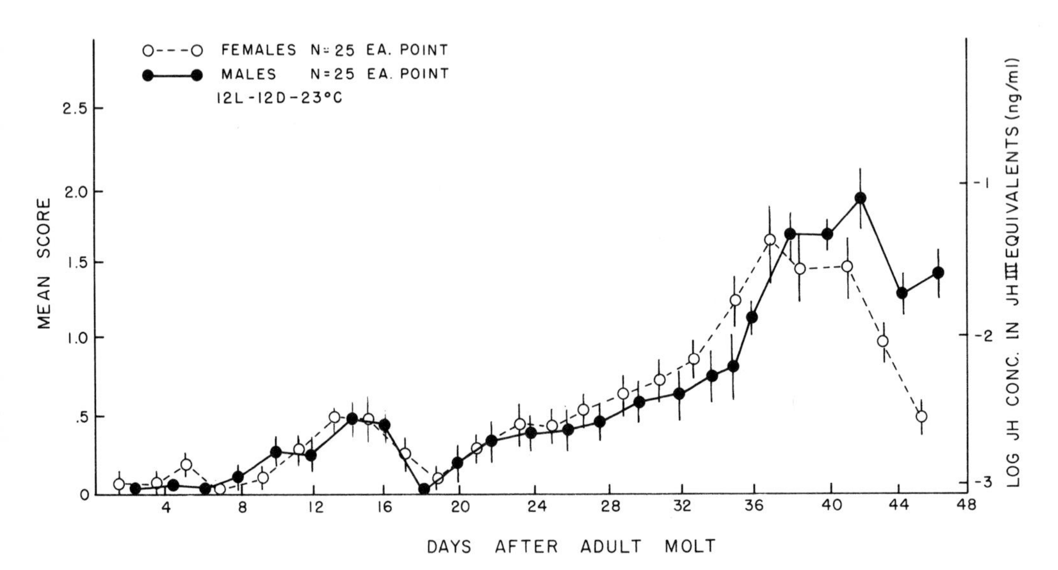

Fig. 12. JH titer determinations on hemolymph taken from animals selected for four generations for late onset of flight, reared under short-day conditions (12L:12D, 24° C). Hemolymph from 25 animals was pooled for each determination. Titer determinations were done on both males (●——●) and females (O----O). Bars indicate ± one standard deviation among assay animals (from Rankin and Riddiford, 1977c).

implicated in the decrease in JH titer during the last larval instar and the pupal stage (Weirich et al., 1973; Sandburg et al., 1975; Weirich and Wren, 1976). Transport and metabolism of JH in *Oncopeltus* seems to be somewhat different.

JH esterase determinations were done on *Oncopeltus* at several stages of development and under various treatment regimes. Determinations of JH specific esterase activity were done on hemolymph obtained from groups of *Oncopeltus* according to the method of Weirich and Wren (1973). Thirty microliters of hemolymph were collected from the front legs and antennae, added to 50 μl buffer (10/10/1 TKM, pH 7.5), and incubated with ^{3}H *Cecropia* JH (NEN) for one hour at 25°C. The reaction was quenched with 1.5 ml ethyl acetate, and the organic fraction was eluted and reduced under nitrogen. The residue was chromatographed on thin-layer silica gel (Eastman) with 35:65 ethyl acetate: hexane plus a small amount of glacial acetic acid. The chromatogram was cut into 1-cm pieces and each piece was put into 5 ml scintillation fluid and counted. The extent of conversion of JH to JH acid (i.e., the ratio of JH acid:JH) was taken as an indication of esterase activity. JH acid standard was obtained by alkaline hydrolysis of ^{3}H-JH.

Esterase determinations were done following starvation to ascertain that the observed decrease in JH titer was due to a decline in allatal activity rather than to an increase in esterase activity. Similarly, we determined JH specific esterase activity in groups of 1-, 2-, 5-, 8-, 11-, 14-, 17-, and 20-day-old adults maintained under either short-day (12L:12D, 23°C) or long-day (16L:8D, 24°C) conditions as well as late third instar larva just prior to the molt to the fourth instar. Further esterase determinations were also done on fourth instar larvae (during each day of the fourth instar) and fifth instar larva (during every day of the fifth instar) in long- and short-day conditions. In *none* of these determinations did we find a significant level of JH esterase activity (Rankin et al., 1977). Furthermore, duplicates of each of the above treatment groups were given 1 μg of ZR512 each or 5 μg JH I each in an attempt to induce the appearance of JH esterase in the hemolymph (Whitmore et al., 1973). Again, no significant conversion of JH to JH acid was observed in any treatment group (Rankin et al., 1977). In order to verify that our incubation procedures were correct, fifth instar *Danaus plexippus* larvae and fifth instar *Manduca sexta* larvae were also assayed for hemolymph JH specific esterase activity. In both experiments appreciable esterase activity was detected. We must conclude from these results therefore that JH metabolism in *Oncopeltus* does not procede via hemolymph esterase activity as it has been shown to do in many other insects, nor do hemolymph esterases play any role in maintaining low titers of hemolymph JH in starved *Oncopeltus*.

Ajami and Riddiford (1973) and Rankin et al. (1977) have shown, however, that when whole *Oncopeltus* were given topical applications of labeled *Cecropia* JH or injected with this hormone, significant JH metabolism to JH acid and lower metabolites could be detected.

In an attempt to ascertain the source of this esterase activity, tissues from adult *Oncopeltus* were cultured for three days in Marks medium (GIBCo) with ^{3}H labeled *Cecropia* JH and the esterase activity was determined as above. Hemolymph, fat body, flight muscle (from ovipositing females), testis, immature ovary, and mature ovary were cultured in this way. Incorporation of labeled methionine and leucine into protein was also measured on similar cultures to insure that such cultured tissues were metabolically active. As Fig. 13b shows, only the mature ovary, of the cultured tissues, was active in the metabolism of JH under these experimental conditions although all tissues were active in protein synthesis (Rankin et al., 1977). It would appear, therefore, that the mature ovary is a primary site of JH metabolism in this species. It is possible that at least part of the inhibitory effect of the mature ovary on flight involves the metabolism of JH to the inactive JH acid by this tissue. Previous JH titer determinations consistently show lower JH titer in ovipositing females than in males of the same age or in preoviposition females (Figs. 7, 8, and 12). Further JH titer determinations are now in progress on older reproductive females and ovariectomized females to determine more exactly ovarian effects on JH titers.

Ovarian inhibition of migratory flight is very likely brought about in several ways. In addition to its action in lowering JH titer, the mature ovary may trigger abdominal stretch receptors (Rankin, 1972), may alter circulating levels of flight fuels, or may secrete a substance which acts as an inhibitory hormone on the flight system. Since JH titers are apparently not reduced to preflight levels, we must at least postulate a second mechanism of flight suppression or a gradual increase in the JH threshold for migratory flight. All of these possibilities are currently being investigated.

Summary and Conclusions

In *Oncopeltus* then, the CA controls the timing and coordination of flight and reproduction through its response to three environmental cues: photoperiod, temperature, and food quality (Fig. 14). Short-day photoperiods induce adult reproductive diapause as long as the temperature remains below 25°C (Dingle, 1974; Caldwell, 1974). Under these conditions the CA initially begins secreting JH, then is turned off, and only gradually is reactivated beginning about day 10. The change in behavior from diapause and migratory flight to mating and oviposition corresponds well with the rising titer of JH. Under long-day photoperiods or at temperatures above 25°C, reproductive diapause is averted with mating beginning about day 6 to 7, soon after the cuticle has hardened, followed by oviposition by day 15 to 16. The CA maintains a low secretion rate during the cuticle hardening and feeding stage, then dramatically increases its output beginning about day 6. This high JH titer leads immediately to egg maturation. At any time during the reproductive phase the lack of food or merely inadequate nutrient supplies will cause inhibition of JH secretion so that the titer falls below the threshold necessary for ovarian development and migratory flight is stimulated. Clearly the stimulation of migratory behavior in response to intermediate JH titers allows dispersal at two critical times (1) Migration before reproduction in the young adult when reproductive potential is at its highest allows for rapid colonization and maximizes utilization of the temporary milkweed habitat (Dingle, 1974). The inhibitory effect of oogenesis on flight behavior effectively partitions the energy investment between flight and egg production so that as oogenesis begins, no further energy expenditure in migratory flight occurs unless adverse conditions develop. (2) In response to food shortage, JH is stimulated for several days thereby maximizing the chance of finding a more favorable environment. As JH titers continue to decline during starvation, even the energy expenditure in flight is curtailed and the individual settles down to survive on its reserves as long as possible (Rankin and Riddiford, 1977a,b).

The use of JH to coordinate various behaviors in *Oncopeltus* is an extremely efficient physiological strategy and may be employed to some extent by other species as well. The monarch butterfly, *D. plexippus*

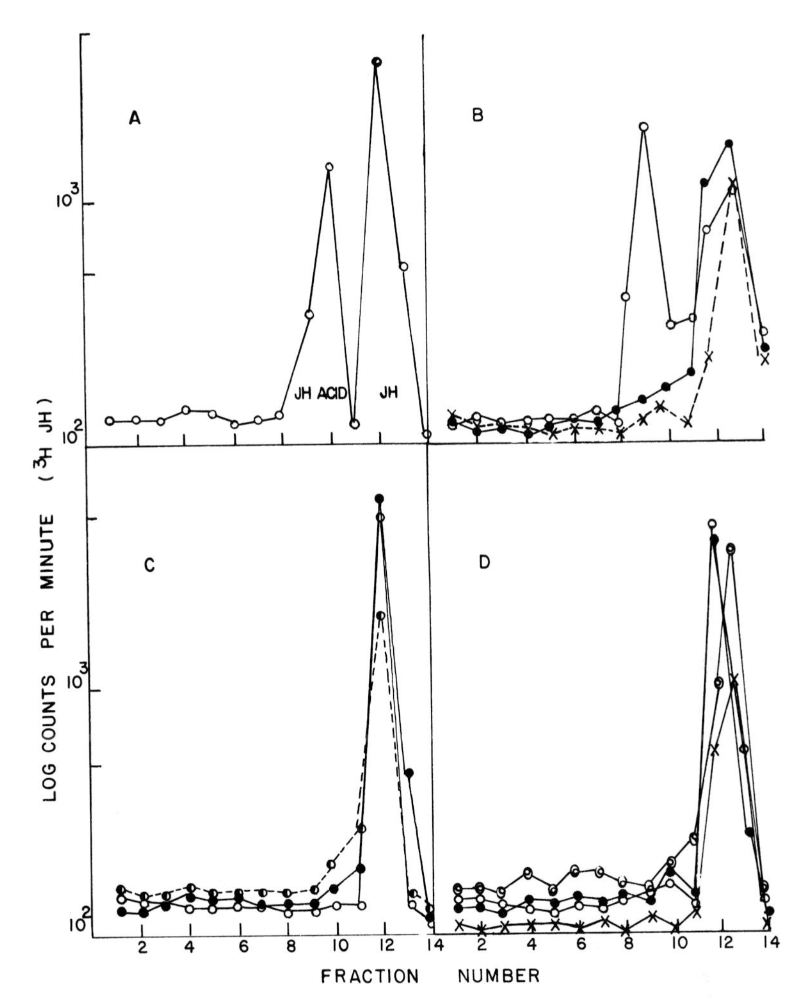

Fig. 13. JH specific esterase activity in *Oncopeltus*.
A. TLC radioactivity profile for JH acid standard obtained by alkaline hydrolysis of ^{3}H-JH and JH standard.
B. TLC radioactivity profile of JH specific esterase activity in mature ovary (O——O), immature ovary (●——●), and fat body (X----X) cultured in Mark's medium (Gibco) for three days with ^{3}H-JH.
C. TLC radioactivity profile of JH specific esterase activity in hemolymph (●——●), flight muscle (●----●), and testis (O——O) cultured in Mark's medium for three days with ^{3}H-JH.
D. TLC radioactivity profile showing JH specific esterase activity in starved adults (X——X), fifth instar (⊚——⊚), fourth instar (●——●), and third instar (O——O) hemolymph incubated with labelled JH for 1 hour at 25°C (from Rankin and Riddiford, 1977a and Rankin, et al., 1977).

and the Colorado potato beetle, *Leptinotarsa decemlineata*, for example, may use similar strategies. *Melolontha* also may use changing titers of JH to stimulate changes in orientation associated with its dispersal flights. The role of the CA in locust flight is still somewhat unclear. However, it may well be involved in the development of the flight muscles (Lee and Goldsworthy, 1976) and the normal timing of migratory flight during the life cycle; other hormones such as the

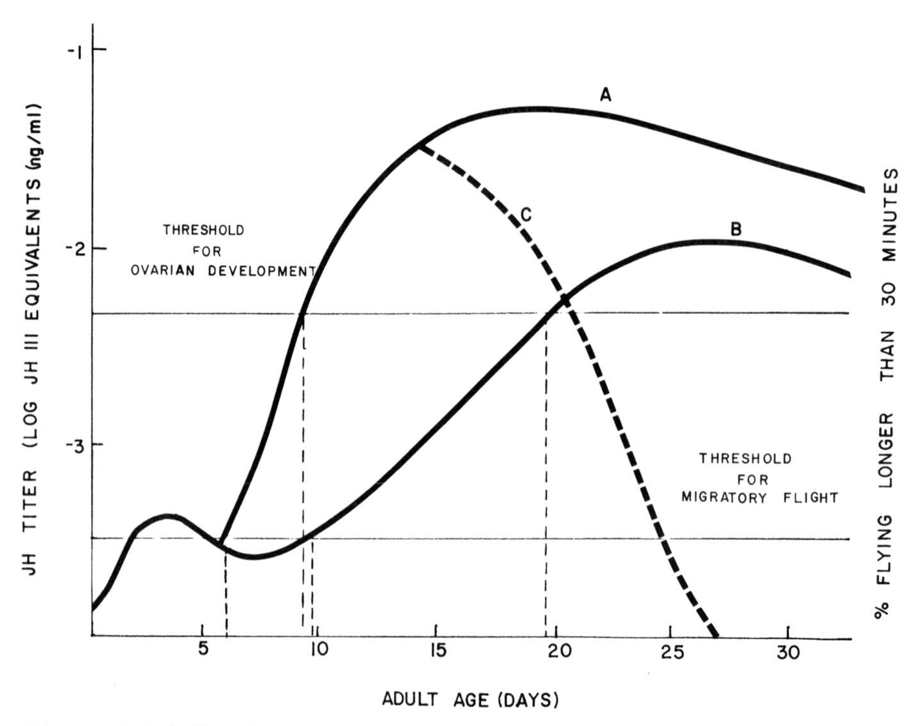

Fig. 14. A model indicating the proposed response of the corpus allatum to the environmental stimuli of photoperiod, temperature, and food deprivation; the relationship of JH titers to ovarian development and migratory behavior. High temperatures and long photoperiods (A) stimulate high JH titers and immediate reproduction, lower temperatures and short photoperiods (B) result in migratory flight in response to intermediate titers, presumably below the threshold for ovarian development, while poor quality food or starvation (C) results in lower JH titers and flight activity followed by a decrease in JH titer and cessation of flight if lack of food persists (from Rankin and Riddiford, 1977b).

adipokinetic hormone and possibly ecdysone are also involved in the control of locust migration -- possibly to a greater extent than JH. Some insects, such as *Lygaeus kalmii* (Caldwell, 1974), *Dysdercus intermedius* (Edwards, 1971), *Dysdercus fulvoniger* (Davis, 1975), and *Ips confusus* (Borden and Slater, 1968) with different migratory and life-history strategies appear to use somewhat different mechanisms for controlling flight behavior.

Wigglesworth has theorized (1970) that juvenile hormone was originally a morphogenetic hormone concerned with maintaining the juvenile morph and inhibiting gonadal development until the adult stage. He suggests that in the course of insect evolution the hormone has acquired a secondary function in many insects, namely, stimulating some aspect of oogenesis after the adult molt in response to environmental cues, probably via control by the central nervous system or brain neurosecretory cells. I would like to suggest that in some insect species which have highly complex life histories and rapid response to changes in environmental parameters, juvenile hormone has acquired a set of tertiary functions such as the control of migration, control of wingless morphs, or caste determination. It would seem that such functions have evolved relatively recently, probably many times in response to specific selection pressures. They would be most likely to occur where there was an advantage to maintaining the option of response to environmental changes after the adult molt in terms of alternative reproductive or life-history strategies. It seems we may predict, therefore, that a particular insect is likely to have evolved some hormonal mechanism for coordinating migratory behavior and reproduction, but we cannot predict *a priori* exactly what that control may be. The particular physiological strategy which has evolved would depend upon the basic physiology of the species, its particular life-history strategy, and the particular selection pressures it had experienced. The mechanism in *Oncopeltus* which utilizes the output of the CA to trigger and closely coordinate dispersal and reproduction is simply one such highly adaptive system.

References

Ajami, A.M., Riddiford, L.M.: Comparative metabolism of the cecropia juvenile hormone. J. Insect Physiol. *19*, 635-349 (1973).

Barker, J.F., Herman, W.S.: Effect of photoperiod and temperature on reproduction of the monarch butterfly, *Danaus plexippus*. J. Insect Physiol. *22*, 1565-1568 (1976).

Beenakkers, A.M.T.: The influence of corpus allatum and corpus cardiacum on lipid metabolism in *Locusta migratoria*. Gen. Comp. Endocr. *13*, Abstract 12 (1969).

Borden, J.H., Slater, C.E.: Induction of flight muscle degeneration by synthetic juvenile hormone in *Ips confusus* (Coleoptera: Scolytidae). Z. Vergl. Physiol. *61*, 366-368 (1968).

Brower, L.: Monarch migration. Natural History, *86*, 40-53 (1977).

Brower, L.: Evidence for interspecific competition in natural populations of the monarch and queen butterflies, *Danaus plexippus* and *D. gilippus berenice* in south central Florida. Ecology 43, 549-552 (1962).

Brower, L.: Studies on the migration of the monarch butterfly I. Breeding populations of *Danaus plexippus* and *D. gilippus berenice* in south central Florida. Ecology *42*, 76-83 (1961).

Caldwell, R.L.: A comparison of dispersal strategies in two species of milkweed bugs, *Oncopeltus fasciatus* and *Lygaeus kalmii*. Ph.D. thesis, University of Iowa (1969.)

Caldwell, R.L.: A comparison of the migratory strategies of two milkweed bugs. In: Experimental Analysis of Insect Behaviour. Barton Browne, L. (ed.). Berlin-Heidelberg- New York: Springer 1974.

Caldwell, R.L., Rankin, M.A.: Effects of a juvenile hormone mimic on flight in the milkweed bug, *Oncopeltus fasciatus*. Gen. Comp. Endocrinol. *19*, 601-605 (1972).

Caldwell, R.L., Rankin, M.A.: Separation of migratory from feeding and reproductive behavior in *Oncopeltus fasciatus*. J. Comp. Physiol. *88*, 313-324 (1974).

Carlisle, D.B., Ellis, P.E.: La persistance des glands ventrales cephaliques chez les Criquets solitaires. C.r. Acad. Sci. Paris *249*, 1059-1060 (1959).

Carlisle, D.B., Ellis, P.E.: Prothoracic gland and gregarious behaviour in locusts. Nature *200*, 603-604 (1963).

Cassier, P.: Action des implantations de corps allates sur la reactivite phototropique de *Locusta migratoria migratorioides* (R. et F.), phase gregaria. C.R. Hebd. Seanc. Acad. Sci., Paris *257*, 4048-4049 (1963).

Cassier, P.: Etude et interpretation des effets a long terme, des implantations abdominales de corps allates, sur la reactivite phototropique de *Locusta migratoria migratorioides* (R. et F.), phase gregaire. C.R. Hebd. Seanc. Acad. Sci., Paris *258*, 723-725 (1964).

Chudakova, I.V., O.M. Bocharova-Messner: Endocrine regulation of the condition of the wing musculature in the imago of the house cricket (*Acheta domestica* L.). Akad. Nauk S.S.S.R. Doklady Biol. Sci. *179*, 157-159 (1968).

Davis, N.T.: Hormonal control of flight muscle histolysis in *Dysdercus fulvoniger*. Ann. Ent. Soc. *68*, 710-714 (1975).

Dahm, K.H., Bhaskaran, G., Peter, M.G., Shink, P.D., Seshan, K.R., Roller, H.: On the identity of the juvenile hormones in insects. In: The Juvenile Hormones. L.I. Gilbert, (ed.). New York: Plenum Press 1976, pp. 19-47.

Dingle, H.: The relation between age and flight activity in the milkweed bug, *Oncopeltus*. J. Exp. Biol. *42,* 269-283 (1965).

Dingle, H.: Some factors affecting flight activity in individual milkweed bugs *(Oncopeltus)*. J. Exp. Biol. *44,* 335-343.

Dingle, H.: Life history and population consequences of density photoperiod and temperature in a migrant insect, the milkweed bug, *Oncopeltus*. Amer. Natur. *102,* 149-163 (1968).

Dingle, H.: Migration strategies of insects. Science N.Y. *175,* 1327-1335 (1972).

Dingle, H.: Diapause in a migrant insect, the milkweed bug, *Oncopeltus fasciatus* (Dallas) (Hemiptera; Lygaeidae). Oecologia *17,* 1-10 (1974).

Dingle, H., Arora, G.K.: Experimental studies of migration in bugs of the genus *Dysdercus*. Oecologia *12,* 119-140 (1973).

Edwards, F.J.: Endocrine control of flight muscle histolysis in *Dysdercus intermedius*. J. Insect Physiol. *16,* 2027-2031 (1970).

Fain, M.J., Riddiford, L.M.: Juvenile hormone titers in the hemolymph during late larval development of the tobacco hornworm, *Manduca sexta* (L.). Biol. Bull. *149,* 506-521 (1975).

Ferkovich, S.M., Silhacek, D.L., Rutter, R.R.: Juvenile hormone binding proteins in the haemolymph of the indian meal moth. Insect Biochem. *5,* 141-150 (1975).

Goldsworthy, G.J., Coupland, A.J., Mordue, W.: The effects of corpora cardiaca on tethered flight in the locust. J. Comp. Physiol. *82,* 339-346 (1973).

Goldsworthy, G.J., Johnson, R.A., Mordue, W.: *In vivo* studies on the release of hormones from the corpora cardiaca of locusts. J. Comp. Physiol. *79,* 85-96 (1972).

Gordon, B.R., Gordon H.T.: Sperm storage and depletion in the spermatheca of *Oncopeltus fasciatus*. Ent. Exp. & Appl. *14,* 425-433 (1971).

Haskell, P.T., Moorhouse, J.E.: A blood-borne factor influencing the activity of the central nervous system of the desert locust. Nature *197,* 56-58 (1963).

Herman, W.S.: The endocrine basis of reproductive inactivity in monarch butterflies overwintering in central California. J. Insect Physiol. *19,* 1883-1887 (1973).

Johansson, A.S.: Relation of nutrition to endocrine-reproductive functions in the milkweed bug, *Oncopeltus fasciatus* (Dallas). Nytt Mag. Zool. *7,* 1-132 (1958).

Johnson, C.G.: Migration and Dispersal of Insects by Flight. London: Methuen, 1969.

Johnson, R.A., Hill, L.: Quantitative studies on the activity of the corpora allata in adult male *Locusta and Schistocerca*. J. Insect Physiol. *19,* 2459-2467 (1973).

Judy, K.J., Schooley, D.A., Hall, M.A., Bergot, B.J., Siddal, J.B.: Chemical structure and absolute configuration of a juvenile hormone from grasshopper corpora allata in vitro. Life Sci. *13,* 1511-1516 (1973).

Kennedy, J.S.: Phase transformation in locust biology. Biol. Rev. *31,* 349-370 (1956).

Kramer, K.J., Dunn, P.E., Peterson, R.C., Hidelisa, L.S., Sanburg, L.L., Law, J.H.: Purification and characterization of the carrier protein for juvenile hormone from the hemolymph of the tobacco hornworm *Manduca sexta* Johannson (Lepidoptera: Sphingidae). J. Biol. Chem. *251,* 4979-4985 (1976).

Lebrun, D.: Corps allates et instinct genesique de *Calotermes flavicolis* Fabr. Le declenchement de l'activite sexuaelle des jeunes imagos ailees de *Calotermes flavicollis* Fabr necessite la presence dans l'organisme d'un taux eleve d'hormone juvenile. C.R. Hebd. Sci., Paris *269,* 632-634 (1969).

Lee, S.S., Goldsworthy, G.J.: Allatectomy and flight performance in male *Locusta migratoria*. J. Comp. Physiol. *100,* 351-359 (1975).

Lee, S.S., Goldsworthy, G.J.: The effect of allatectomy and ovariectomy on flight performance in female *Locusta migratoria migratorioides* (R. & F.) Acrida *5,* 169-180 (1976).

Mayer, R.J., Candy, D.J.: Control of haemolymph lipid concentration during locust flight. An adipokinetic hormone from the corpora cardiaca. J. Insect Physiol. *15,* 598-611 (1969).

Michel, R.: Influence des corpora cardiaca sur la tendance au vol soutenu de criquet pelerin *Schistocerca gregaria* (Forsk.). J. Insect Physiol. *18*, 1811-1827 (1972a).
Michel, R.: Etude experimentale de l'influence des glandes prothoraciques sur l'activite de vol du Criquet Pelerin *Schistocerca gregaria*. Gen. Comp. Endocrin. *19*, 96-101 (1972b).
Michel, R.: Variations de la tendance au vol soutenu du criquet pelerin *Schistocerca gregaria* apres implantations de corpora cardiaca. J. Insect Physiol. *19*, 1317-1325 (1973).
Michel, R., Bernard, A.: Influence de la pars intercerebralis sur l'induction au vol soutenu chez le criquet pelerin *Schistocerca gregaria*. Acrida *2*, 139-149 (1973).
Mordue, W., Goldsworthy, G.J.: The physiological effects of corpus cardiacum extracts in locusts. Gen and Comp. Endocrin. *12*, 360-369 (1969).
Muller, P.J., Masner, P., Trautmann, K.H.: The isolation and identification of juvenile hormone from cockroach corpora allata *in vitro*. Life Sci. *15*, 915-921 (1974).
Odhiambo, T.R.: The metabolic effects of the corpus allatum hormone in the male desert locust. J. Exp. Biol. *45*, 51-63 (1966).
Urquhart, F.A.: The Monarch Butterfly. Toronto: Univ. Toronto Press 1960, 361pp.
Urquhart, F.A., Urquhart, N.R.: A study of the peninsular Florida population of the monarch butterfly (*Danaus p. plexippus*; Danaidae), J. Lepid. Soc. *30*, 73-87 (1976).
Rankin, M.A.: An experimental analysis of the physiological control of flight and reproduction in *Oncopeltus fasciatus* (Heteroptera: Lygaeidae). Ph.D. Thesis, University of Iowa (1972).
Rankin, M.A.: The hormonal control of flight in the milkweed bug, *Oncopeltus fasciatus*. In: Experimental Analysis of Insect Behavior. Barton Browne, L. (ed.) Berlin-Heidelberg-New York: Springer 1974.
Rankin, M.A., Gandy, W., Jordan, R. Juvenile hormone specific esterase activity in *Oncopeltus*: A mechanism of ovarian regulation of flight behavior. (Ms. in preparation.)
Rankin, M., Riddiford, L.M.: The hormonal control of migratory flight in *Oncopeltus fasciatus*: The effects of the corpus cardiacum, corpus allatum and starvation on migration and reproduction. Gen. and Comp. Endocrin. (in press).
Rankin, M., Riddiford, L.M.: The significance of hemolymph juvenile hormone titer changes in the timing of migration and reproduction in adult *Oncopeltus fasciatus*. (J. Insect Physiol. in press) (1977b).
Rankin, D., Riddiford, L.M.: The influence of artificial selection on hemolymph juvenile hormone titers and the timing of reproduction and flight in adult *Oncopeltus fasciatus*. (Ms. in preparation) (1977c).
Robinson, N.L., Goldsworthy, G.J.: The effects of locust adipokinetic hormone on flight muscle metabolism *in vivo* and *in vitro*. J. Comp. Physiol. *89*, 369-377 (1974).
Sandburg, L.L., Kramer, K.J., Kexdy, F.J., Law, J.H.: Juvenile hormone-specific esterases in the haemolymph of the tobacco hornworm, *Manduca sexta*. J. Insect Physiol. *21*, 873-887 (1975).
Stengel, M.: Migratory behaviour of the female of the common cockchafer *Melolontha melolontha* L. and its neuroendocrine regulation. In: *Experimental Analysis of Insect Behaviour*. Barton Browne, L. (ed.). Berlin-Heidelberg-New York: Springer 1974.
Stengel, M., Schubert, G.: Role des Corpora allata dans le comportement migrateur de la femelle de *Melolontha melolontha* L. (Coleoptere Scarabidae). C.R. Hebd. Seanc. Sci., Paris (D) *270*, 181-184 (1970).
Stengel, M., Schubert, G.: Influence des Corpora allata de la femelle pondeuse de *Melolontha melolontha* L. (Coleoptere Scarabidae) sur l'ovogenese de la femelle prealimentaire. C. R. Hebd. Seanc. Acad. Sci., Paris (D) *274*, 426-428 (1972a).
Stengel, M., Schubert, G.: Influence des Corpora allata de la femelle pondeuse de *Melolontha melolontha* L. (Coleoptere Scarabidae) sur le comportement migrateur du male. C.R. Hebd. Seanc. Acad. Sci., Paris (D) *274*, 568-570 (1972b).
Stengel, M., Schubert, G.: Influence de la Pars intercerebralis et des Corpora cardiace de la femelle pondeuse sur l'ovogenese de la femelle prealimentaire de *Melolontha melolontha* L. (Coleoptere Scarabidae). C.R. Hebd. Seanc. Acad. Sci.,

Paris (D) 275, 1653-1654 (1972c).
Stengel, M., Schubert, G.: Role de la Pars intercerebralis et des Corpora cardiaca de la femelle pondeuse de *Melolontha melolontha* L. (Coleoptere Scarabidae) dans le comportement migratoire de la femelle prealimentaire. C.R. Hebd. Seanc. Acad. Sci., Paris (D) *275*, 2161-2162 (1972d).
Strong, L.: The effect of enforced locomotor activity on lipid content in allatectomized males of *Locusta migratoria migratorioides*. J. Exp. Biol. *48*, 625-630 (1968a).
Strong, L.: Locomotor activity, sexual behavior, and the corpus allatum hormone in males of *Locusta migratoria migratorioides*. J. Insect Physiol. *14*, 1685-1692 (1968b).
Trautmann, K.H., Suchy, M., Masner, P., Wipf, H.K., Schuler, A.: Isolation and identification of juvenile hormones by means of a radioactive isotope dilution method: Evidence for JH III in eight species from four orders. In: The Juvenile Hormones. Gilbert, L.I. (ed.). New York: Plenum Press 1976, pp. 118-130.
Truman, J.W., Riddiford, L.M., Safranek, L.: Hormonal control of cuticle coloration in the tobacco hornworm, *Manduca Sexta:* Basis of an ultrasensitive bioassay for juvenile hormone. J. Insect Physiol. *19*, 195-205 (1973).
Wajc, E.: The effect of the corpora allata on flight activity of *Locusta migratoria migratorioides* (R. & F.). Ph.D. Thesis, University of London (1973).
Wajc, E., Pener, M.P.: The effect of the corpora allata on the flight activity of the male African migratory locust, *Locusta migratoria migratorioides* (R. & F.) Gen. Comp. Endocr. *17*, 327-333 (1971).
Weirich, G., Wren, J.: The substrate specificity of juvenile hormone esterase from *Manduca sexta* hemolymph. Life Sci. *13*, 213-226 (1973).
Weirich, G., Wren, J.: Juvenile hormone esterase in insect development. A comparative study. Physiol. Zool. *49*, 341-350 (1976).
Weirich, G., Wren, J., Siddall, J.B.: Development stages of the juvenile hormone esterase activity in hemolymph of the tobacco hornworm, *Manduca sexta*. Insect Biochem. *3*, 397-407 (1973).
Weis-Fogh, T.: Fat combustion and metabolic rate of flying locusts (*Schistocerca gregaria* F.). Phil. Trans. B *237*, 1-36 (1952).
Whitmore, D., Gilbert, L.I., Ittycheriah, P.I.: The origin of hemolymph carboxylesterases 'induced' by the insect juvenile hormone. Mol. Cell. Endocrin. *1*, 37-54 (1974).
Wigglesworth, V.B.: Insect Hormones, San Francisco: W.H. Freeman, 1970.
Wilde, J. de; Boer, J.A. de: Physiology of diapause in the adult Colorado potato beetle. II. Diapause as a case of pseudoallatectomy. J. Insect Physiol. *6*, 152-161 (1961).
Wilde, J. de; Boer, J.A. de: Humoral and nervous pathways in photoperiodic induction of diapause in *Leptinotarsa decemlineata*. J. Insect Physiol. *15*, 661-675 (1969).
Wilde, J. de, Staal, G., Kort, C. de, DeLoof, A., Baard, G.: Juvenile hormone titer in the hemolymph as a function of photoperiodic treatment in the adult Colorado beetle (*Leptinotarsa decemlineata* Say). Proc. K. Ned. Akad. Wet. (C) *71*, 321-326 (1968).

The Evolution and Ecology of Flight: the "Oceanographic" Approach

R. C. RAINEY

Introduction

Our theme is "Evolution of escape," by contrasted strategies of migration and of diapause. Migration is characterized not only by escape from adverse conditions, but also by the active exploitation of temporary habitats - as Southwood (1962) has emphasized. We now know that far back in geological time (and long before the first birds) there were already wind systems available of a kind likely to have enabled the early insects to locate and exploit new food supplies following rains - as Desert Locusts (*Schistocerca gregaria* Forsk.) can so spectacularly do to this day. Thus the extent and consistency of orientation of fossil sand dunes in the western United States provide evidence of prevailing easterly tradewinds in Permo-Carboniferous times, with the dunes, by continental drift, then in appropriately lower latitudes than their present position (Runcorn, 1961).

Ecology and Mechanism of Locust Migration

The story of Desert Locust migration is introduced by Figure 1 (Rainey, 1969), which presents the spatial distribution of the probability of breeding during the two contrasted months of May and September. It illustrates the geographical extent of seasonal migration needed between successive generations, and gives some indication of the completeness and the degree of regularity of these movements. Now these areas and seasons of breeding are in fact areas and seasons of rain, with the summer rainfall regime south of the Sahara contrasting with the winter and spring rains around the Mediterranean and the Persian Gulf; and it has long been known that the migrations of Desert Locusts take them out of areas which are drying up and into areas of new rain. To outline how this happens, we begin with Figure 2, looking upwards during the passage of a locust swarm flying across what are now the Tsavo National Parks in Kenya; the photograph records the striking uniformity of orientation which is characteristic of such flying locusts. This uniform orientation, giving an impression of purposeful navigation, irresistibly suggests itself as in some way enabling the locusts to reach a far-distant goal of new food supplies (as they so often do). Just how misleading is this impression of purposeful orientation became clear only when aircraft observations made it possible for the first time to follow the hour-to-hour and day-to-day movements of individual swarms and to relate these movements to the wind in which each swarm was flying. In Figure 2 the long arrows show the direction of displacement of this swarm as a whole, as given by aircraft fixes of its position an hour before and an hour after the photograph was taken. The swarm as a whole was in fact traveling directly with the wind, which is shown by the small arrow. Figure 2 shows what the swarm looked like; Figure 3 (Rainey, 1976) shows what it did. The inset of Figure 3 was provided by a series of photographs like Figure 2, taken throughout the passage of the swarm overhead (Waloff, 1972), and shows that the swarm comprised large numbers of such groups of flying locusts, but with a wide diversity of orientation between different groups, so that the swarm as a whole traveled with the direction of the wind. At both the leading and trailing edges of the swarm the locusts were heading into the swarm and so contributing to its cohesion. The displacement of the swarm recorded during the three days it was followed by the aircraft is shown on the map in Figure 3. The next three figures present further aircraft observations of swarm movements, in wind fields of increasing complexity. First, the near-parallel tracks of swarms traveling in company in near-uniform winds are shown in Figure 4 (Rainey, 1976), which also shows the maintenance of cohesion and indeed of plan area of individual swarms - with gregarious behavior operating to offset the disruptive effects of the vigorous atmospheric turbulence of fair weather in these arid regions. Next,

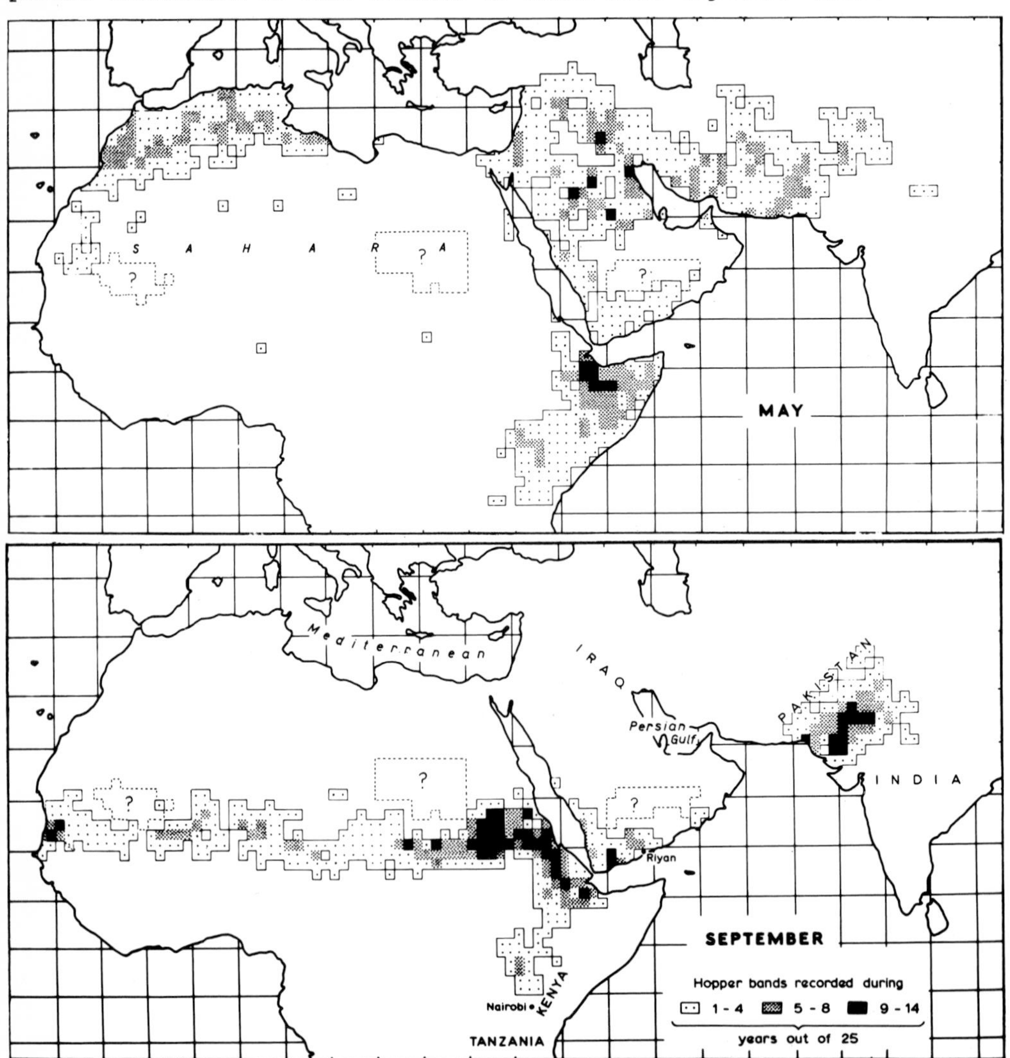

Fig. 1. Contrasted seasonal breeding areas of the Desert Locust illustrated by the number of years during 1939-63 in which bands of nymphs were recorded in each degree-square of latitude and longitude during the months of May and September; queries indicate areas from which reports are largely or wholly lacking.

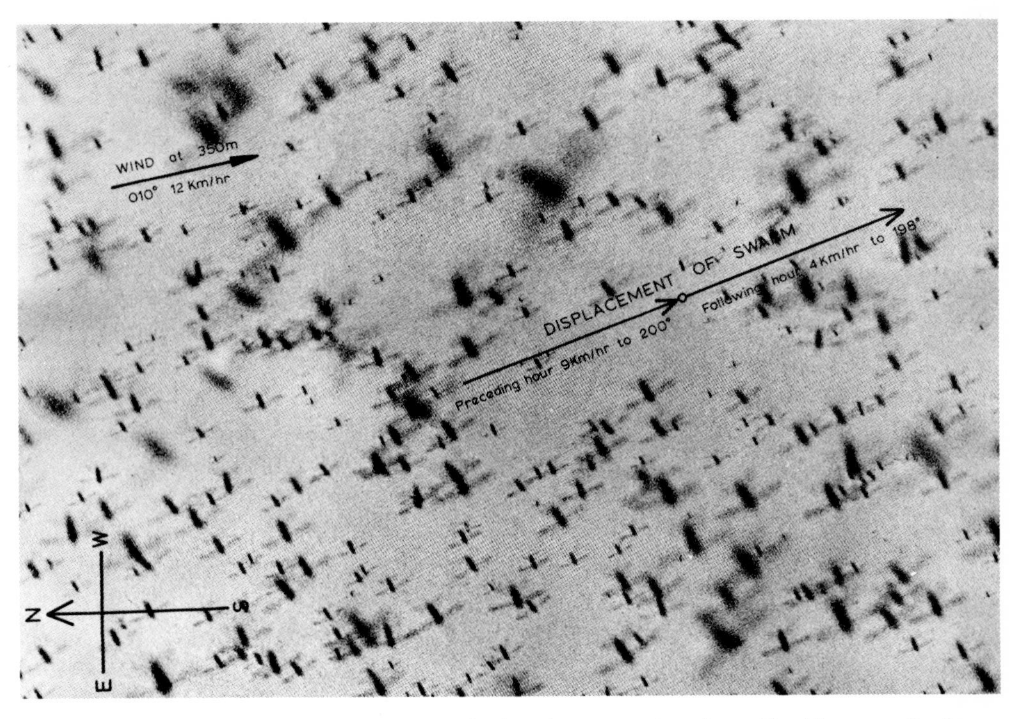

Fig. 2. Group of uniformly orientated flying locusts viewed vertically upwards from the ground near Mtito Andei in Kenya at 1625 on 10 February 1955 during the passage of a swarm of which the track is shown in Figure 3. The topmost locusts were recorded by aircraft observation at 720 m above the ground, and the corresponding wind is shown at a representative height of 350 m (Rainey 1963, 1976).

Figure 5 (Rainey, 1976) shows the movements of a single swarm in changing winds, with morning westerlies and afternoon northeasterlies. Lastly, Figure 6 (Rainey, 1963) shows the track of a swarm which encountered a complete reversal of wind direction, and in consequence moved in a complete loop - not exactly purposeful navigation! Figure 7 (Rainey, 1963) summarizes the results of the 42 fully documented occasions of such aircraft observations of swarm tracks with instrumental wind observations at the appropriate time, place, and height. The top of this diagram is not north, but is the direction from which the wind is coming in each case, and the circles indicate the ground speed of each swarm as a percentage of the speed of the wind in which it flies. These swarms were in fact all traveling directly downwind, within the established limits of accuracy of the observations.

Suggested Evolutionary Origin of Flight

Downwind movement necessarily means movements on balance toward and with zones of wind convergence, which are areas across whose perimeter at any one time there is a net excess of inflowing air. Wind convergence (at relatively low levels, i.e., within the lowest few kilometres of the atmosphere) necessarily means ascent of air, causing cooling by expansion, condensation, and rain if the process continues long enough. Wind convergence is indeed *essential for the production of rain* ; hence the immense survival value of downwind movement to an insect like the Desert Locust (Rainey, 1951), inhabiting as it does arid regions but with an egg stage needing free soil-water for its successful development. This behavior thus enables the locusts to use the kinetic energy of the atmospheric circulation to locate and exploit the temporary but very extensive ephemeral vegetation which follows rains in these arid regions (Rainey, 1977) in a manner which may perhaps illustrate the evolutionary origin of flight itself (Wigglesworth, 1963). Thus the first accidentally airborne insects

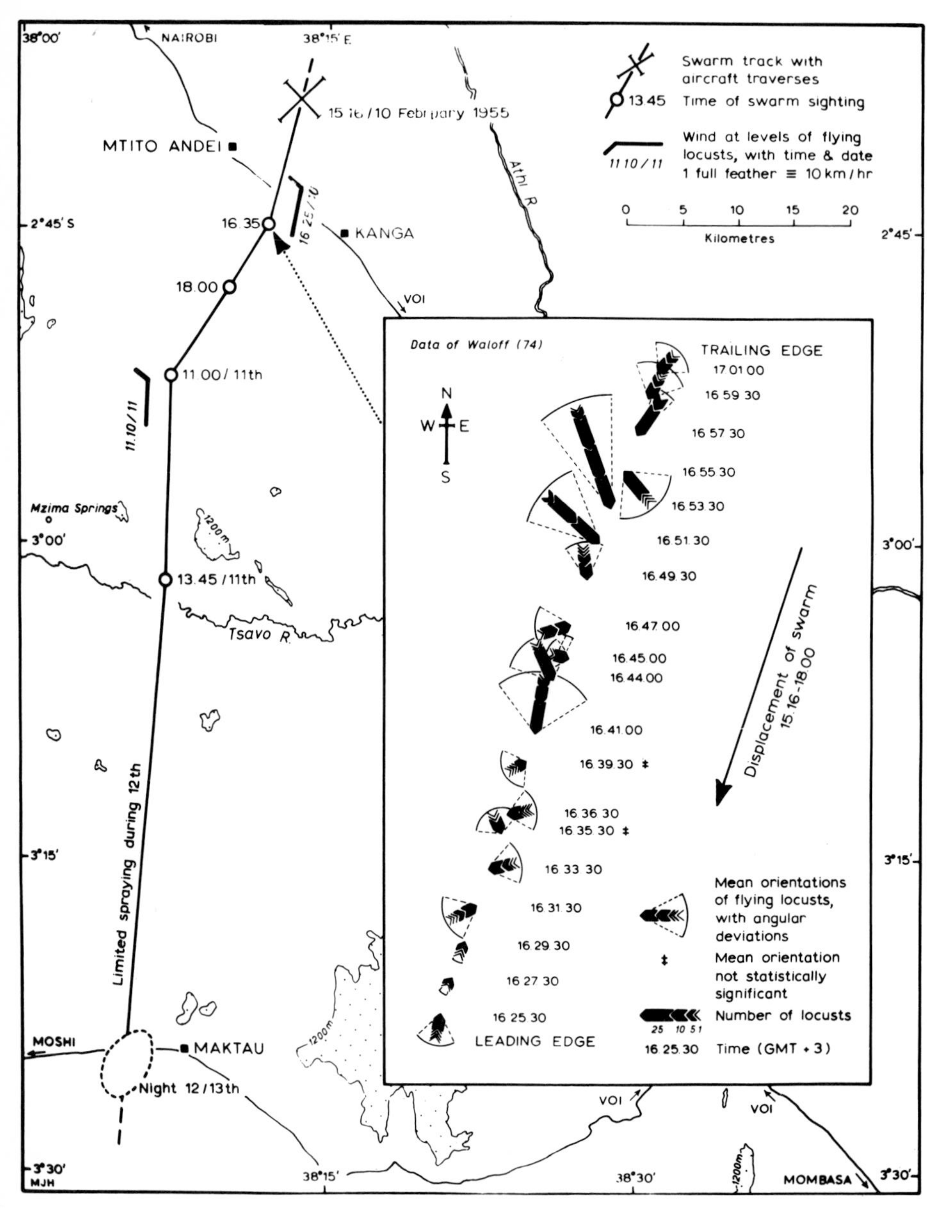

Fig. 3. Swarm track and orientations of flying locusts: orientations (data of Waloff, 1972) recorded by series of photographs similar to Figure 2 taken at an adjacent fixed point at about two-minute intervals throughout the passage of the swarm.

may be envisaged as carried up in powerful thermal up-currents like dust devils, in the kind of semidesert environment in which downwind displacement (toward zones of wind convergence with their rain) could have had an immediate survival value (Rainey, 1965). It has recently been suggested that an early aquatic environment may have been involved in evolution along these general lines (Wigglesworth, 1976).

Patterns of Intercontinental Locust Migration

Figure 8 (Rainey, 1951) presents an early specific example, showing the accumulation of swarms in the Inter-Tropical Convergence Zone (ITCZ), a feature of the global atmospheric circulation within which winds (tradewinds and monsoons) originating from the northern hemisphere meet others originating from the southern hemisphere. Utilizing these general findings in interpreting the extensive but inevitably incomplete reports of Desert Locusts available from all sources for one particular year, with the help of correspondingly comprehensive analyses of relevant current meteorological data, it has been possible to deduce for that year the source areas, routes, times, and termination areas of the main population movements involved, for example (Fig. 9) in connecting the two contrasted seasonal breeding areas shown in generalized form in Figure 1. Incidentally in October 1954 at least nine Desert Locusts were taken in the British Isles (the first such record since 1869), but again (in 1954) moving down-

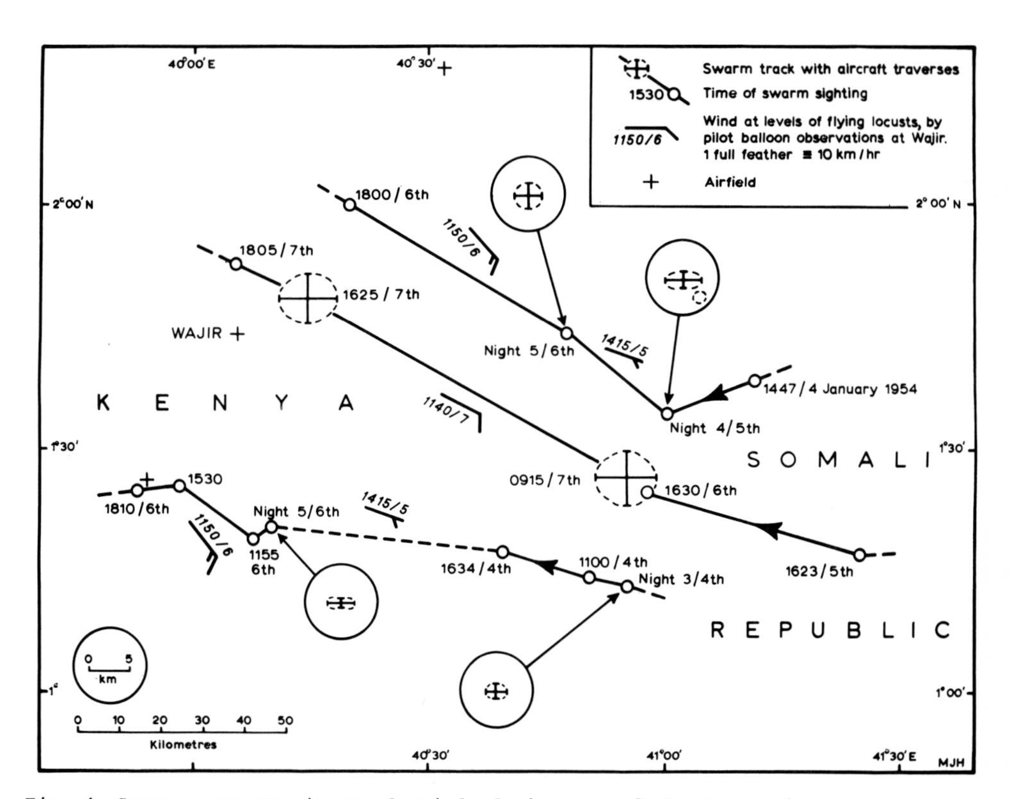

Fig. 4. Swarm movements in steady winds during one of the last major locust invasions of Kenya.

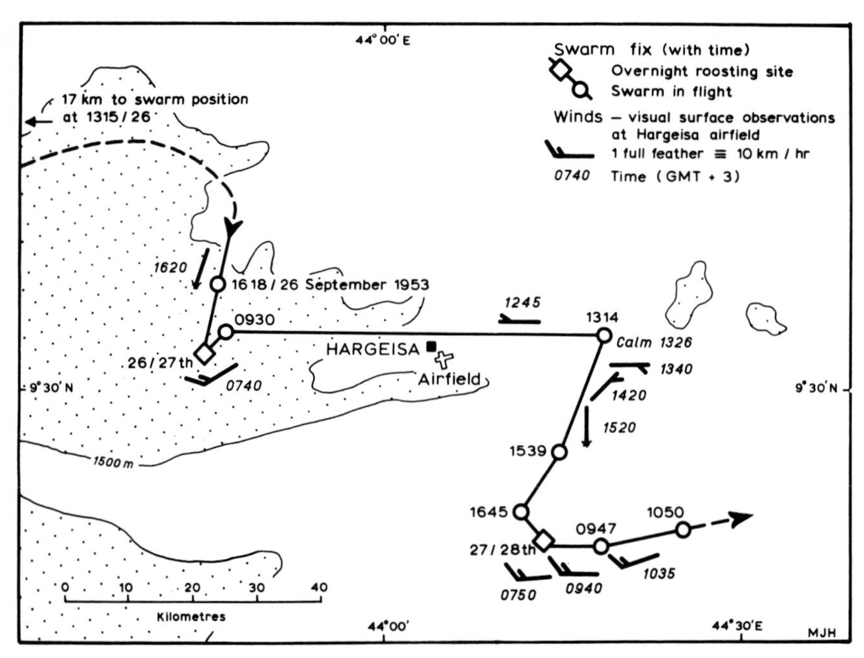

Fig. 5. Movements of a swarm with alternating winds in the Inter-Tropical Convergence Zone over the Somali Republic.

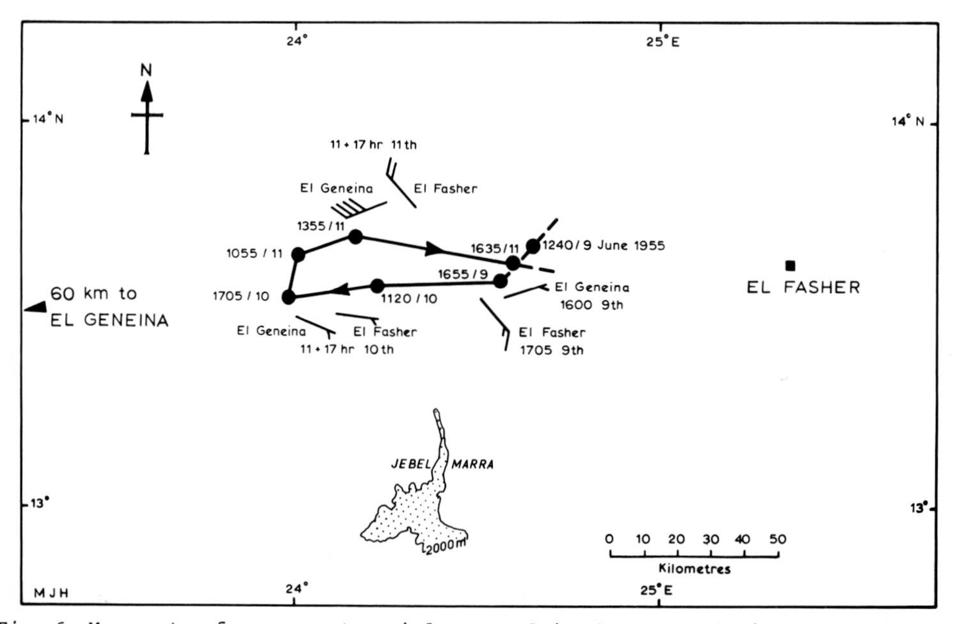

Fig. 6. Movements of a swarm at a wind reversal in the Inter-Tropical Convergence Zone over the western Sudan. Winds by pilot balloon ascents at El Fasher and El Geneina; one full feather represents 10 km/hr.

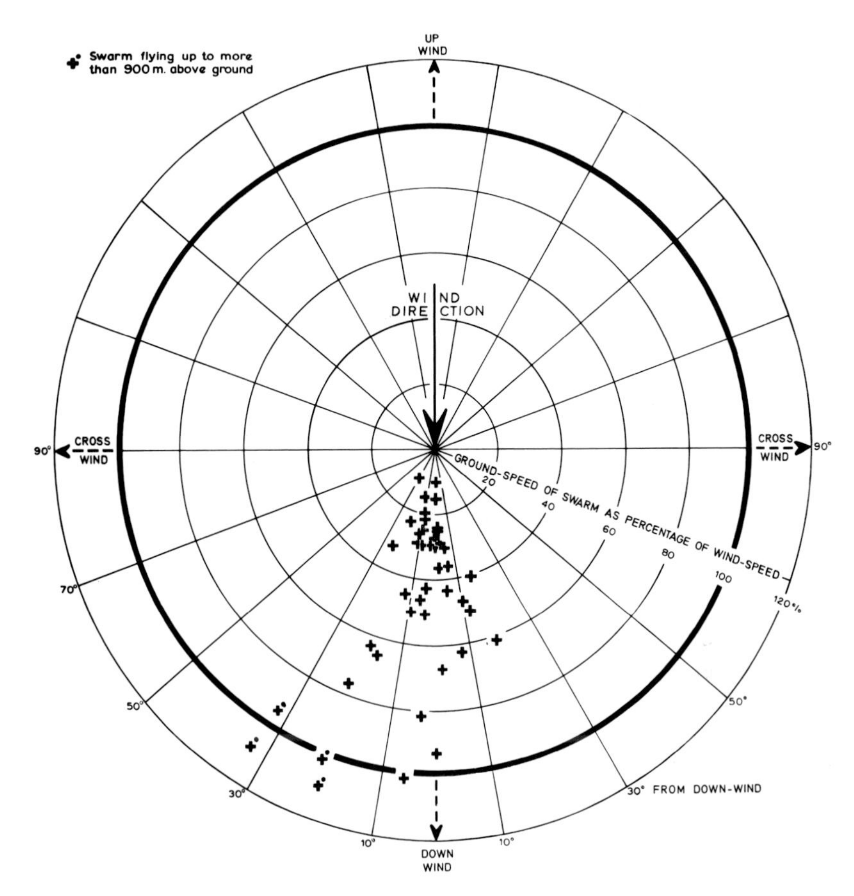

Fig. 7. Direction and speed of displacement of individual swarms in relation to wind, from aircraft observations of swarm tracks: Kenya, Tanzania, Somali Republic 1951-57.

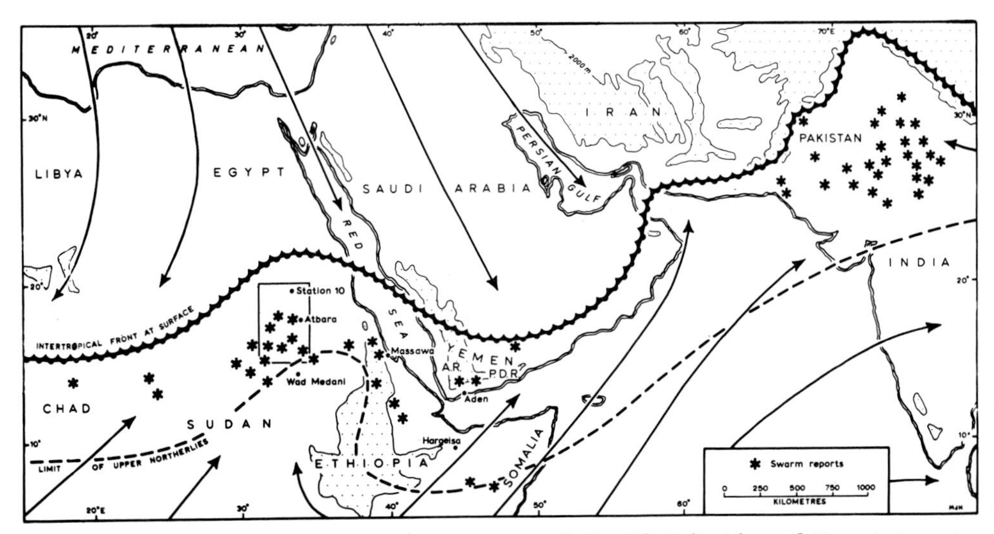

Fig. 8. The Inter-Tropical Convergence Zone and the distribution of Desert Locust swarms with the ITCZ temporarily near stationary: swarm reports and winds 12-31 July 1950. Small rectangle shows area of Sudan covered by Figure 10.

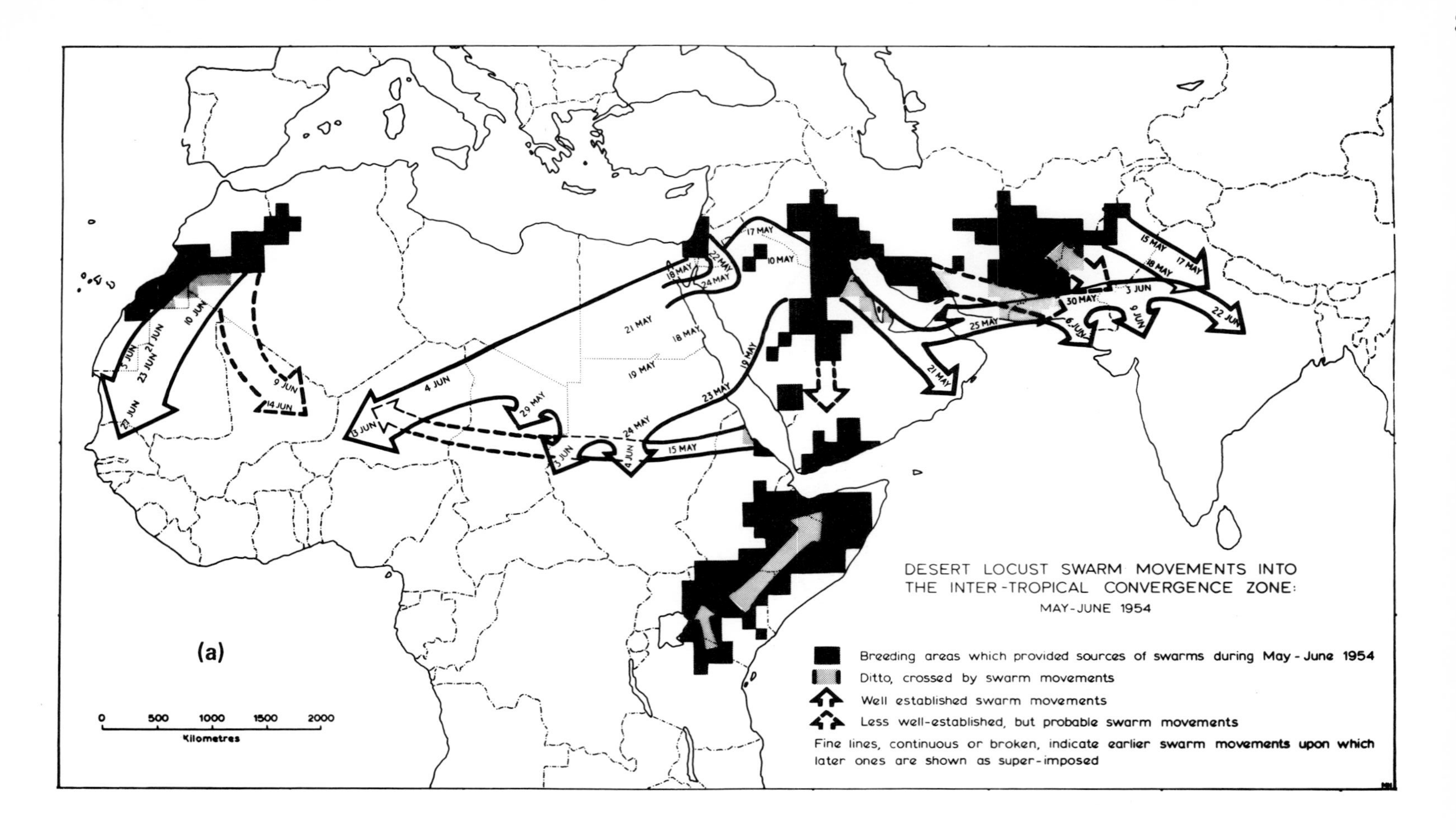
DESERT LOCUST SWARM MOVEMENTS INTO THE INTER-TROPICAL CONVERGENCE ZONE:
MAY-JUNE 1954
Breeding areas which provided sources of swarms during May - June 1954
Ditto, crossed by swarm movements
Well established swarm movements
Less well-established, but probable swarm movements
Fine lines, continuous or broken, indicate earlier swarm movements upon which later ones are shown as super-imposed
(a)
0
500
1000
1500
2000
Kilometres
15 MAY
17 MAY
18 MAY
3 JUN
30 MAY
9 JUN
22 JUN
6 JUN
25 MAY
21 MAY
17 MAY
10 MAY
22 MAY
24 MAY
18 MAY
21 MAY
18 MAY
19 MAY
19 MAY
23 MAY
24 MAY
15 MAY
4 JUN
3 JUN
29 MAY
4 JUN
13 JUN
10 JUN
21 JUN
23 JUN
9 JUN
14 JUN

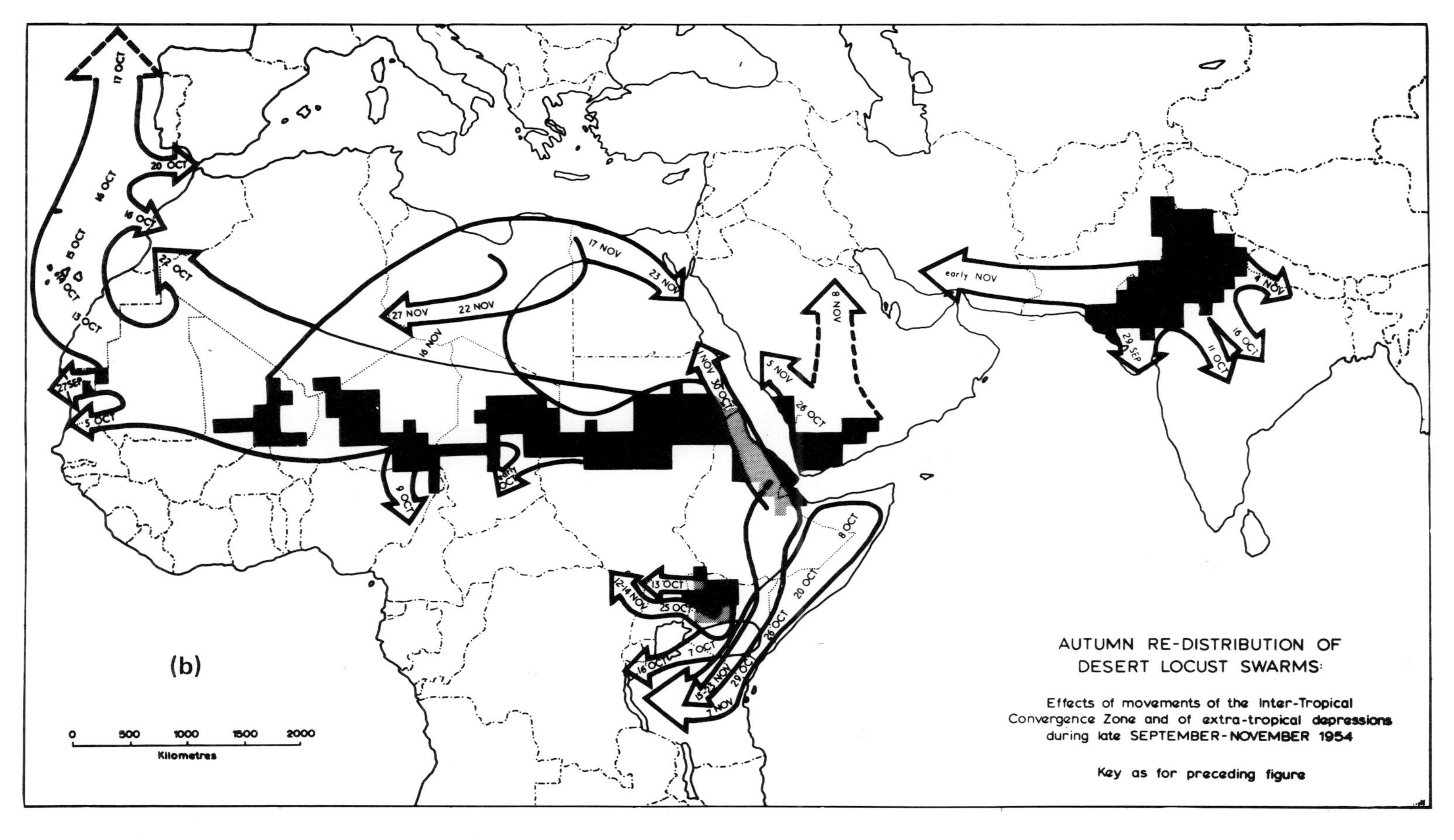

Fig. 9. Major Desert Locust migrations 1954: deductions from a uniquely comprehensive analysis of locust records and corresponding meteorological data for a year of widespread infestations. As in most years, there were two main periods of extensive migration, as shown, with the intervening months characterized by extended periods of effectively static locust distribution, attributable to effects of wind systems and of limiting temperatures (Rainey, 1963, 1969).

wind toward a zone of active convergence (which gave the worst rainstorm in living memory in southern Scotland!) - an encouraging example of an apparent exception substantiating a rule (Rainey, 1963).

Exploring Wind Currents for Concentrations of Other Airborne Insects

Figure 10 shows the first traverse of the ITCZ by an aircraft specially equipped for precision wind finding and for insect sampling, and demonstrates the location within the Zone of the sharply defined air-mass boundary termed the Inter-Tropical Front (ITF). We have been immensely indebted to Professor R.J.V. Joyce and to Ciba-Pilatus Ltd for opportunities of participating in a most rewarding series of expeditions with this aircraft. We were indeed working with it on the migrations of *Simulium* in West Africa, for the World Health Organization's Onchocerciasis Control Programme, over the period of the Washington Congress (at which this paper had accordingly to be presented *in absentia*). These expeditions have provided much new and unexpected information on atmospheric features as well as on their airborne fauna (e.g., Rainey and Joyce, 1972; Rainey, 1976). Thus, for example, after the ITF had been located as in Figure 10, it was found possible to fly box patterns to provide more detailed information on the winds in the vicinity of the front (Rainey, 1974), and in particular to provide quantitative evidence on the intensity of the wind covergence and in turn on the rate of concentration which is imposed by this convergence on airborne insects constrained (e.g., by air temperature) against unlimited ascent. Much of this concentrating effect is localized close to the front, as first shown by hourly catches using suction traps on a 15-m tower during the passage of the ITF (Bowden and Gibbs, 1973) - and shown still more strikingly by radar, as in Figure 12. It was noted that the largest of the suction-trap catches was taken in the top trap, at 15 m, consistent with attributing this high density to a process operating well above ground and crop levels.

Figure 11 shows the ITCZ during a major night flight of Sudan Plague Grasshoppers - a flight of which the scale and especially the height would have entirely escaped notice without radar. From his radar observations, Professor Schaefer inferred immediately that "medium-sized grasshoppers" were flying up to 1200 m, and at a density at 450 m which he estimated would give a catch of five grasshoppers during a 20-minute trapping run with the aircraft net. Aircraft trapping was immediately undertaken to test these inferences; four medium-sized grasshoppers (3 *Aiolopus simulatrix* and 1 *Catantops axillaris*) were in fact taken during a 20-minute trapping run at 450 m, and one further *Aiolopus* during a subsequent run at 1200 m.

Figure 12 is one of Professor Schaefer's radar photographs showing an insect concentration (on this occasion probably mainly moths) providing a line-echo at the Inter-Tropical Front. The plan-position-indicator photo was taken at 3° elevation of the radar beam, giving a map form of presentation with the range rings at a spacing of 450 m and with a triple exposure to record the individual insects' echoes three times. Disregarding the ground clutter in the center of the photograph and the vertical lines which are radar echoes from the edges of sorghum rotation crops in the standard rectangular 42-hectare irrigated fields of the Sudan Gezira, the ITF is represented by the line-echo which is orientated SW/NE; it was moving to the northwest and had passed overhead within a minute of the abrupt corresponding wind shift, to SE, recorded by a sensitive anemometer. Upper wind observations by pilot balloon, which we were able to make very shortly

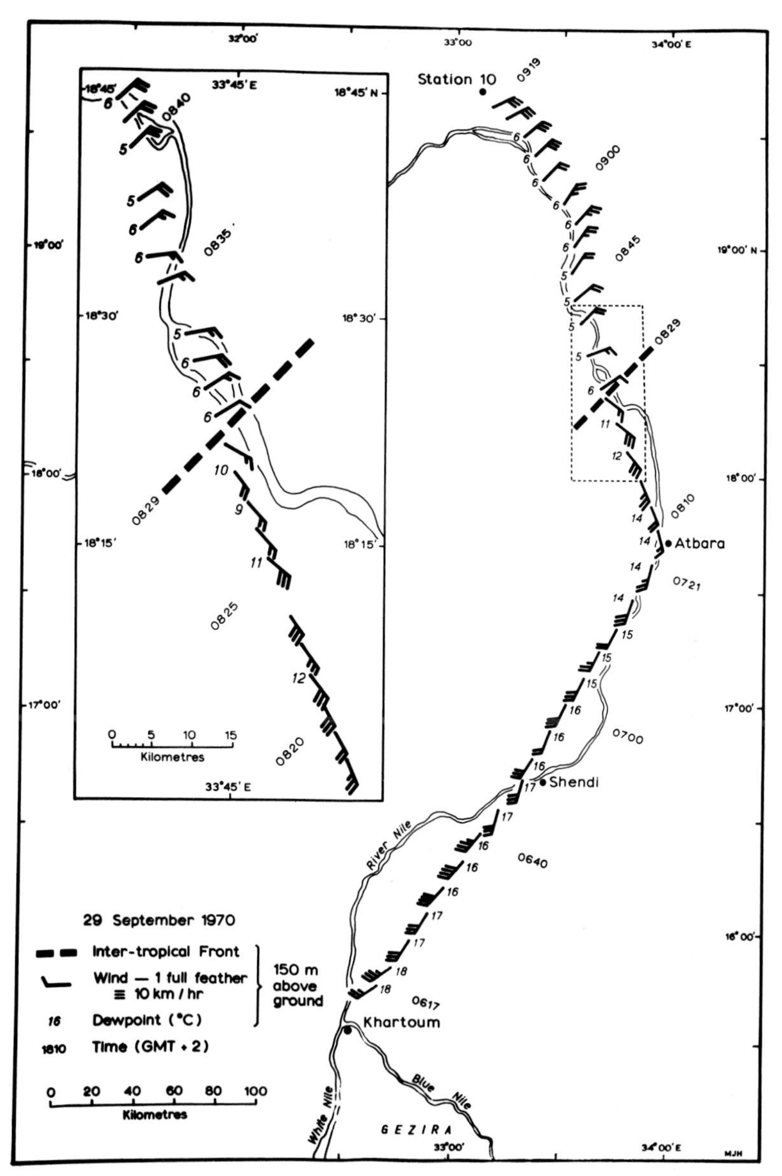

Fig. 10. Wind-finding aircraft traverse through the Inter-Tropical Convergence Zone, Northern Sudan - area outlined in Figure 8: Pilatus Turbo-Porter research aircraft with Decca Doppler navigation and wind-finding system. Note the Inter-Tropical Front with its characteristically abrupt wind shift and contrast in humidity between dry NE tradewinds and moist southerly monsoon: 29 September 1970 (Rainey, 1973).

after the passage of the front, showed overriding northerly winds still present above the undercutting southeasterlies, in the same way as shown in Figure 11 for the occasion of our first grasshopper catches.

During 1973-76 we have also been greatly privileged to participate in a research program on the flight of spruce budworm moths (*Choristoneura fumiferana* - Tortricidae) undertaken by the Canadian Forestry Service

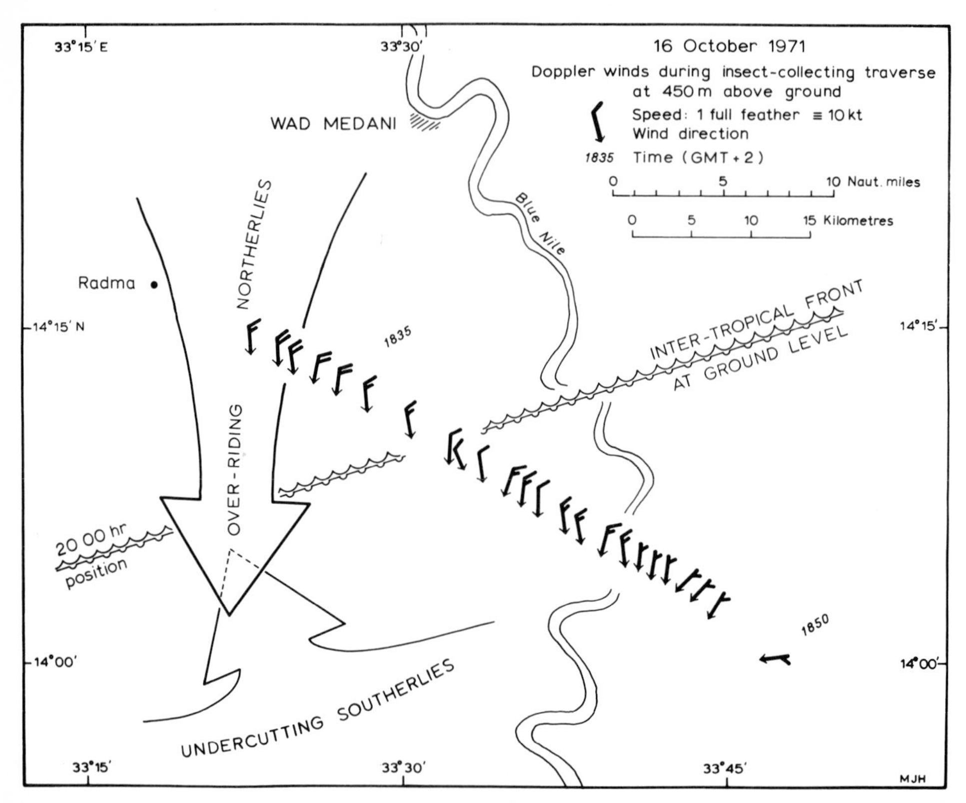

Fig. 11. Inter-Tropical Convergence Zone during major flight of grasshoppers: 16 October 1971. The high-flying grasshoppers were initially observed and recognized by ground-based radar (Schaefer, 1976) and confirmed by aircraft collections at 450 and 1200 m above the ground, within the overriding northerly winds and above the undercutting southerlies as the two wind currents converged at the sloping Inter-Tropical Front (Rainey, 1974a).

and the New Brunswick Department of Natural Resources, making similarly coordinated use of radar, specially-instrumented aircraft, and ground observations.

Figure 13 (Rainey, 1975) shows how, with a DC-3 aircraft using Bendix Doppler navigation and wind-finding equipment, it was possible to locate and explore a Canadian front, not revealed by routine synoptic weather analysis, but marked by a very sharply defined wind shift, with the main transition zone (on the traverse shown enlarged in the inset) no more than 300 m wide. This was found to be associated with wind convergence of an intensity implying, on the simplest of assumptions, concentration at rates giving more than ten-fold hourly increases in the area densities of flying moths. At Chipman a radar line-echo, representing in all probability another part of the same feature, aligned NNE/SSW and visible on the ground radar for a length of 30 km, was recognized independently by Professor Schaefer (to whom I am indebted for the use of these unpublished observations) and confirmed as due to spruce budworm moths at high density by aircraft trapping by a Cessna from Chipman at 90-300 m from 2033 to 0025 the same night. This same concentration of moths, envisaged as having settled later in the night not far from the line of the front as found from the DC-3 and seen on the radar, may also be suggested as perhaps having accounted for the specially high moth kills (up to 60 moths/m^2) which were recorded in the particular block indicated by the rectangle

SW of Bathurst, following large-scale aircraft spraying operations against settled moths next day. Figure 13 also summarizes the detailed evidence available on the source areas of the moths in the airborne concentration, provided by the routine CFS pupal survey counts at their thousand sampling points in New Brunswick (weighted by the area of susceptible forest within each 200 km^2 unit), and by the emergence data of the current season for each of the seven phenological zones covering the province.

Another occasion during the same season, on which our program coordinator Dr. Greenbank, Professor Schaefer, and I have recently been making a detailed study, shows particularly clearly just how incomplete and misleading field studies of such insect flight would have been without the use of radar and aircraft. It was near Renous, in the hills of central New Brunswick, toward midnight on a cold evening, with no moth flight at all to be seen from the ground and at a temperature at ground level of 10°C (some 5° below the threshold for flight activity), at which no moth flight would have been envisaged. But the Renous radar showed an extremely dense layer of moths in flight at approximately 100 m above the ground; aircraft trapping not only confirmed them as spruce budworm moths but gave densities again consistent with the radar estimates; and the wind-finding DC-3 again recorded a front at an appropriate time and place to account for this concentration.

There is increasing evidence that insects in flight at initially low densities can at times be concentrated by wind convergence to within

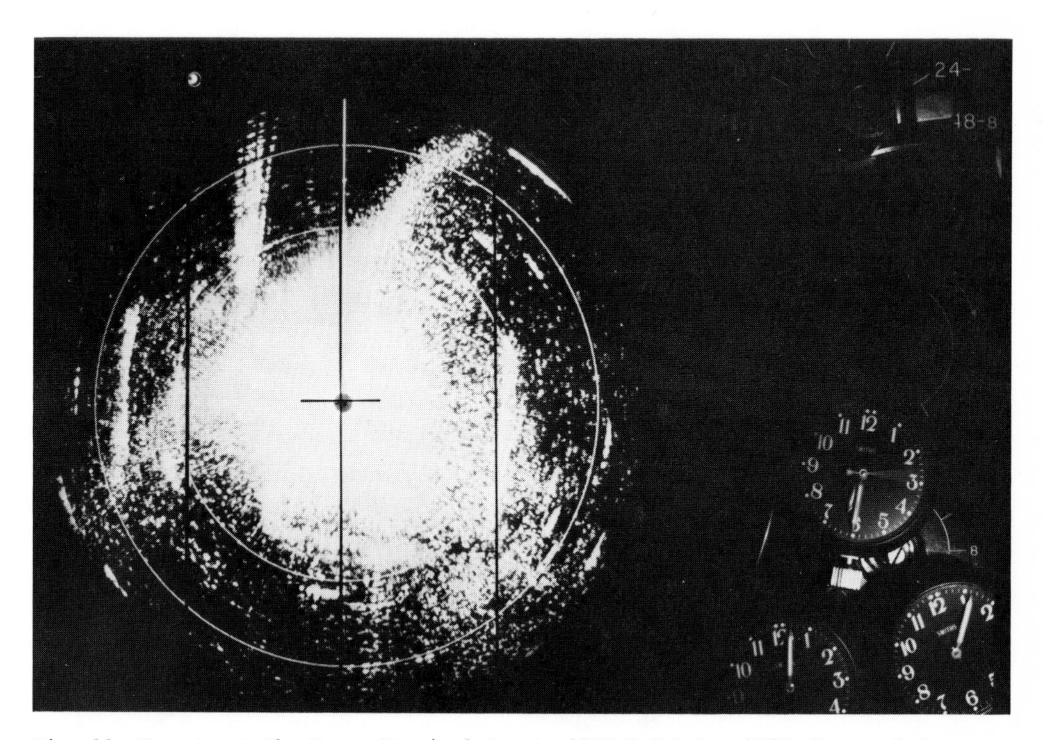

Fig. 12. Insects at the Inter-Tropical Front: 1830 9 October 1973; Kumor, Sudan (Schaefer, 1976). Many moths, off Gezira crops, at high density along front which had passed overhead from SE two minutes previously; to be distinguished from permanent echoes, e.g., from nearer edges of sorghum fields aligned N/S.

range of mutual perception, and it has been suggested that, for some insects, even more biologically important than long-range transport may be such concentrating effects of convergent wind systems in facilitating the meeting of the sexes (Rainey, 1976).

Some Strategies of Migration in North America

Finally, Figure 14 (Rainey, 1973) shows the geographical displacement achieved by Urquhart's classical Monarch (marked and released near Toronto in September 1957 and recaptured 3000 km away in Mexico four months later), together with two contrasted strategies of migration by other North American organisms. At one extreme, evidence of the astonishing navigational success of the migration of the minute hummingbird is provided by the size of the egg clutch of this species, which is only two eggs. At the other extreme, airborne organisms even as completely passive as fungus spores can nevertheless undertake regular transcontinental seasonal migrations [the plant pathologists' phrase (Gregory, 1961), not mine] if they are produced in sufficiently immense numbers; the survival value of these fungus migrations arises because these wheat rusts are unable to survive either the winter in the north or the summer in the south. Thus long-range seasonal displacements of small windborne insects also may perhaps have been an early evolutionary development. On the other hand, the perma-

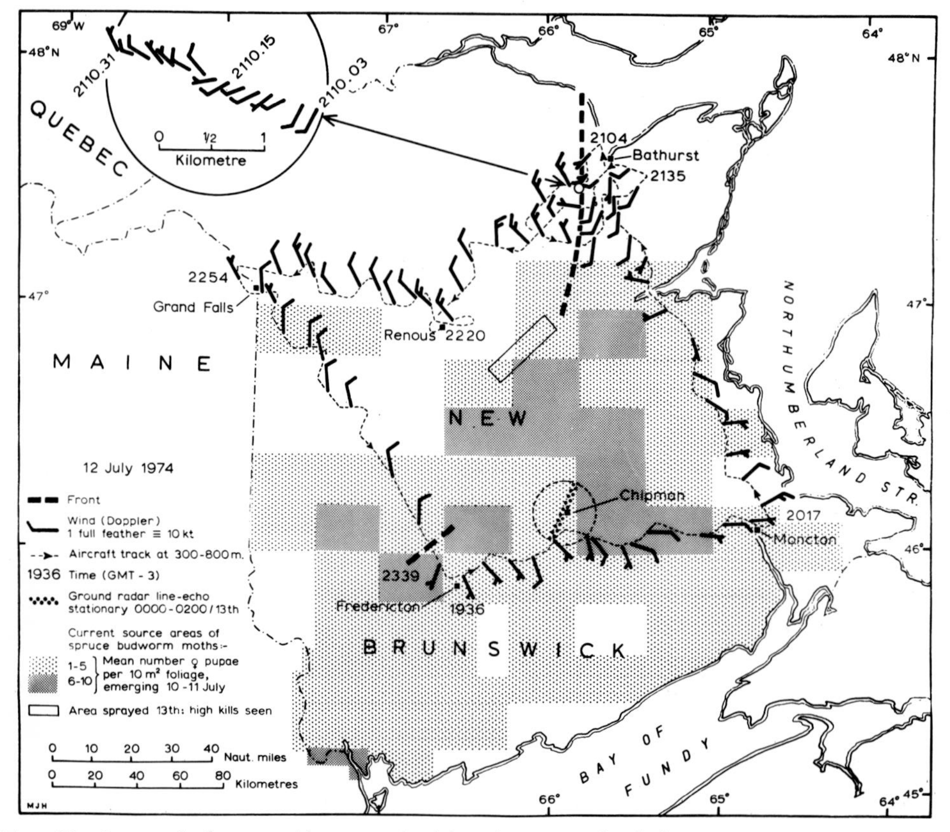

Fig. 13. Spruce budworm moth concentration in zone of wind convergence at a Canadian front. Integrated findings from specially instrumented aircraft, ground-based radar, province-wide field survey, and assessment of large-scale spray trials.

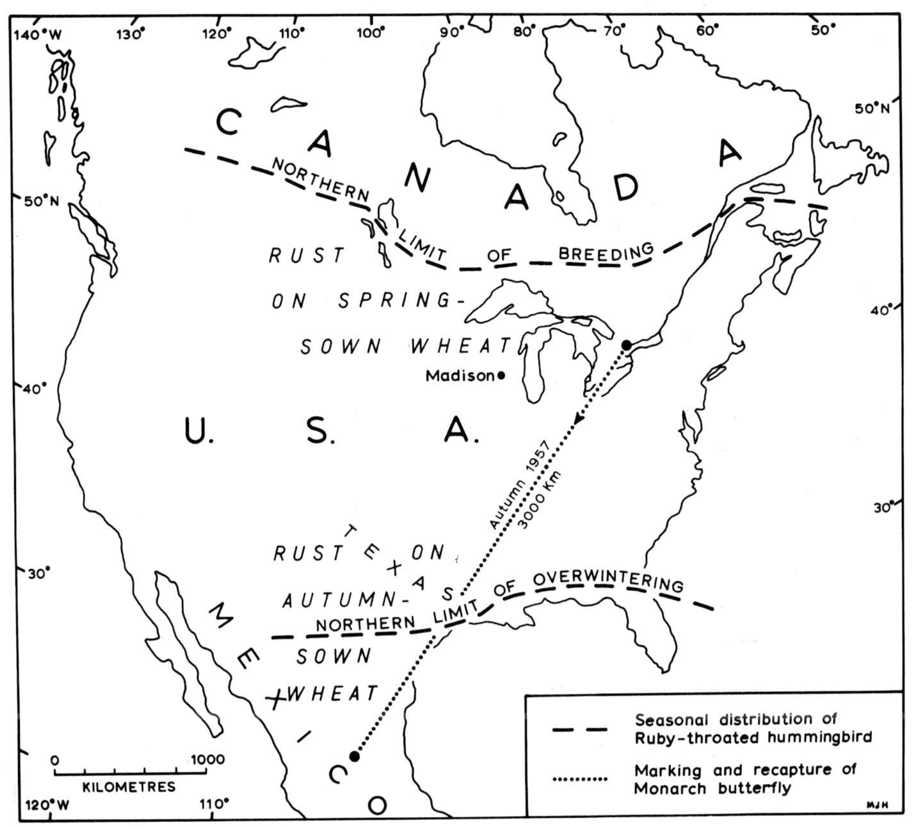

Fig. 14. Contrasted strategies of migration. Mark-and-recapture record for Monarch butterfly (*Danaus plexippus* L.; Urquhart, 1960), with seasonal population displacements of the Ruby-throated Hummingbird (*Archilochus colubris* L.; Godfrey, 1966) and of the wheat-rust fungi *Puccinia graminis* and *P. triticina*; spores of *P. graminis* from a source in Texas have been recorded as being deposited in Madison, some 1000 km from their source, at a rate of the order of 1000 spores per m^2 per 24 hours.

nent occupation of localized ecological niches by individual flying insects, may require accurately "keeping station" on landmarks by predominantly upwind orientation, and may often involve hovering flight which is aerodynamically more exacting than fast forward flight (Weis-Fogh, 1976). This appears to represent more specialized behavior and may accordingly be a later evolutionary development than long-range migration (Rainey, 1976).

Concluding Suggestion from Oceanography

In conclusion, it is suggested that the physical, atmospheric environment of airborne insects requires of serious insect ecologists study as thorough as that which marine biologists give as a matter of course to the physical environment of the ocean.

Acknowledgments The writer is greatly indebted both to Frank Walsh, of the WHO Onchocerciasis Control Programme, for undertaking at short notice to present this paper at the Congress, and to Margaret Haggis, of the Centre for Overseas Pest Research, for helping to complete the text in Bobo Dioulasso in time for the Congress, as well as for her many contributions to the results of this fieldwork over the past seven years. The writer is also indebted to the World Meteorological Organization for the use of Figures 6 and 7; to the Royal Entomological Society and to Blackwell Scientific Publications for Figures 2, 3, 4, and 5 and, with Prof. G.W. Schaefer, for Figure 12; to the Royal Meteorological Society for Figures 1, 8, 9, 10, and 14; to the British Crop Protection Council for Figure 11; and to the International Agricultural Aviation Centre for Figure 13.

References

Bowden, J., Gibbs, D.G.: Light-trap and suction-trap catches of insects in the northern Gezira, Sudan, in the season of southward movement of the Inter-Tropical Front. Bull. Ent. Res. *62*, 571-596 (1973).

Godfrey, W.E.: The birds of Canada. Bull. Natn. Mus. Can. (Biol.) *73* (1966), 428 pp.

Gregory, P.H.: The Microbiology of the Atmosphere. Plant Science Monographs: Leonard Hill London/Interscience New York (1961).

Rainey, R.C.: Weather and the movements of locust swarms: a new hypothesis. Nature, Lond. *168*, 1057-1060 (1951).

Rainey, R.C.: Meterology and the migration of Desert Locusts: applications of synoptic meteorology in locust control. Tech. Notes World Met. Org. *54* (1963); 115 pp. (also as Anti-Locust Mem. *7*).

Rainey, R.C.: The origin of insect flight: some implications of recent findings from palaeoclimatology and locust migration. 12th Int. Congr. Ent., London, 134 (1965).

Rainey, R.C.: Effects of atmospheric conditions on insect movements. Quart. J. Roy. Met. Soc. *95*, 424-434 (1969).

Rainey, R.C.: Airborne pests and the atmospheric environment. Weather, Lond. *28*, 224-239 (1973).

Rainey, R.C.: Biometeorology and insect flight: some aspects of energy exchange. Ann. Rev. Ent. *19*, 407-439 (1974).

Rainey, R.C.: Flying insects as ultra-low-volume spray targets. Br. Crop Prot. Counc. Monogr. *11*, 20-28 (1974a).

Rainey, R.C.: New prospects for the use of aircraft in the control of flying insects and in the development of semi-arid regions. 5th Int. Agric. Aviat. Congr., Kenilworth, 229-233 (1975).

Rainey, R.C.: Flight behaviour and features of the atmospheric environment. Symp. R. Ent. Soc. Lond. *7*, 75-112 and 272-273 (1976).

Rainey, R.C.: Rainfall: scarce resource in 'opportunity country.' Phil. Trans. R. Soc. (B) *278*, 439-455 (1977).

Rainey, R.C., Joyce, R.J.V.: The use of airborne Doppler equipment in monitoring wind-fields for airborne insects. 7th Int. Aerospace Instrum. Symp. Cranfield 8.1-8.4 (1972).

Runcorn, S.K.: Climatic change through geological time in the light of the palaeomagnetic evidence for polar wandering and continental drift. Quart. J. Roy. Met. Soc. *87*, 282-313 (1961).

Schaefer, G.W.S.: Radar observations of insect flight. Symp. R. Ent. Soc. Lond. *7*, 157-197 (1976).

Southwood, T.R.E.: Migration of terrestrial arthropods in relation to habitat. Biol. Rev. *37*, 171-214 (1962).

Urquhart, F.A.: The Monarch butterfly. University of Toronto Press (1960).

Waloff, Z.: Orientation of flying locusts (*Schistocerca gregaria* Forsk.) in migrating swarms. Bull. Ent. Res. *62*, 1-72 (1972).

Weis-Fogh, T.: Energetics and aerodynamics of flapping flight: a synthesis. Symp. R. Ent. Soc. Lond. *7*, 48-72 (1976).

Wigglesworth, V.B.: The origin of flight in insects. Proc. R. Ent. Soc. Lond. (C) *28*, 23-32 (1963).

Wigglesworth, V.B.: The evolution of insect flight. Symp. R. Ent. Soc. Lond. *7*, 255-273 (1976).

2 Diapause, Development, and Phenology

Diapause has two broad roles to play in insect life histories. It permits escape in time, but it also interacts with development rates to synchronize life cycles, determine patterns of voltinism, and generally regulate seasonal phenologies. The papers in this section discuss both these roles. They also examine the genetic and environmental variance within and between species that influences the depth and duration of diapause and the stimuli which induce and maintain it.

The Taubers and S. Masaki use comparative studies of closely related species to analyze adaptations controlling the seasonal timing of life cycles and the evolution of phenological strategies as they involve diapause. The Taubers' work on chrysopids shows that although timing may be similar in different species, the underlying mechanisms may be quite different. On the other hand, organisms as unrelated as insects and birds can show strong similarities in seasonal timing and responses to stimuli. Phenological diversification in some chrysopids has resulted in sufficient genetic divergence to produce speciation. Similar phenological mechanisms may have resulted in speciation in some of the crickets studied by Masaki.

Masaki makes use of the unique features of the Japanese islands, extending from cold temperate to subtropical regions, to study latitudinal influences on size and voltinism patterns in crickets. Genetic variability has permitted adaptation to a range of environmental circumstances with resultant clinal variation in diapause, development, and voltinism. Interaction between length of growing season and voltinism patterns leads to interesting latitudinal discontinuities in body size. Considerable genetic flexibility is suggested by the rapid loss of diapause in a cricket recently introduced from temperate Japan to the subtropical Bonin Islands.

In the third paper in this section, M. A. Hoy reviews work on both inter- and intrapopulation variation in diapause and associated traits emphasizing her own research on the gypsy moth, *Lymantria dispar*, and its parasitoid *Apanteles melanoscelus*. She discusses the rapid responses to artificial selection, often inadvertent, in many species and their potential importance in applied entomology. Finally she discusses the techniques available for the genetic analysis of diapause and some of the problems inherent in "genetic dissection" because of the diverse components of the diapause response.

The final three papers in this section concentrate on diapause and development in specific insects. G. P. Waldbauer considers discontinuous variation in the spring emergence times of saturniid moths and a clinal gradient in this polymorphism. After analyzing the genetic background, he considers the adaptive significance of the variation in terms of a "bet-hedging" phenological strategy. J. Lumme reviews information concerning the photoperiodic induction of diapause in various species of *Drosophila* inhabiting northern and subartic areas. As in other groups, there is much clinal variation in critical photoperiod. Lumme then uses classic crossing and backcrossing techniques with different geographical races of *D. littoralis* to demonstrate that in this species a single mendelian autosomal unit exerts the major genetic influence on the photoperiodic reaction. The longer critical daylength was in general dominant over the shorter although dominance was incomplete. Because *Drosophila* has long been a favorite of geneticists, it should provide excellent material to study the genetic basis of phenological variation.

In his paper C. A. Istock considers some global questions concerning the evolution of fitness. He considers first Fisher's fundamental theorem of natural selection and suggests that the requirement of directional selection may pose severe problems in the use of the theorem as a general predictor of the microevolutionary process. Rather variation in selection may result in genetic variance for fitness. Istock examines development time and diapause tendency in the pitcher plant mosquito *Wyeomyia smithii* as two characters which have a profound bearing on fitness. The developmental characters have a "quasi-stable" genetic variance and influence the evolution of adaptation through the control they exert over the number and timing of generations each year. Sexual reproduction is important in maintaining the mixed life-history strategy.

The importance of genetic and environmental variance in the timing of phenologies through diapause and development is apparent in all these papers. It is equally obvious that their roles have been barely adumbrated. The evolution of phenological strategies should therefore be an important and exciting subject for future research.

Evolution of Phenological Strategies in Insects: a Comparative Approach with Eco-Physiological and Genetic Considerations

MAURICE J. TAUBER AND CATHERINE A. TAUBER

The variability that characterizes natural environments confronts organisms with numerous vital problems. To overcome these problems, organisms have evolved diverse ecological, behavioral, physiological, and genetic adaptations that comprise their adaptive strategies (Levins, 1968; see also Valentine, 1976). Probably the most important and all-encompassing environmental variable that organisms face is seasonality, and thus one of the most basic and most unifying components of each species' overall adaptive strategy is its *phenological strategy*, i.e., the timing of the periods of its activity and dormancy in relation to annually cycling environmental factors - both biotic and abiotic. Thus, knowledge of the evolution of seasonal adaptations is a fundamental and unifying component in our understanding of the evolution of overall adaptive strategies of both plants and animals. This knowledge is obtained through comparative studies of phenology and environmental physiology.

Among the Insecta, for example, the life cycle of each species encompasses four main phases: reproduction, growth and development, dormancy, and (to a greater or lesser degree) movement and migration. The seasonal timing of each of these phases is adapted, in part, to circumvent unfavorable physical factors and, in part, to synchronize activity with favorable physical conditions and seasonally available, biotic requisites (such as food, mates, oviposition sites, periods free from natural enemies). As such, phenological strategies govern not only interactions with the physical environment, but also interactions between individuals within a species and interactions between members of different species. Thus, phenological strategies play a dominant role in determining the seasonal activity and seasonal abundance of each species, and they are also a basic component underlying intra- and interspecific interactions at the population and community levels (see Lieth, 1974).

As a result, phenological investigations are not only of theoretical importance to evolutionary biologists and ecologists, but they also have a central place in applied ecology. For example, a crucial aspect of modern insect pest management is the modeling of the dominant interacting subsystems within agroecosystems (Haynes et al., 1973; Ruesink, 1976; Tauber and Tauber, 1976). These model systems are designed to predict crop yields, numerical changes in pest and beneficial species populations, and the outcome of interactions among the crop, pests, and beneficial species under actual and projected environmental conditions. The degree of synchrony in the seasonal timing of various phases in the life histories of key species is a major factor that profoundly influences the interactions of the various biotic components in agroecosystems. Consequently, reliable, predictive crop modeling requires comprehensive, quantitative analyses of the cause-effect relationship between relevant, seasonally variable, environmental factors and the reactions of the biotic elements -- host plants, pests, and beneficial species -- to these factors.

Among insects, there is great diversity in the adaptations that serve to synchronize the various phases in their life histories with appropriate seasons. Given this diversity, we will discuss two important topics. First, precisely what are the adaptations that control the seasonal timing of various phases in insect life cycles? Can these adaptations be categorized? Do adaptive seasonal patterns follow phylogenetic lines? Second, how do seasonal strategies evolve? Is their evolution subject only to relatively slow change that confers ever-increasing adaptation of the population or the species, or can they diverge quickly and thus presumably have a pivotal role in the diversification of phylads? These two topics are dealt with in sections II and III below, and our responses to the above questions are derived mainly from aspects of work in our laboratory.

To begin answering these questions, it is first necessary to make comparative studies, not only of insects' seasonal activity patterns, but also of the underlying eco-physiological responses that determine the overt expression of seasonality. Such investigations benefit greatly when they are based on natural field populations, because ecologically meaningful data are thus more likely to result. And, since insects' patterns of response to the various physical factors change drastically throughout the year (e.g., Danilevsky et al., 1970; Tauber and Tauber, 1973d), analyses which are conducted periodically throughout the entire annual cycle can reveal the full range of the insects' changing responses (Tauber and Tauber, 1973a).

Most comparative phenological studies of insects are concerned with intraspecific variation, and they have demonstrated clines of variability in phenological adaptations, e.g., latitudinal variation in critical photoperiod, diapause depth, and temperature responses (e.g., Masaki, 1961, 1965, 1967a, this volume; Danilevsky, 1965; Tauber and Tauber, 1972) and altitudinal variation in critical photoperiod (Bradshaw, 1976). Selection experiments have shown the polygenic basis underlying such inter- and intrapopulation variability (e.g., Dingle, 1974). Recent experimental evidence indicates that natural selection maintains the variability in these seasonal life-history traits in natural populations (Istock et al., 1976). In contrast, few studies have revealed patterns of interspecific divergence in phenological adaptations and their associated seasonal strategies. These studies have dealt primarily with crickets (e.g., Alexander and Bigelow, 1960; Masaki, 1967b, 1973, this volume; Alexander, 1968; Fontana and Hogan, 1969), mosquitoes (in Danilevsky, 1965), sawflies (Knerer and Atwood, 1973), and green lacewings (Tauber and Tauber, 1973; Tauber and Tauber, 1976b), and they have pointed to the role of phenological divergence in the diversification and speciation of populations (both allopatrically and sympatrically). At this point, a full understanding of both the evolution of phenological adaptations and its role in speciation (see Bush, 1975; Tauber and Tauber, 1977) is dependent on a more thorough knowledge of the genetic mechanisms involved. This is an important consideration in our work, and we discuss some of our recent findings in later sections.

I. Comparative Analyses and Classifications of Dormancy

I.A. Limitations of a Broad Comparative Approach

All evolutionary studies, whether they deal with morphology, physiology, behavior, etc., include categorization and classification as essential components. At present, there are several classifications of dormancy among insects (e.g., Müller, 1970; Mansingh, 1971; Thiele,

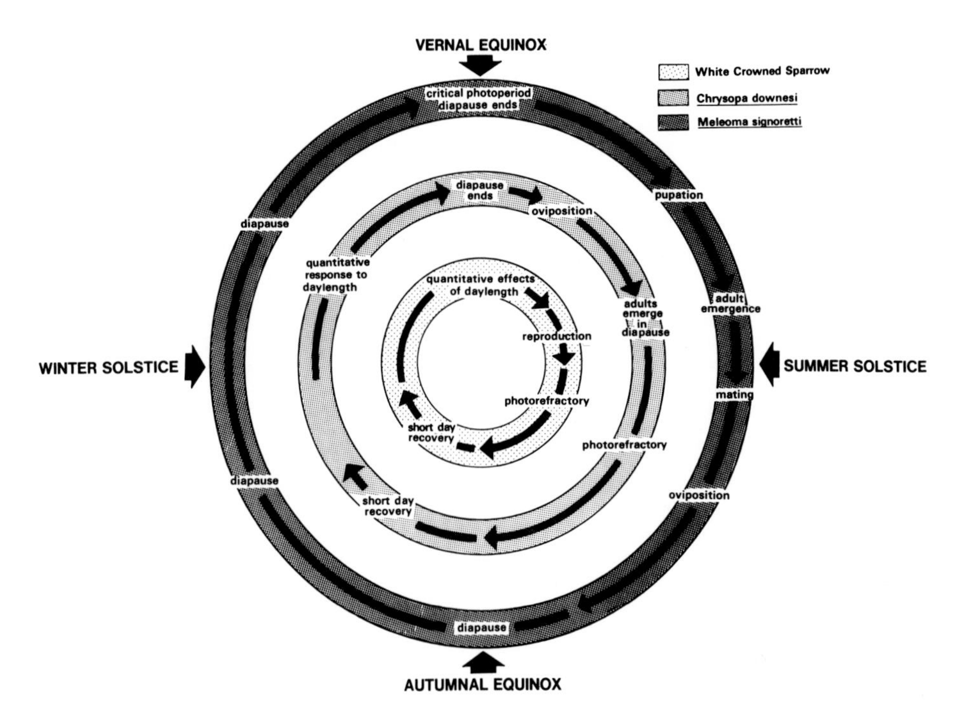

Fig. 1. The photoperiodic control of annual seasonal cycles in three species. Each circle represents the annual cycle of one species. Although diapause in the two univoltine insect species, *Chrysopa downesi* and *Meleoma signoretti* (compare two outside circles) overlaps during late summer, autumn, and winter and ends about the time of the vernal equinox, the underlying photoperiodic responses differ in the two species. The white-crowned sparrow (data from Farner and co-workers) and the insect *C. downesi* (compare two inner circles) show remarkable overlap in (1) the features of their reproductive cycles (e.g., photorefractory phase, short-day recovery, quantitative response to daylength), (2) the annual timing of these events, and (3) the photoperiodic control of these events.

1973). While these schemes contain useful concepts and are based on some excellent experimental work, they are not useful for an evolutionary analysis of phenological strategies in insects. They are artificial classifications that do not solve the problem of whether the various mechanisms involved in controlling dormancy are homologous or analogous phenomena. Our investigations show that similar seasonal activity patterns in different species can be controlled by very different stimulus-response patterns; i.e., there has been considerable convergent evolution in phenological adaptations and seasonal strategies among unrelated animals. Admittedly, this problem is not easily solved, and our following examples, based on several comparisons, will illustrate the difficulties.

Our first example presents a comparative study of two insect species, *Chrysopa downesi* Banks and *Meleoma signoretti* Fitch, both of which are green lacewings (Neuroptera: Chrysopidae). *C. downesi* has a life history pattern typical of an early spring breeder (Fig. 1). Unmated adults overwinter in a state of reproductive diapause. Oviposition occurs in early spring, and larvae are present from the beginning of May through the first part of June. Adults eclose at the end of June and immediately enter an aestival-autumnal-hibernal diapause. Thus, *C. downesi* has one generation per year.

Photoperiod is the primary environmental factor controlling *C. downesi*'s seasonal cycle as follows. Firstly, unless a sensitive stage (i.e., the third instar larva or the pupa) experiences an increase in daylength, from short day to long day, the resulting adult enters diapause (Tauber and Tauber, 1976a). Therefore, since adults emerge around the end of June, under natural conditions diapause ensues in all emerging adults because their larval and pupal stages occurred when the rate of increase in daylength was small (Tauber and Tauber, 1976b). Secondly, during summer the diapausing adults are refractory to an increase in daylength, i.e., an artificial increase in daylength at this time will not terminate diapause. However, if naturally aestivating adults are subjected to a period of short days in the laboratory and subsequently to a period of long days, diapause terminates and reproduction begins. In nature, this short-day requirement is fulfilled sometime around October (Tauber and Tauber, 1976b). Thirdly, after the short-day requirement has been completed, field-collected adults, when moved into the laboratory, break diapause and begin to reproduce in response to daylength. The rate at which they break diapause is quantitatively related to the daylength they experience (Tauber and Tauber, 1975b). Thus, in nature, diapause is maintained throughout late autumn and early winter because the short days during this period retard the rate of diapause development. Then, as the days grow longer in late winter (end of February to end of March), the rate of diapause development is accelerated in direct relationship to the actual length of day, and diapause ends by the time of the vernal equinox (Tauber and Tauber, 1975b, 1976b).

Now, let's contrast this unicyclic seasonal pattern and the ecophysiological responses that underlie it with those of another unicyclic species of Chrysopidae, *M. signoretti* (see Fig. 1). This species overwinters as a third instar larva within the cocoon. Both pupation and adult emergence occur in spring, and reproduction takes place in summer. Toward the end of summer the larvae spin cocoons and enter a late summer-autumnal-hibernal diapause. Like *C. downesi, M. signoretti* has one generation per year and an extensive diapause period that is controlled primarily by photoperiod and that ends at the time of the vernal equinox. However, the photoperiodic responses in this species are very different from those in *C. downesi*. For example, diapause in *M. signoretti* is induced and maintained through the action of daylengths shorter than a critical photoperiod (Tauber and Tauber, in press); any daylength longer than the critical photoperiod prevents diapause. Thus, *M. signoretti* has no short-day/long-day requirement for diapause prevention. There is also no photorefractory phase during diapause, i.e., daylengths longer than the critical photoperiod for diapause termination cause diapause termination. In nature diapause is maintained during late summer, autumn, and winter because daylengths are below the critical photoperiod. Diapause terminates at the time of the vernal equinox through the action of daylengths longer than a critical daylength. This species, unlike *C. downesi,* does not show a quantitative response to the actual duration of daylength (Tauber and Tauber, 1975b); it has an all-or-none response to long vs. short days (Table 1).

From our comparison between *C. downesi* and *M. signoretti,* we conclude that in different species, similar seasonal patterns of diapause can be based on very different sets of responses to environmental stimuli. Because of this and because these phenotypic differences probably reflect relatively large genetic differences, we cannot assume that the seasonal adaptations of species such as *C. downesi* and *M. signoretti* are either homologous or closely analogous. Thus, the types of dormancy in these two species cannot be included within the same category in either an evolutionary or an artificial system of classification. Not only do their physiological response patterns show very little rela-

Table 1. Types of photoperiodic stimulus-response patterns that regulate insect diapause maintenance and termination in nature

Stimulus	Response
1. Long vs. short days	All-or-none: A. Critical photoperiod for diapause maintenance does not change during diapause; therefore short days maintain and long days terminate diapause, e.g. *Meleoma signoretti*. B. Critical photoperiod decreases as diapause progresses; therefore short days maintain diapause and photoperiod has no active role in terminating diapause, e.g. *Chrysopa harrisii*.
2. Absolute duration of daylength	Graded: A. Decreasing daylengths decelerate diapause development, e.g. *Chrysopa carnea*. B. Increasing daylengths accelerate diapause development, e.g. *Chrysopa downesi*.
3. Direction of change in daylength	All-or-none (probably)[a]

[a]An all-or-none (diapause induction vs. diapause avoidance) response to lengthening, long days occurs in *Gerris* (Vepsäläinen 1971, 1974), but we have no examples where the direction of daylength change influences diapause maintenance or termination.

tionship to each other (e.g., quantitative vs. all-or-none response to daylength, presence vs. absence of a short-day/long-day response and a photorefractory phase), but the seasonal photoperiodic cues that the two species perceive are different. For example, in the winter, *C. downesi* measures and responds to the duration of the photophase (or scotophase), whereas *M. signoretti* does not show any overt response until after the late winter daylengths exceed a certain critical value. Moreover, from a practical viewpoint, e.g., for purposes of ecosystem modeling, the dormancies of the two species must be treated differently. Taking these facts into account, we need to reexamine, for both theoretical and practical reasons, the current classifications of dormancy, such as Müller's (1970) and Mansingh's (1971), because both of these classifications would place the two species into the same category.

Our second comparison involves two phylogenetically very different organisms -- an insect, the lacewing *C. downesi*, and a bird, the white-crowned sparrow, *Zonotrichia leucophrys gambelii*. The data on the white-crowned sparrow that we present are taken largely from the investigations of Farner and his co-workers (see Farner and Follett, 1966; Farner and Lewis, 1971; Farner, 1975).

The white-crowned sparrow is a single-brooded bird (Fig. 1); migration and reproduction occur in the spring. As in *C. downesi*, reproduction is followed by an aestival photorefractory phase during which an artificial increase in daylength does not stimulate reproduction. Just as in *C. downesi*, only after the birds have experienced the relatively short days of the autumnal equinox will transfer to long-day conditions initiate reproduction. And again similar to *C. downesi*, after the short-day recovery from the photorefractory phase, the rate at which reproduction is initiated (i.e., the rate of gonadal growth, prenuptual moult, and certain premigratory events) is quantitatively related to the absolute duration of daylength.

From the comparison between *C. downesi* and the white-crowned sparrow, we conclude that phylogenetically unrelated organisms can exhibit remarkable similarities both in their seasonal strategies and in the photoperiodic responses that underlie them. These similarities reflect analogous, and obviously not homologous phenomena, and from this we conclude that temperate-zone animals have evolved relatively few general seasonal patterns of response to photoperiod. Further evidence to support this generalization is found, for example, in the types of photoperiodic stimuli and related overt response patterns that are known to influence diapause maintenance and termination in insects (Tauber and Tauber, 1976c). Specifically, three aspects of naturally changing daylengths provide organisms with reliable photoperiodic stimuli that indicate or forecast seasonal changes, and diapausing insects show only a few types of responses to these stimuli (Table 1). From this we conclude that convergent evolution in seasonal life-history strategies and in patterns of response to environmental cues is common.

If the above generalizations are correct, then broad comparisons based on phenological response patterns will not prove useful for generalizing on the origin and evolution of phenological strategies. Nor will these broad comparisons be helpful in developing an evolutionary classification of dormancy. Therefore, future research in this area should be aimed at both inter- and intraspecific comparative studies of the responses and the genetic bases for the seasonal strategies in closely related groups of animals. In this way the homologous components of the patterns can be identified and the comparisons will have evolutionary meaning. [See Lofts and Murton (1968) and Murton (1975) for discussions of evolutionary trends in the seasonal cycles of avian reproduction.]

I.B. Value of a Broad Approach

The inapplicability of the broad comparative approach to specific phenological problems that we discussed above should not be interpreted to mean that these types of studies are not valuable. Although their usefulness to the analysis of the evolution of phenological strategies is presently very limited, they do have an important place in several other areas, as follows.

1. *Comparative animal physiology:* Notable examples are found in the analysis of hormonal regulation of diapause (e.g., Wigglesworth, 1965; de Wilde, 1970; Chippendale, 1977) and in the analysis of the photoperiodic clock (e.g., Lees, 1972; Bünning, 1973).

2. *Systematics:* Certain phenological characteristics vary consistently between taxonomic categories; broad (between genera) comparison of these characteristics often provide useful and very important taxonomic information. For example, in the Chrysopidae, the diapausing stage -- but not the stage sensitive to diapause-inducing stimuli -- is an excellent taxonomic character (Table 2).

3. *Ecosystem modeling:* Modern ecosystem models incorporate seasonality as a basic component in determining the interactions of the various plant and animal species (Lieth, 1974). These models require several types of phenological data, e.g., information on the requirements and responses of individual species to prevailing environmental parameters.

Table 2. Characteristic overwintering states for selected genera of North American Chrysopidae

Genus	Overwintering Stage
Chrysopa oculata group (=*Chrysopa*)	third instar in cocoon
Chrysopa perfecta group (=*Anisochrysa*)	free-living larva
Chrysopa carnea group (=*Chrysoperla*)	adult
Meleoma	most = third instar in cocoon
Suarius	third instar in cocoon
Nodita	free-living third instar

Therefore, in ecosystem modeling comparative studies of the mechanisms controlling the seasonality of various interacting species in the ecosystem have an important role (Tauber and Tauber, 1976d).

II. Evolution of Phenological Strategies Within a Species-Group

This part of our discussion concentrates on the evolutionary development of phenological strategies. We address the problem of whether the evolution of phenological strategies occurs only through small, adaptive changes or whether large, divergent changes can occur. And, we base our analysis on inter- and intraspecific comparisons within a single, wide-ranging species group.

The subject of our study was the *Chrysopa carnea* species group. Members of this group characteristically overwinter in the adult stage (Table 2); they undergo reproductive diapause. The three species that we concentrate on here are *Chrysopa harrisii* Banks, *C. downesi*, and *C. carnea* Stephens. Within *C. carnea* we will discuss two strains -- the *carnea* strain and the *mohave* strain. *C. downesi* and *C. carnea* are very closely related to each other, whereas *C. harrisii* is more closely related to other species within the *C. carnea* group which are not discussed here. First, we give a brief account of each species' seasonal cycle and then we compare the eco-physiological (thermal, photoperiodic, and dietary) and genetic mechanisms involved in controlling the seasonal timing of activity and dormancy in these insects. Finally, using these comparisons as a basis, we discuss the evolution of the phenological strategies.

C. harrisii seasonal cycle: *C. harrisii* is a multivoltine species (Fig. 2) that occurs on conifers throughout northern and eastern North America. Two generations (with perhaps a very small third generation) are produced each summer in the Ithaca, New York, area; during this time, photoperiod plays no role in regulating the rate of growth and development, and temperature is the primary physical factor controlling these processes. Adults that emerge during the latter part of August or during September enter diapause in response to daylengths which, at that time of year, are shorter than the species' critical photoperiod. Once

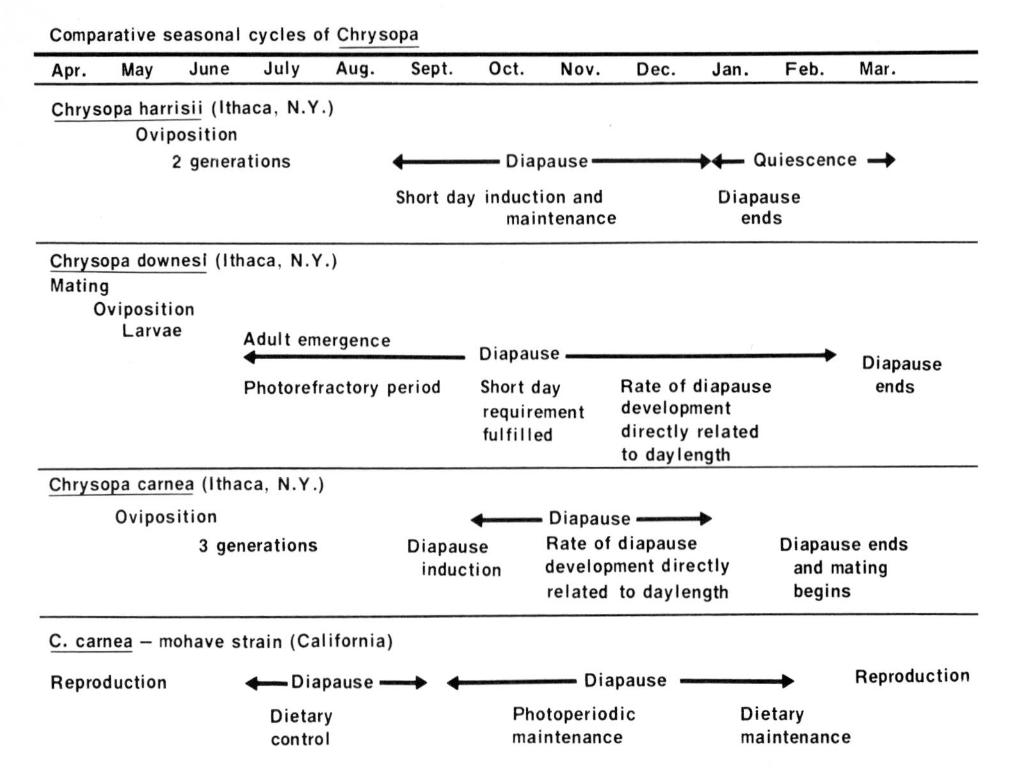

Fig. 2. Annual seasonal cycles of three species in the *Chrysopa carnea* species group. *Chrysopa harrisii* and *Chrysopa downesi* occur sympatrically on conifers; they differ, however, in their annual cycles: *C. Downesi* is univoltine whereas *C. harrisii* is multivoltine. Diapause in both the *carnea* and the *mohave* strains of *C. carnea* is photoperiodically controlled; however, diapause in the *mohave* strain also has a food-mediated component. *C. downesi* and *C. carnea,* which are sibling species, show divergent life cycles, whereas *C. harrisii,* which is the most distantly related member of this group shows considerable convergence in its life cycle with *C. carnea.*

initiated, diapause is maintained in *C. harrisii* adults through their response to short days, i.e., diapause is maintained because the autumnal days are shorter than the critical photoperiod, not because they are decreasing in length (Tauber and Tauber, 1974). During autumn, the critical photoperiod for diapause maintenance gradually decreases until sometime between the end of December and the end of January when all response to daylength ceases. Thus, diapause ends at this time without the intervention of a particular stimulus.

After diapause termination, the insects remain quiescent until temperatures exceed the thermal threshold for postdiapause development. In relation to other species in the *carnea* species-group, *C. harrissi*'s threshold for development is quite high, and this high threshold probably serves as the main mechanism for preventing premature oviposition by overwintering females. Mating by posthibernating adults usually occurs in midspring; we have observed oviposition by this species in nature during mid-May.

C. carnea seasonal cycle. *C. carnea* occurs throughout most of the Holarctic region in a variety of habitats and also on a variety of agricultural crops. Thus far, two morphologically and ecologically distinct strains have been described -- the *carnea* strain throughout most of North America and the *mohave* strain in specific areas of the far west (Tauber and Tauber, 1973; Tauber and Tauber, 1975a).

The *carnea* strain occurs in Ithaca, New York, where three generations are produced each year (Fig. 2). During summer, the adults feed on honeydew and pollen (Hagen, 1950; Sheldon and MacLeod, 1971). There are no dietary requirements for mating in this strain (Tauber and Tauber, 1975a), and reproduction continues as long as daylengths are permissive. Temperature is the main physical factor regulating the rate of growth, non diapause development, and oviposition. Diapause begins at the end of summer when daylengths fall below the critical photoperiod, and it is maintained during autumn and early winter through the species' quantitative response to decreasing daylengths, i.e., as autumnal daylengths get shorter, the rate of diapause development is decelerated and diapause persists (Tauber and Tauber, 1973b). Diapause ends around midwinter without the intervention of a particular stimulus, and after diapause ends, the rate of postdiapause development (leading to oviposition) is regulated by temperature accumulations above the lower thermal threshold for development. Thus, after diapause ends the timing of vernal reproduction is mainly under the control of temperature.

C. downesi seasonal cycle. We described *C. downesi*'s seasonal cycle in Section I.A of this paper, and it is summarized in Figure 2.

II.A. Thermal Adaptations and Their Significance

Interspecific comparisons. In all three species, *C. harrisii*, *C. downesi*, and *C. carnea*, photoperiod has not been shown to have any effect on nondiapause development and growth; and assuming that the food supply is sufficient, the rate of preimaginal growth and development is primarily dependent on temperature. In comparing the thermal requirements for growth in these three species (Table 3) we see some large differences and some similarities. For example, the average lower thermal threshold for development (t) is significantly different (0.05 level) between each of the three species. *C. harrisii* has the highest threshold, *C. carnea* has the lowest, and *C. downesi* is intermediate. The K values (heat degree days above the threshold that are required for completion of development) for the three species are not grossly different except in the pupal-pharate adult stage. In this stage *C. harrisii* has a considerably lower K value than the other two species. Although all our tests were run under diapause averting conditions, this large difference may be related to differences in the stages that are sensitive to diapause inducing stimuli in *C. harrisii* and in the other two species. In *C. harrisii*, the adult is the primary sensitive stage, and therefore all physiological changes related to diapause initiation or avoidance probably occur after adult emergence. However, in *C. carnea* and *C. downesi*, the preadult stages are the primary ones sensitive to diapause inducing stimuli, and many of the physiological changes related to diapause induction or avoidance occur prior to emergence from the cocoon (Tauber and Tauber, unpublished). These differences in K values for the pupa-pharate adult probably evolved in association with differences in diapause characteristics between the relatively distantly related species.

The thermal adaptations of the three species have important roles in fine-tuning the species' overall life-history strategies. For ex-

Table 3. Interspecific comparison of thermal requirements (°C) for nondiapause growth and development of species from Tompkins County, NY[a]

Stage[b]	*Chrysopa harrisii* t[c]	*Chrysopa harrisii* K[d]	*Chrysopa downesi* t	*Chrysopa downesi* K	*Chrysopa carnea* t	*Chrysopa carnea* K
Egg	11.9°	60.4	11.4°	67.0	9.4°	75.5
First instar	11.4°	51.8	10.8°	52.6	9.2°	56.4
Second instar	10.3°	55.2	10.3°	46.9	9.7°	44.2
Free-living third instar	13.2°	44.5	10.7°	51.2	9.2°	48.6
Third instar in cocoon	11.9°	49.2	11.1°	49.2	9.7°	48.1
Pupa	13.6°	78.3	10.3°	116.5	9.3°	104.8
$\overline{x}$ ± s.d.	12.1 ± 1.2°		10.8 ± 0.4°		9.4 ± 0.2°	

[a]All tests were made under LD 16:8 with first-generation offspring of field-collected females.

[b]Ecdysis to ecdysis (e.g., "pupa" includes the unemerged pharate adult).

[c]t = lower threshold for development, based on linear regression method.

[d]K = heat degree days above t required for completion of stage.

ample, in the unicyclic species, *C. downesi*, the relatively high t values play a very important role in that they decelerate egg and larval development so that the subsequent stages that are sensitive to diapause averting stimuli do not coincide with the occurrence of the diapause averting stimuli in nature. This ensures that diapause will not be avoided and that the *C. downesi* unicyclic mode of life will be preserved. Such precise seasonal timing is important from two standpoints: first it serves to synchronize the occurrence of active adult and larval stages with the annual occurrence of food on conifers (Tauber and Tauber, in manuscript), and it serves as a barrier to hybridization and possible swamping by its sympatric sibling species, *C. carnea* (Tauber and Tauber, 1976b).

The differences in t values of the multivoltine *C. harrisii* and *C. carnea* are reflected in seasonal differences in the timing of their post-diapause oviposition and in the number of generations they produce each year. *C. harrisii,* although it ends diapause approximately 1 1/2 months before *C. carnea,* does not begin oviposition until slightly after *C. carnea* does; subsequently, *C. harrisii* produces two generations (with perhaps a very small third generation), and *C. carnea* usually produces a full three generations in Ithaca.

Intraspecific comparisons.

Although, Danilevsky (1965) stressed that, "All physiological mechanisms that regulate the synchronization of the life cycle with the seasonal rhythm of climate are subject to variation," he did not provide examples of adaptive variation in thermal requirements for non-diapause or postdiapause development. [Danilevsky did give examples of geographic variation in thermal reactions during diapause; see Tauber and Tauber (1976d) for a recent review.] We will not discuss here intraspecific variation in temperature requirements for non-diapause or diapause development; however, we will compare the temperature requirements for postdiapause development of several North American populations of *C. carnea*. The reproductive diapause in this species ends in midwinter and postdiapause development encompasses subsequent development leading to oviposition. The rate of postdiapause development determines when, after diapause ends, oviposition will begin in the spring. Presumably selection pressure results in considerable geographic adjustment in this characteristic.

Our comparisons of six geographic populations of *C. carnea* (Table 4) show distinct trends in both their t and K values for postdiapause development. The two populations which experience the more rigorous and least variable winter conditions [the Brownstown (Wash.) and the Portage la Prairie (Manitoba) populations] have the lowest thresholds (t) for development. This adaptation allows the two populations to begin development while temperatures are still relatively low. On the other hand, considerably higher t values occur in the populations from the Corcoran (Ca.), Sedona (Ariz.), Quincy (Fla.), and Ithaca (NY) populations, which experience less rigorous and more variable winter conditions than the Brownstown and Manitoba populations. This characteristic probably prevents premature postdiapause oviposition during periods of warm weather in late winter and early spring when prey for the larvae would not be available. The differences in the K values between the six populations probably reflect differences in the onset of the growing seasons at the various localities. For

Table 4. Intraspecific comparison of thermal requirements (°C) for postdiapause development[a] by various geographic populations of *Chrysopa carnea*[b]

Locality	t[c]	k[d]
6 miles E. Portage la Prairie, Manitoba, Canada	6.6°	89.3
Brownstown, Yakima Co., Wash.	7.2°	61.9
8 miles S. Corcoran, Kings Co., Ca.	9.2°	67.3
9 miles N. Sedona, Coconino Co., Ariz.	9.1°	69.1
Quincy, Gadsden Co., Fla.	9.9°	68.1
Ithaca, Tompkins Co., NY	8.6°	80.5

[a]Period from end of diapause to first oviposition (see Tauber and Tauber, 1973c, for methods).

[b]All tests were run under LD 16:8 with F_1 or F_2 generation offspring of field-collected females.

[c]t = lower threshold for development, based on linear regression method.

[d]K = heat degree days above t required for completion of development.

example, the growing season in Brownstown (Wash.) begins earlier than in Manitoba, and this condition is associated with the fewer heat degree days that are required by the Brownstown population to begin its vernal oviposition. Similarly, the Ithaca (NY) growing season begins considerably later than those of either Quincy (Fla.), Corcoran (Ca.), or Sedona (Ariz.), and this difference is reflected in the higher K value of the Ithaca population.

II.B. Photoperiodic Responses: Genetic Control and Evolution

Interspecific Comparisons.

In the three species, *C. harrisii*, *C. downesi*, and *C. carnea*, photoperiod plays the major role in inducing and maintaining diapause, and in *C. downesi*, it also plays an important role in diapause termination (Tauber and Tauber, 1973d, 1974, 1976a,b). Various types of photoperiodic stimuli can influence diapause in these three species, both in nature and in the laboratory (Table 5), and in comparing the three species we see similarities as well as differences in their photoperiodic response patterns. The degree of similarity between the species coincides with the degree of phylogenetic relationship (based on morphology) between the species. For example, our evidence indicates that, unlike *C. carnea* and *C. downesi*, the more distant relative, *C. harrisii* responds in an all-or-none fashion to daylengths longer or shorter than the critical photoperiod, and that *C. harrisii* adults are much more sensitive to photoperiod than adults of the other two species. Thus, although *C. carnea* and *C. harrisii* have almost identical seasonal life cycles (Fig. 2), these cycles are based on very different photoperiodic response patterns. On the other hand, the sibling species *C. downesi* and *C. carnea*, which have very different seasonal cycles, show many similarities in their patterns of response to photoperiod that we consider homologous. That is, they can perceive and react to three aspects of photoperiod: the absolute duration of daylength, an actual change in daylength, and a critical photoperiod (Table 5). Also, the developmental stages that are most sensitive to photoperiodic stimuli are similar in these two species (Table 5). These data, in combination with results from our hybridization tests (Tauber et al., 1977), support our hypothesis that the univoltinism of *C. downesi* evolved directly from *C. carnea*'s multivoltinism.

The periods during which mating in nature takes place in the two sibling species are separated temporally (Fig. 2). *C. downesi* mates only during April, after *C. carnea*'s postdiapause mating period (February-early March) but before the mating time of *C. carnea*'s first summer generation. Differences in the diapause characteristics and photoperiodic responses of the two species are responsible for the temporal asynchrony of mating. The example, *C. downesi* has a short-day/long-day requirement for diapause prevention; it undergoes an aestival "photorefractory" phase, and responds quantitatively to daylength during winter rather than autumn. Hybridization tests between *C. carnea* and *C. downesi*, show that these differences in photoperiodic responses are based on single allele differences at two autosomal loci (Tauber et al., 1977).

Thus, in comparing these two species, we see a direct evolutionary line, with no intermediates, between their grossly different seasonal patterns (e.g., a multivoltine, spring and summer breeder to a univoltine, early spring breeder), and we also see that these different seasonal patterns, which reflect large phenotypic differences in response to photoperiod, are based on very small genetic differences.

Table 5. Interspecific comparison of photoperiodic stimuli controlling diapause in the *Chrysopa carnea* species group from Tompkins County, NY.

	Chrysopa harrisii	*Chrysopa downesi*	*Chrysopa carnea*
In Nature:			
Diapause inducing stimulus	daylengths shorter than critical photoperiod (late summer)	any slowly increasing, constant, or decreasing daylengths (late spring, early summer)	daylengths shorter than critical photoperiod (very late summer, early autumn)
Diapause maintaining stimulus	autumn daylengths shorter than critical photoperiod for diapause maintenance	summer and early autumn: long daylengths; late autumn and early winter: absolute duration of daylength	absolute duration of autumnal and early hibernal daylengths
Diapause terminating stimulus	none(end of Dec. to end of Jan.)	absolute duration of late winter daylengths (mid-March)	none (end of Jan. to end of Feb.)
In Laboratory:			
Number of critical photoperiods involved in diapause prevention	one (long day)	two (short day/long day)	one (long day)
Quantitative response to daylength	none apparent	yes	yes
Sensitivity to daylength changes	none apparent	yes	yes
Stage sensitive to diapause inducing (or preventing) stimuli	mainly adult	mainly third instar and pupa	mainly third instar and pupa

Table 6. Intraspecific comparison of critical photoperiod and diapause intensity in

	N.Y. Ithaca, Tompkins Co.		Fla. Quincy, Gadsden Co.		Wash. Brownstown, Yakima Co.	
	% diapause	Diapause duration[a]	% diapause	Diapause duration[a]	% diapause	Diapause duration[a]
LD 10:14	100	87 ± 8	100	92 ± 20	100	138 ± 25
LD 11:13	100	63 ± 12	100	64 ± 6	-	-
LD 12:12	100	40 ± 9	100	54 ± 1	100	96 ± 23
LD 13:11	100	29 ± 6	100	42 ± 8	100	48 ± 11
LD 14:10	0	-	0	-	100	54 ± 13
LD 15:9	0	-	-	-	0	-
LD 16:8	0	-	0	-	0	-

[a]In days, $\bar{x}$ ± s.d. (No. ≅ 10 pairs each condition).

Intraspecific Comparisons

The sharp difference in photoperiodic response between *C. carnea* and *C. downesi* is in marked contrast to the clinal variation found in diverse geographic populations of *C. carnea* (*carnea* strain). For example, in the populations listed in Table 6 (with the exception of the Ithaca and Quincy populations), we see a pattern similar to that in populations of other geographically widespread species (see Masaki, 1961, 1965, 1967a; Danilevsky, 1965), i.e., a longer critical photoperiod for diapause induction generally characterizes populations from more northern regions. Also, the six populations show a corresponding geographic variation in the intensity (duration) of diapause (Table 6). Experiments (Tauber and Tauber, unpublished) indicate that crosses between populations result in offspring with critical photoperiods and diapause depths that are intermediate to the two parental populations. In addition, the quantitative value of these characteristics is highly sensitive to directional selection (e.g., in selection experiments we reduced and lengthened both the diapause duration and the critical photoperiod) thus indicating that these characters have high additive genetic variance (i.e., they are under polygenic control). The high degree of genetic variation in these diapause characteristics probably has a large part in determining the expansive geographic distribution of this species; less flexibility would result in a more restricted distribution.

II.C. Diet as an Adaptive Stimulus

As mentioned above, *C. carnea* has two recognized strains (Tauber and Tauber, 1973), the widespread *carnea* strain discussed above, and the *mohave* strain, which we will discuss here. In contrast to the *carnea* various geographic populations of *Chrysopa carnea*

Ca. Corcoran, 8 miles S., Kings Co.		Ariz. Sedona, 9 miles N., Coconino Co.		Ariz. Chandler, Pinal Co.	
% diapause	Diapause duration[a]	% diapause	Diapause duration[a]	% diapause	Diapause duration[a]
100	96 ± 13	100	45 ± 14	91	29 ± 13
100	60 ± 5	100	37 ± 15	100	28 ± 2
100	39 ± 7	92	25 ± 10	80	16 ± 7
100	24 ± 10	0	-	0	-
50	11 ± 4	0	-	0	-
0	-	-	-	-	-
0	-	0	-	0	-

strain, the *mohave* strain occurs in restricted areas of western United States that are characterized by hot, dry summers during which prey abundance is variable. The *carnea* and *mohave* strains readily interbreed both in nature and in the laboratory.

Like the *carnea* strain, the *mohave* strain has an autumnal-early hibernal diapause that is controlled by photoperiod and that we consider to be homologous to the *carnea* strain diapause. However, in the *mohave* strain, reproduction in spring is initiated after the photoperiodic maintenance of this diapause ends *and* after their prey become abundant (Tauber and Tauber, 1973b). Therefore, *mohave* has both a photoperiodic and a dietary component to its overwintering diapause. In addition, the *mohave* strain can enter an aestival diapause if prey are scarce. This aestival diapause is under dietary control -- the presence or absence of prey regulates its induction and termination (Tauber and Tauber, 1973b).

Thus, in its evolution, the *mohave* strain has kept the basic "*carnea* type" diapause that is induced and maintained by photoperiod, but it has added a responsiveness to prey as a second, essential component in the control of diapause. We therefore conclude that there is an evolutionary progression from the autumnal-early hibernal diapause regulated by a single major factor (photoperiod) to an aestival-autumnal-hibernal diapause with two controlling factors (photoperiod and prey). Our hybridization tests and our analysis of natural hybrid populations show that *carnea* is dominant over *mohave* (Tauber and Tauber, 1973). We are currently studying the genetics involved in the shift from the *carnea* to the *mohave* seasonal strategy.

II.D. Evolutionary Trends

Our comparative studies of seasonal strategies in the *Chrysopa carnea* species group indicate that the evolution of phenological adaptations in these insects has included two processes -- anagenesis (phyletic evolution) and cladagenesis (splitting) (see Avise and Ayala, 1975; Ayala 1975, 1976).

a. The anagenetic process of phenological adaptation. This course involves evolutionary alterations in the quantitative aspects of dormancy; these alterations are exemplified by the geographic variation in the quantitative value of the critical photoperiod, depth (duration) of diapause, and thermal requirements for development (t and K values) found in *C. carnea* populations (Tables 4 and 6). In selection and hybridization experiments, these characteristics show a polygenic basis of inheritance and high additive genetic variance (high heritability). As a result, they appear to be highly responsive to natural selection, and thus the geographic variation in these characteristics allows each population to adapt to the seasonal conditions of its locality (latitude, altitude, etc.).

We conclude that *C. carnea* with its large genetic variance in the quantitative aspects of its phenological characteristics has considerable flexibility in adjusting to diverse geographic areas. This plasticity accounts for its expansive geographic distribution; more limited variance would probably result in a more restricted distribution of the species. We also conclude that the evolution of phenological adaptations by the anagenetic method contributes to the allopatric diversification of populations, and that if these populations become geographically isolated, it would contribute to the process of speciation.

b. The cladogenetic process of phenological diversification. This course of evolution in phenological characteristics of the *carnea* group involves qualitative changes in the dormancy-controlling mechanisms, and it results in the divergence of phenological patterns. These changes include major alterations in the diapause controlling factor (e.g., changes in the photoperiodic responses in *C. downesi*), and/or the acquisition of an additional diapause controlling factor (e.g., the dietary requirement in the *mohave* strain of *C. carnea*). Such changes result in large phenotypic differences in seasonal strategies, but they are based on relatively small genetic changes. Moreover, they appear to evolve both sympatrically as in *C. downesi* (Tauber and Tauber, 1977), and allopatrically as in the *mohave* strain (Tauber and Tauber, 1973).

This cladogenetic method of phenological evolution appears to have a direct role in the evolutionary diversification of populations. For example, the rapid evolution of a unicyclic seasonal pattern appears to have been a significant aspect in the sympatric speciation of *C. downesi* from *C. carnea* (Tauber and Tauber, 1977), and the asynchrony in their reproductive periods is a major factor that prevents these two species from hybridizing under natural conditions (Tauber and Tauber, 1976b). Also, the morphologically distinct *mohave* strain of *C. carnea*, with its response to dietary stimuli and its aestival diapause, probably evolved allopatrically from the *carnea* strain as a result of strong selection for reduced reproduction and increased survival during the dry, relatively preyless summer in certain areas of western United States. It is likely that this strain would have remained distinct had its seasonal environment not been altered by man's horticultural practices (Tauber and Tauber, 1973; Tauber and Tauber, 1975a).

III. Summary

An analysis of the evolution of phenological strategies in insects leads us to three generalizations.

1. Among animals, there is considerable convergence both in seasonal strategies and in the underlying responses to environmental factors. This convergence makes identification of homologous adaptations difficult, and one result of this problem is that current classifications of insect dormancy are artificial systems that are of relatively little value in unraveling the evolution of phenological strategies in insects.

2. The development of more useful schemes and the elucidation of evolutionary trends in phenological adaptations requires research with closely related species that compares not only the seasonal cycles of activity, but also the environmental and genetic factors that control the seasonal cycles. The full range of each species' pattern of response to physical and biotic stimuli must be shown and compared with those of closely related species so that homologous components can be identified and their evolution traced.

3. The evolution of phenological strategies involves two pathways: (a) the anagenetic process which provides adaptation to local conditions through the action of natural selection on quantitative aspects of phenological characteristics, and (b) the cladogenetic process which produces divergent phenological strategies through large phenotypic alterations. The anagenetic pathway includes changes in the quantitative value of the critical photoperiods, the depths (duration) of diapause, the thermal requirements for postdiapause development, etc. This pathway is based on the high heritability of these polygenically controlled characteristics (see Dingle, 1974) and on the continuous action of natural selection (see Istock et al., 1976; Istock, this volume), and it results in the evolutionary adaptation of geographic populations of widespread species to their diverse seasonal conditions. In contrast, the cladogenetic method includes qualitative changes in eco-physiological responses to environmental factors, and it results in the diversification of phenological adaptations and seasonal strategies. This diversification is based on relatively few gene changes, and it can occur either allopatrically or sympatrically. As a result, phenological diversification can play a major role in both allopatric and sympatric speciation (Tauber and Tauber, 1977).

Acknowledgments We thank J.R. Nechols and J.J. Obrycki for their help during the preparation of this manuscript. We also thank L.A. Falcon, J.G. Franclemont, C.D. Johnson, M.K. Kennedy, G. Tamaki, and W.H. Whitcomb for their help in collecting specimens. We are especially grateful to R.L. Ridgway, P.D. Lingren, and D.L. Bull for their sustained cooperation.

References

Alexander, R.D.: Life cycle origins, speciation, and related phenomena in crickets. Quart. Rev. Biol. *43*, 1-41 (1968).

Alexander, R.D., Bigelow, R.S.: Allochronic speciation in field crickets, and a new species, *Acheta veletis*. Evol. 14, 334-346 (1960).

Avise, J.C., Ayala, F.J.: Genetic change and rates of cladogenesis. Genetics *81*, 757-773 (1975).

Ayala, F.J.: Genetic differentiation during the speciation process. In: Evolutionary Biology. Dobzhansky, T., Hecht, M.K., Steere, W.C. (eds.). New York: Plenum Press 1975 pp. 1-78.

Ayala, F.J.: Molecular genetics and evolution. In: Molecular Evolution. Ayala, F.J. (ed.). Mass.: Sinauer Assoc. 1976, pp. 1-20.

Bradshaw, W.E.: Geography of photoperiodic response in diapausing mosquito. Nature, Lond. *262*, 384-385 (1976).

Bünning, E.: The Physiological Clock. London: English Univ. Press 1973.

Bush, G.L.: Modes of animal speciation. Ann. Rev. Ecol. & Systematics *6*, 339-364 (1975).

Chippendale, G.M.: Hormonal regulation of larval diapause. Ann. Rev. Ent. *22*, 121-138 (1977).

Danilevsky, A.S.: Photoperiodism and Seasonal Development of Insects. London: Oliver & Boyd 1965.

Danilevsky, A.S., Goryshin, N.I., Tyshchenko, V.P.: Biological rhythms in terrestrial arthropods. Ann. Rev. Ent. *15*, 201-44 (1970).

Dingle, H.: The experimental analysis of migration and life-history strategies in insects. In: Experimental Analysis of Insect Behaviour. Barton Browne, L. (ed.). New York: Springer. 1974.

Farner, D.S.: Photoperiodic controls in the secretion of gonadotropins in birds. Amer. Zool. *15* (Suppl.), 117-135 (1975).

Farner, D.S., Follette, B.K.: Light and other environmental factors affecting avian reproduction. J. Anim. Sci. *25* (Suppl.), 90-115 (1966).

Farner, D.S., Lewis, R.A.: Photoperiodism and reproductive cycles in birds. In: Photophysiology. Giese, A.C. (ed.). New York: Academic Press 1971, pp. 325-370.

Fontana, P.G., Hogan, T.W.: Cytogenetic and hybridization studies of geographic population of *Teleogryllus commodus* (Walker) and *T. oceanicus* (Le Guillou) (Orthoptera: Gryllidae). Aust. J. Zool *17*, 13-35 (1969).

Hagen, K.S.: Fecundity of *Chrysopa californica* as affected by synthetic foods. J. Econ. Ent. *43*, 101-104 (1950).

Haynes, D.L., Bradenburg, R.K., Fisher, P.D.: Environmental monitoring network for pest management systems. Environ. Ent. *2*, 889-899 (1973).

Istock, C.A., Zisfein, J., Vavra, K.J.: Ecology and evolution of the pitcher-plant mosquito. 2. The substructure of fitness. Evol. *30*, 535-547 (1976).

Knerer, G., Atwood, C. E.: Diprionid sawflies: polymorphism and speciation. Science *179*, 1090-1099 (1973).

Lees, A. D.: The role of circadian rhythmicity in photoperiodic induction in animals. In: Circadian Rhythmicity. Proc. Int. Symp. Circadian Rhythmicity 1970, pp. 87-110.

Levins, R.: Evolution in Changing Environments. New Jersey: Princeton Univ. Press. 1968.

Lieth, H.: Phenology and Seasonality Modeling. New York: Springer-Verlag. 1974.

Lofts, B., Murton, R. K.: Photoperiodic and physiological adaptations regulating avian breeding cycles and their ecological significance. J. Zool., Lond. *155*, *327-394* (1968).

Mansingh, A.: Physiological classification of dormancies in insects. Can. Entomol. *103*, 983-1009 (1971)

Müller, H.J.: Formen der Dormanz bei Insekten. Nova Acta Leopoldina *35*, 1-27 (1970).

Masaki, S.: Geographic variation of diapause in insects. Bull. Fac. Agric. Hirosaki Univ. *7*, 66-98 (1961).

Masaki, S.: Geographic variation in the intrinsic incubation period: a physiological cline in the Emma field cricket (*Teleogryllus*). Bull. Fac. Agric. Hirosaki Univ. *11*, 59-90 (1965).

Masaki, S.: Geographic variation and climatic adaptation in a field cricket (Orthoptera: Gryllidae). Evol. *21*, 725-741 (1967).

Masaki, S.: Climatic adaptation and photoperiodic response in the band-legged ground cricket. Evol. *26*, 587-600 (1973).

Masaki, S., Ohmachi, F.: Divergence of photoperiodic response and hybrid development in *Teleogryllus* (Orthoptera: Gryllidae). Kontyu *35*, 85-105 (1967).

Murton, R.K.: Ecological adaptation in avian reproductive physiology In: Avian Physiology, Peaker, M. (ed.). London: Academic Press 1975, pp. 149-175.
Ruesink, W.G.: Status of the systems approach to pest management. Ann. Rev. Ent. *21*, 27-44 (1976).
Sheldon, J.K., MacLeod, E.G.: Studies on the biology of the Chrysopidae II. The feeding behavior of the adult of *Chrysopa carnea* (Neuroptera). Psyche *78*, 107-121 (1971).
Tauber, C.A., Tauber, M.J.: Diversification and secondary intergradation of two *Chrysopa carnea* strains. Can. Ent. *105*, 1153-1167 (1973).
Tauber, C.A., Tauber, M.J.: A genetic model for sympatric speciation through habitat diversification and seasonal isolation. Nature, Lond. *268*, 702-705 (1977).
Tauber, C.A., Tauber, M.J., Nechols, J.R.: Two genes control seasonal isolation in sibling species. Science *197*, 592-593 (1977).
Tauber, M.J., Tauber, C.A.: Geographic variation in critical photoperiod and in diapause intensity of *Chrysopa carnea*. J. Insect Physiol. *18*, 25-29 (1972).
Tauber, M.J., Tauber, C.A.: Insect phenology: criteria for analyzing dormancy and for forecasting postdiapause development and reproduction in the field. Search (Agric.), Cornell Univ. Agric. Exp. Sta., Ithaca, N.Y. *3*, 1-16 (1973a).
Tauber, M.J., Tauber, C.A.: Nutritional and photoperiodic control of the seasonal reproductive cycle in *Chysopa mohave*. J. Insect Physiol. *19*, 729-736 (1973b).
Tauber, M.J., Tauber, C.A.: Seasonal regulation of dormancy in *Chrysopa carnea*. J. Insect Physiol. *19*, 1455-1463 (1973c).
Tauber, M.J., Tauber, C.A.: Quantitative response to daylength during diapause in insects. Nature, Lond. *224*, 296-297 (1973d).
Tauber, M.J., Tauber, C.A.: Thermal accumulations, diapause, and oviposition in a conifer-inhabiting predator, *Chrysopa harrisii*. Can. Ent. *106*, 969-978 (1974).
Tauber, M.J., Tauber, C.A.: Criteria for selecting *Chrysopa carnea* biotypes for biological control: adult dietary requirements. Can. Ent. *107*, 589-595 (1975a).
Tauber, M.J., Tauber, C.A.: Natural daylengths regulate insect seasonality by two mechanisms. Nature, Lond. *258*, 711-712 (1975b).
Tauber, M.J., Tauber, C.A.: Developmental requirements of the univoltine species *Chrysopa downesi*: photoperiodic stimuli and sensitive stages. J. Insect Physiol. *22*, 331-335 (1976a).
Tauber, M.J., Tauber, C.A.: Environmental control of univoltinism and its evolution in an insect species. Can. J. Zool. *54*, 260-265 (1976b).
Tauber, M.J., Tauber, C.A.: Insect seasonality: diapause maintenance, termination, and postdiapause development. Ann. Rev. Ent. *21*, 81-107 (1976c).
Tauber, M.J., Tauber, C.A.: Physiological responses underlying the timing of vernal activities in insects. Int. J. Biometeor. *20*, 218-222 (1976d).
Thiele, H.U.: Remarks about Mansingh's and Müller's classifications of dormancies in insects. Can. Ent. *105*, 925-928 (1973).
Valentine, J.W.: Genetic strategies of adaptation. In: Molecular Evolution. Ayala, F.J. (ed.). Mass.: Sinauer Assoc. 1976, pp. 78-94.
Vepsäläinen, K.: The role of gradually changing daylength in determination of wing length, alary dimorphism and diapause in a *Gerris odontogaster* (Zett.) population (Gerridae, Heteroptera) in South Finland. Ann. Acad. Sci. Fenn., A. IV, *183*, 1-25 (1971).
Vepsäläinen, K.: Lengthening of illumination period is a factor in averting diapause. Nature, Lond. *247*, 385-386 (1974).
Wigglesworth, V.B.: The Principles of Insect Physiology. London: Methuen. (1965).
Wilde, J. de: Hormones and insect diapause. Mem. Soc. Endocrinol. *18*, 487-514 (1970).

Seasonal and Latitudinal Adaptations in the Life Cycles of Crickets[1, 2]

SINZO MASAKI

1. Introduction

In the temperate regions, crickets usually have well-defined seasonal life cycles. Their developmental stages occur every year in a regular sequence in relation to the seasonal changes, and the individual life cycles in each species are more or less synchronized. Among those species inhabiting the main part of the Japanese islands to the north of 30° N, only *Gryllodes sigillatus* and *Myrmecophilus sapporoensis* can be found at various stages throughout the year. The fact that these exceptional species are confined to special habitats protected from violent fluctuations of weather conditions, houses and ant nests, respectively, suggests that adaptation to temperate climates requires a definite pattern of seasonal life cycle. The latter is, of course, due to the existence in the life cycle of a particular stage highly resistant to a long period of arrested development. Adaptation to wide seasonal fluctuations in the environment is therefore not attained by a general increase in the tolerance range, but by the physiological division of the developmental cycle into the active and diapause phases with different optima of temperature, roughly corresponding to the summer and winter conditions, respectively (Browning, 1952a,b; Hogan, 1960a,b; Masaki, 1960, 1962, 1963, 1965; Rakshpal, 1962a,b,c, 1964). In view of the principle of allocation (Levins, 1968), this specialization seems to give a higher adaptability to temperate climates. It enables crickets to skip the harsh season by the diapause phase without impeding adaptation by the active phase to the favorable season. The synchronization of breeding activity might have been another impetus to this line of evolution, or simply a due consequence of the seasonal development.

If an appropriate seasonal cue is incorporated into the mechanism switching the physiological state from one phase to the other, the adaptive tactic becomes more efficient. Like many other organisms, crickets also are highly sensitive to photoperiod, and show a wide spectrum of responses (Alexander, 1968; Fuzeau-Braesch, 1966; Ismail and Fuzeau-Braesch, 1976; Masaki, 1966, 1967, 1972, 1973; Masaki and Ohmachi, 1967; Masaki and Oyama, 1963; Saeki, 1966; Tanaka et al., 1976). The origins of diapause and associated photoperiodic responses are without doubt one of the most important pivots in the evolution of crickets in the temperate region. It seems, however, obviously impossible to infer this sort of evolution from any fossil materials. Only comparative methods will give a feasible way of approach to the problem.

2. Convergent Evolution of Seasonal Life Cycle

The seasonal adaptations of crickets can be compared at two different levels: interspecific and intraspecific. In order to envisage the life-cycle origins and speciation problems, Alexander and Bigelow (1960) and

Alexander (1968) compared the life cycles of different species of crickets in North America, and found that the egg and late juvenile are the predominant stages of diapause among those insects. They ascribed this to preadaptations to the winter conditions at these particular stages, which may be spent in the soil and therefore more protected from the cold than in any other stage. Similar comparison of the crickets' life cycles in Japan generally supports their arguments (Ohmachi and Matsuura, 1951). Namely, diapause at any stage other than the egg or late juvenile is not found among those species living in more or less exposed situations. Owing to its insulating effect, the soil indeed offers the best winter refuges for insects, and many species pass the winter at the subterranean stage in their life, or actively burrow into the soil for the sole purpose of hibernation.

The behavioral preadaptation does not, however, seem to be the whole story. Various species of field crickets such as *Teleogryllus emma*, *Velarifictorus aspersus* in Japan (Masaki, 1960; Umeya, 1946) and *Teleogryllus commodus* in Australia (Brookes, 1952; Hogan 1960a) enter an egg diapause at similar stages of embryogenesis, after water absorption and just before segmentation. *Gryllus pennsylvanicus* in North America does so at a slightly advanced stage (Rakshpal, 1962a). On the other hand, the eggs of *Pteronemobius* species enter diapause before water absorption, possibly at a younger stage than the above species (Masaki, 1960). Diapause does not, therefore, occur randomly at any stage of embryogenesis among different species. A particular stage may have a higher physiological potentiality to acquire resistance to arrested conditions than do others. A similar potentiality may also exist in the late nymphal stage. Overwintering at this stage is found not only in those species which live in a burrow, but also in others such as *Pteronemobius* and *Trigonidium* which do not show any excavating activity. Diapause is most likely to evolve at those stages which are both behaviorally and physiologically preadapted to hibernation.

Another important inference emerging from the interspecific comparison is the recurrent and independent origin of diapause in various phylogenetic groups. Many of the Japanese species of crickets are more closely related to tropical forms than to one another. Some of them, in fact, have been regarded as conspecific with tropical forms (Shiraki, 1930; Chopard, 1967, 1969). In the ground cricket genus *Pteronemobius*, eight morphologically distinct entities have been recognized in the main island of Japan (Honsyu), and all but one, *P. nitidus*, diapause as eggs. The two most common species have been identified as *P. taprobanensis* and *P. fascipes*, respectively, both originally described from southeastern Asia, and their tropical populations virtually lack the ability to diapause (unpublished observation). Even in the same genus, therefore, the egg diapause might have been evolved in each species quite independently of one another. The similar seasonal strategies adopted by different members of the same genus in the temperate climate may therefore be a result of convergent evolution, but not due to the inheritance of diapause genes from their common ancestor.

Convergent evolution of the seasonal life cycle is also suggested by crossing between the closely related field crickets, *T. emma* and *T. yezoemma*, both of which obligatorily enter an egg diapause (Ohmachi and Masaki, 1964). Quite unexpectedly, diapause was abbreviated in the reciprocal hybrid eggs. This might be interpreted as showing that different sets of genes were responsible for inducing diapause in the parental species, and that they failed to express themselves fully in the heterozygous conditions. If so, the two species should also have evolved their egg diapause independently of one another, even though they are so closely related that they easily produce F_1 hybrids in the laboratory.

Comparison of cricket life cycles in different parts of the world may give further evidence for convergent evolution. For instance, the egg diapause of antipodean *T. commodus* cannot share a common origin with that of any congeneric species in the northern temperate region. It is completely separated from the northern relatives by the tropical zone, where its nondiapause sibling species, *T. oceanicus*, occurs. The egg diapause of *G. pennsylvanicus* and *G. ovisopsis* (Walker, 1974) has again obviously originated independently from that in the Asian field crickets, for nymph-overwintering is the more common type of life cycle in this genus in North America. As pointed out by Alexander (1968), similar selection pressures on the developmental cycle with similar adaptive potentiality have resulted in parallel evolution in different species in different parts of the world.

The widespread tendency for convergent evolution suggests that the genetic system controlling the developmental program is more or less versatile, and that it can readily be channeled to evolve diapause. The existence of very closely related or sibling species which over-winter at different developmental stages further corroborates this assertion. In such cases, diapause has been evolved at different stages repeatedly in the ancestral form. This fact is important for understanding not only the origin of diapause but also speciation phenomena. There are possibilities for allochronic or parapatric speciation through divergent evolution of diapause in crickets (Alexander, 1968; Alexander and Bigelow, 1960; Alexander and Meral, 1967; Bigelow, 1958, 1960a,b, 1962; Bigelow and Cochaux, 1962; Fontana and Hogan, 1969; Lloyd and Oace, 1975; Masaki, 1961, Masaki and Ohmachi, 1967; Walker, 1964b, 1974) as well as in other groups of arthropods (Ankersmit, 1964; Bush, 1969; Ghent and Wallace, 1958 cited by Alexander, 1968; Tamura and Koseki, 1974; Tauber and Tauber, 1976a,b). Some original data pertaining to this topic will be given in Section 6.

3. Genetic Variability of Seasonal Life Cycles

More direct evidence for the versatility of the genetic system controlling the developmental program has been derived from intraspecific comparisons. Variations in developmental traits have been found in many species of insects, and this suggests that each population is genetically more or less heterogeneous. Therefore, strains with different developmental traits can be selected in the laboratory. Thus, a strain of *Gryllus campestris* averting diapause was established by selection (Ismail and Fuzeau-Braesch, 1976). Continuous laboratory rearings also inadvertently resulted in a decreased incidence or complete elimination of egg diapause in *T. commodus* (Bigelow and Cochaux, 1962; Cousin, 1961). Examples of this kind are rather scarce in crickets, but there is a growing body of literature showing that, in various groups of insects, diapause characteristics can easily be modified by artificial selection through several generations, as cited by other authors in this volume.

In most of these selection experiments, nondiapause strains were established from predominantly diapause populations. A few attempts were also undertaken to select for a stronger tendency to diapause, but no one seems to have ever tried to create a diapause strain from a virtually nondiapause population occurring in the tropical or subtropical region. Obviously, however, it is the latter kind of selection which is important for the evolution of diapause. Selection of a strain with an egg diapause from a predominantly nymph-overwintering, subtropical population of *P. fascipes* is now in progress in our laboratory, and the percentage of egg diapause has increased from about 10%

or less of the original level to more than 70% during eight generations of selection.

Since genetically variable populations are regularly subject to differential climatic selection in different localities, the variability in developmental characteristics should be best illustrated by different geographic populations. Such a variation will provide important information on the evolution of seasonal adaptation, because it must have been established through the historical process of distributional or climatic fluctuations, and therefore reflects not only the geographic but also, to some extent, the temporal sequence of evolutionary changes. As reviewed by several authors (Beck, 1968; Danilevskii, 1961; Danilevsky et al., 1970; Lees, 1955; Masaki, 1961; Saunders, 1976; Tauber and Tauber, 1976a), diapause characters are, in fact, highly variable among geographic populations. A general trend in such variations is the latitudinal increment in the critical length of photoperiod for the induction of diapause, which is a due consequence of selection by the reversed latitudinal gradients of heat quantity and daylength during the growing season. Thus, in the beach ground cricket, *Pteronemobius csikii*, the critical duration of parental photoperiod for the induction of egg diapause varies from about 12 hr at 27°48'N to about 14 hr 30 min at 35°25'N (unpublished observation).

The developmental cycle in any population is, however, defined by several different parameters, and the critical daylength is only one of them, even though it is of special importance in actually or potentially multivoltine populations. Intraspecific variations in other aspects of development have also been known to occur. The incidence and intensity of diapause vary among different geographic populations of *T. commodus* in Australia (Bigelow, 1962; Bigelow and Cochaux, 1962; Hogan, 1965, 1966), and the duration of nymphal development in *Gryllus veletis* and *G. fultoni* in North America (Bigelow, 1960, 1962). Many more examples may be found in other groups of insects to show geographic variations in the incidence, intensity, or termination of diapause. As will be shown later, all these aspects of diapause are closely interrelated or coadapted, and some of them may even be regarded as different facets of a single genetic trait.

A much more instrumental approach will be derived from synthesis of the interspecific and intraspecific levels of comparison. The result of natural selection is obviously determined by a complicated interplay of the external and internal factors. Even when the former remains constant, the latter may change with the mode of life. Under the same climatic conditions, different species may show either similar or dissimilar trends of evolutionary response, depending on the similarity or dissimilarity of their seasonal adaptations. The best way to study this evolutionary interplay is perhaps to compare intraspecific variations of various species along the same climatic gradient. We shall therefore turn to comparisons of variation in certain species of crickets and attempt to clarify, if possible, the rules governing their seasonal and geographic adaptations.

4. Parallel Univoltine Clines

4.1. The Area and Climate

The geographic area available for my study was a long island chain off the east coast of the Eurasian continent: the Japanese archipelago. The area represents only a very small portion of the world, but the

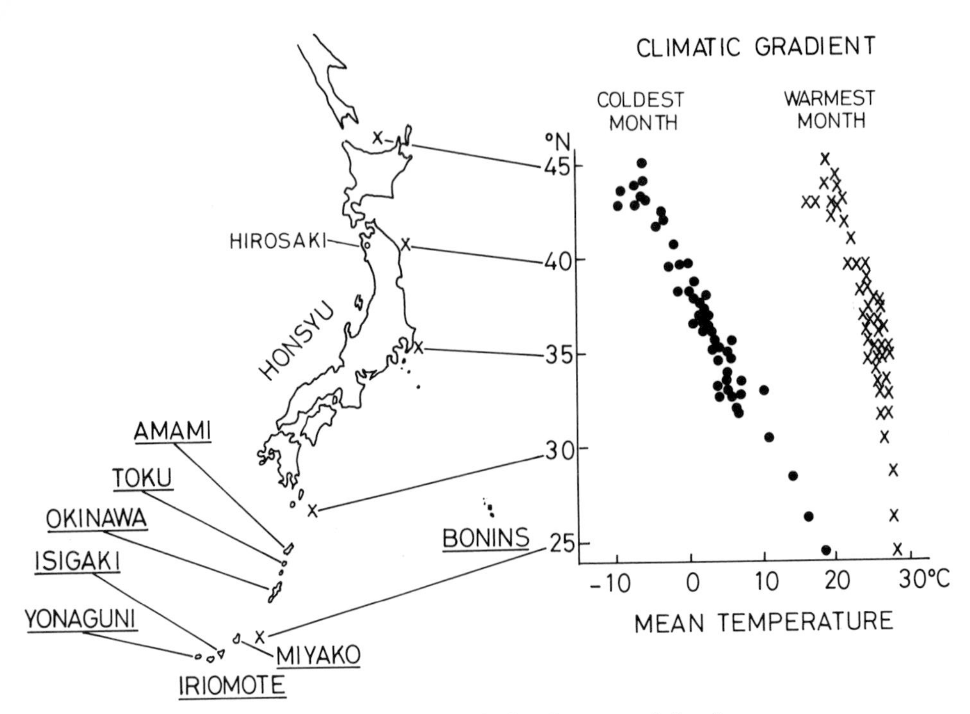

Fig. 1. A map and the climatic gradient of the Japanese islands.

islands scatter widely from north to south over a stretch of about 3000 km between 24 and 45°N. Naturally, there is a very conspicuous climatic gradient, of which temperature is undoubtedly the most significant factor, the annual mean ranging from about 5°C in the northern end to 23°C in the southern end (Fig. 1). The area is generally moist, with the annual precipitation ranging from about 800 to 3000 mm. There is no conspicuous dry period anywhere during the warm season.

These climatic features of the islands, together with the highly omnivorous habit of the crickets studied, would have excluded complicating factors and established relatively simple patterns of geographic variation, which may facilitate our analysis.

4.2. Diapause Intensity

More than half of the cricket species in the Japanese islands are univoltine and overwinter in the egg stage, as typically represented by *T. emma* (Fig. 2). This species occurs in habitats between 43 and 30°N with the annual mean temperature ranging from about 7 to 19°C. In spite of this wide range of climatic conditions, the life cycle is highly stabilized throughout its distributional area; the adult emerges and sings everywhere in the local autumn. This seasonal homeostasis is not simply due to the genetic uniformity. As shown in Figure 2, there was a conspicuous difference in the duration of egg stage between a northern(43°03'N) and a southern(33°34'N) population (Masaki, 1963). The means at 25°C were 93 and 132 days, respectively, and this difference persisted through three successive generations in the laboratory. The genetic control of the duration of diapause was further confirmed by crossing experiments. The reciprocal crosses between the northern

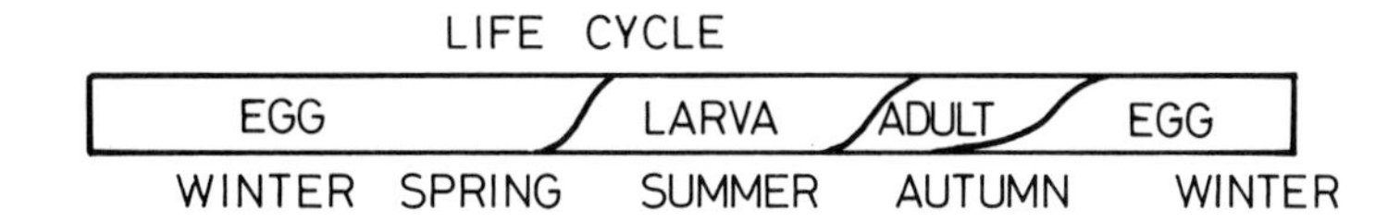

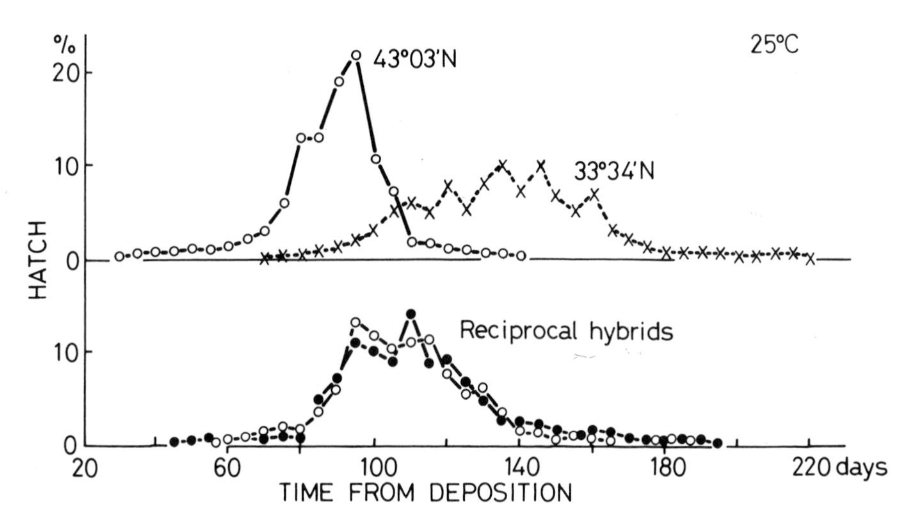

Fig. 2. The seasonal life cycle of the Emma field cricket and the distribution of hatching time in a northern and a southern strain and their hybrids.

and southern stocks yielded similar hybrids intermediate in the duration of diapause between the parents. Both male and female genes were therefore responsible for diapause.

This contrasts with the results of crossing experiments between strains with different diapause characters of *Bombyx mori* (Nagatomo, 1953) or *Atrachya menetriesi* (Ando and Miya, 1968). In those insects, the diapause character of the F_1 hybrid was entirely determined by the female parent. The male parent was, on the other hand, found to be influential on the egg diapause in the interspecific hybrids of Australian (MacFarlane and Drummond, 1970) and Japanese (Ohmachi and Masaki, 1964) species of *Teleogryllus*. This difference is probably related to the diapause mechanisms; in the silkworm or the leaf beetle, the chorion or yolk which is laid down before fertilization may be responsible for diapause, while in the crickets the egg structure formed after fertilization may be more important.

In the field, the hibernating period is of course much longer in the colder north, but contrary to this the northern eggs terminated diapause in a shorter time than the southern ones. This paradoxical result was not due to the particular effect of warm temperatures at which eggs were maintained. As in many other insects, the egg diapause is terminated by incubation at a high temperature after several weeks of chilling at 5-10°C. The median time of chilling for diapause termination was about 50 days in the northern eggs and about 100 days in the southern eggs. Therefore, it was the intensity of diapause which was variable between the local samples.

In order to know the adaptive background of this variation, more than 30 local samples were collected as adults and the diapause intensities of their eggs were compared under identical conditions (Fig. 3, left;

Masaki, 1965). There was clearly a linear increasing tendency from north to south. Similar trends were repeatedly found at different temperatures. The geographic regularity of this variation strongly suggests that selection by the climatic gradient is involved. It was represented with considerable accuracy by the equation

$$Y = 226 - 4.65\ X_N + 0.33\ X_E - 0.016\ X_A$$

Where Y is the mean duration of egg stage at 25°C, and X_N, X_E, and X_A are the latitude, longitude, and altitude, respectively, of the original locality. The variables in this equation could be substitued by certain climatic indices without reducing the predictive power, among which the sum of the monthly means of daily maximum temperature from October to April seemed to be the best predictor. From this analysis, it was inferred that selection is exerted through the risk of untimely termination of diapause, which may be caused by warm termperature before and during hibernation. Such a risk is obviously greater in the warmer south, where a more intense diapause would be favored. On the other hand, an intense diapause is not required in the north where the winter is persistently cold, but it may increase the risk of failure to mature within the short growing season by delaying the spring hatch.

4.3. Nymphal Development Time

Another cline was found in the nymphal development time (Fig. 3, right; Masaki, 1967). The northern nymphs matured faster than the southern

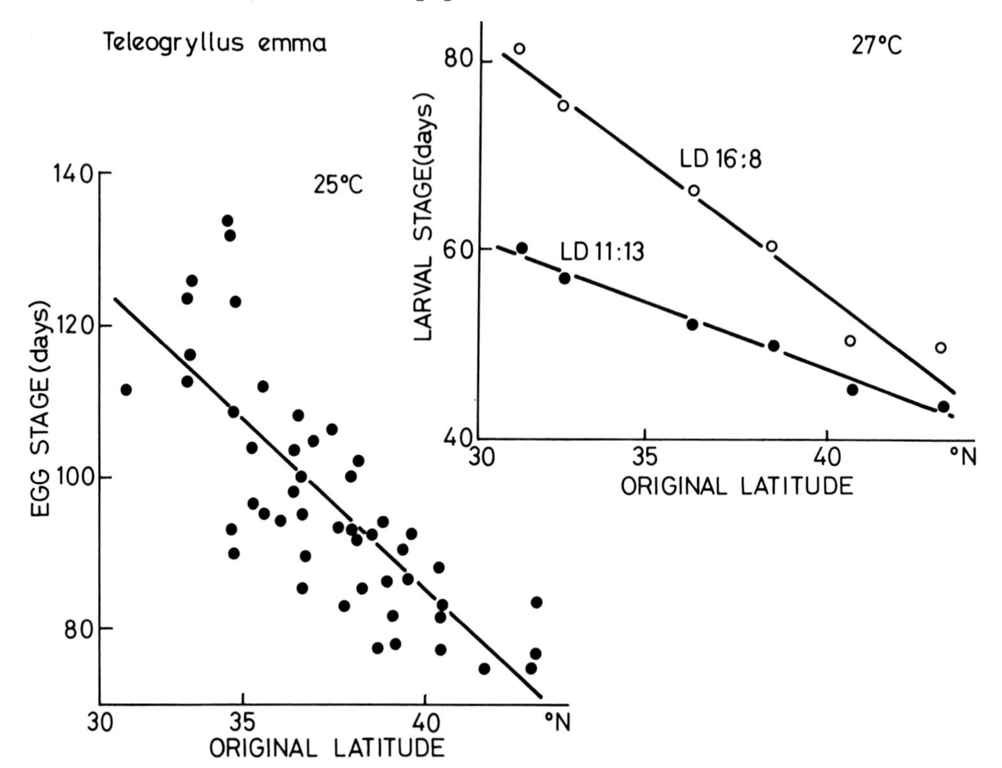

Fig. 3. Physiological clines in the Emma field cricket. Left, the duration of diapause egg stage. Right, the duration of nymphal development at two photoperiods.

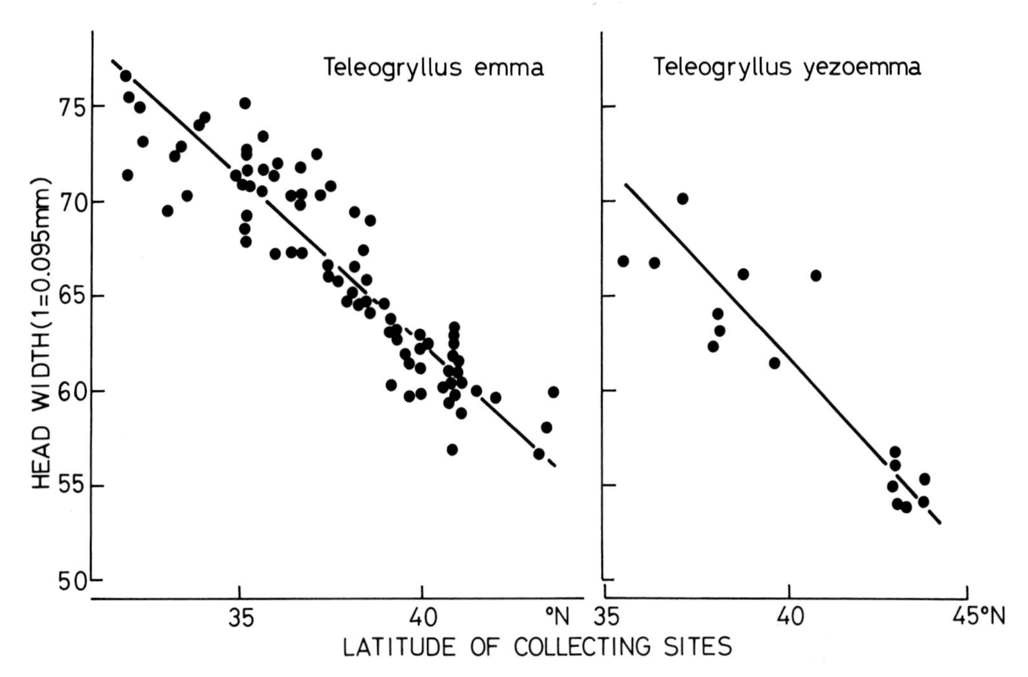

Fig. 4. Adult-size clines in two species of field crickets. Points indicate means of local samples collected in autumn.

ones. Reciprocal crosses between a northern and a southern stock produced F_1 hybrids intermediate in development time between the parents, but they were slightly closer to the mother than to the father (Masaki, 1963), suggesting that cytoplasmic as well as chromosomal factors were involved. The geographic regularity of this variation again leaves little doubt that the amount of heat available in the growing season is the selecting agency.

Figure 3 (right) shows further that the nymph is sensitive to photoperiod. Adults appeared later at a long photoperiod than at a short, and more so in the southern than in the northern strains. This response, together with the genetically controlled duration of development, ensures diapause eggs to be laid in autumn. When laid too early, they may terminate diapause before winter; when laid too late, they may fail to enter diapause. This response is also effective in synchronizing the breeding activity. The short-day type regulation of this kind has scarcely been reported, but it is widespread among univoltine crickets hibernating as eggs (Masaki, 1966, 1972; Saeki, 1966). A similar regulation of development before hibernation occurs also in nymph-overwintering species, *G. campestris* (Ismail and Fuzeau-Braesch, 1976) and *Pteronemobius nitidus* (Tanaka, unpublished). In *T. emma* and other egg-overwinterers, any kind of nymphal diapause is not involved in this response, a long photoperiod simply extending the nymphal growth. Consequently, the longer nymphal stage at a long photoperiod results in a larger adult than at a short photoperiod.

4.4. Adult Size

The adult size of *T. emma* also varies with the geographic variation in the nymphal development time. There is thus a latitudinal cline in adult size (Masaki, 1967; Fig. 4, left). The regularity of this variation was again tested by multiple regression analysis:

$$Y = 126 - 1.98\ X_N + 0.116\ X_E - 0.011\ X_A \text{ for the male,}$$

$$Y = 115 - 1.75\ X_N + 0.131\ X_E - 0.009\ X_A \text{ for the female,}$$

where Y is the head width (1 = 0.095 mm) and X's are as defined before. The partial regression of the head width on either latitude or altitude was highly significant, but that on longitude was not. The partial regression coefficients for these variables indicate that shifts of about 1° in latitude and of 180 m in altitude of the habitat location are almost equal in their effects on the adult size. This is highly suggestive, because in these areas such geographic shifts are equally accompanied by about 1°C change in the annual mean temperature. More than 80% of the variance between the local samples is accounted for by the regression on the annual mean temperature alone. This strongly supports the view that the size cline is related to the climatic selection.

Being just the reverse of what might be expected from Bergmann's rule, however, the size cline rejects the possibility that the retention of body temperature is somehow responsible. A more likely hypothesis is that the cline is mainly a by-product of climatic selection of the development time. If the adult size itself is the target of natural selection, the capacity for egg production in the female, or the mating competition in the male may be involved. In these respects, a larger body size would be selected for, but selection of the development time by a given duration of growing season sets the upper limit for the adult size. The possibility of selection by predation is obscure, since the cricket may be attacked by a variety of predators of different sizes at different growth stages, such as mites, spiders, entomophagous insects, lizards, frogs, snakes, birds, or mammals (Walker, 1964a; Gabbutt, 1957).

This sort of climatic selection may be widespread. Similar size clines would therefore be expected among those insects with similar developmental physiology. Parallel size clines have thus been found in several other univoltine crickets in North America (Alexander and Bigelow, 1960) as well as in Japan (Fig. 4, right). As far as the heat requirement is concerned, a multivoltine life cycle appears to be possible in the south of their distributional areas, where the heat quantity is two or three times as much as in the north. Those crickets, nevertheless, have responded to the climatic gradient by genetically varying their heat requirements in parallel to the available heat. Their life cycle seems to work as an evolutionary feedback system, which mitigates the disturbing effect of climatic variation.

In the Japanese islands, crickets with homeostatic univoltinism, such as *T. emma*, *T. yezoemma*, and the autumn form of *Velarifictorus aspersus*, occur only to the north of 30°N, despite their suspected tropical origin. Possibly, their ancestral populations with a well-established univoltine cycle reached the archipelago from the continent first at middle latitudes. The probability was small for them to slip out of the evolutionary feedback and establish a multivoltine cycle in response to temporally or geographically increasing warmth. As the distributional or climatic changes were usually gradual, they would have encountered only such a slight climatic change in each generation as could be compensated for by a minor genetic alteration within the framework of univoltinism. This situation would have persisted until the limit of adaptation by an obligatory egg diapause was reached. On the other hand, if a species suffers such an abrupt and drastic climatic change that is far beyond the feedback capacity of an established life cycle, the latter would be perturbed and a possibility for a new selective response would arise.

5. Geographic Variation in Voltinism

5.1. Complication of Size Trend

There is no reason to suspect that the genetic correlation between the adult size and development time is restricted to the univoltine crickets. If it exists also in other species with variable voltinism, the geographic variation in adult size would not follow a simple linear trend. If a species varies its life cycle, for example, from bivoltine in the south to univoltine in the north, the decreased number of annual generations in the north would allow a longer time for development of each generation, which in turn may result in a larger adult size than in the southern bivoltine area. Within the area where the same voltinism is maintained, however, the adult size would still tend to be smaller northwards, for the same reason as in a univoltine cycle. A variation in voltinism would therefore complicate the geographic trend of variation in adult size. If this reasoning is correct, it is possible to infer from the geographic size trend what kind of life-cycle change occurs.

Such complicated variations in adult size are found in the two species of ground crickets, *P. fascipes* and *P. taprobanensis* (Fig. 5). They are similar to one another, but quite different from what we have found in

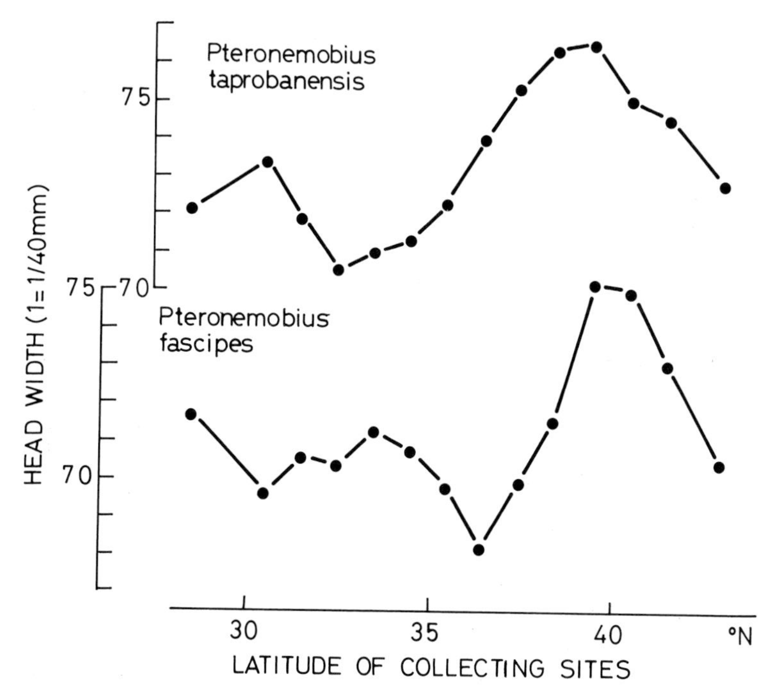

Fig. 5. Latitudinal variation in adult size of two species of ground crickets with variable voltinism. Points indicate latitudinal means of samples collected in autumn.

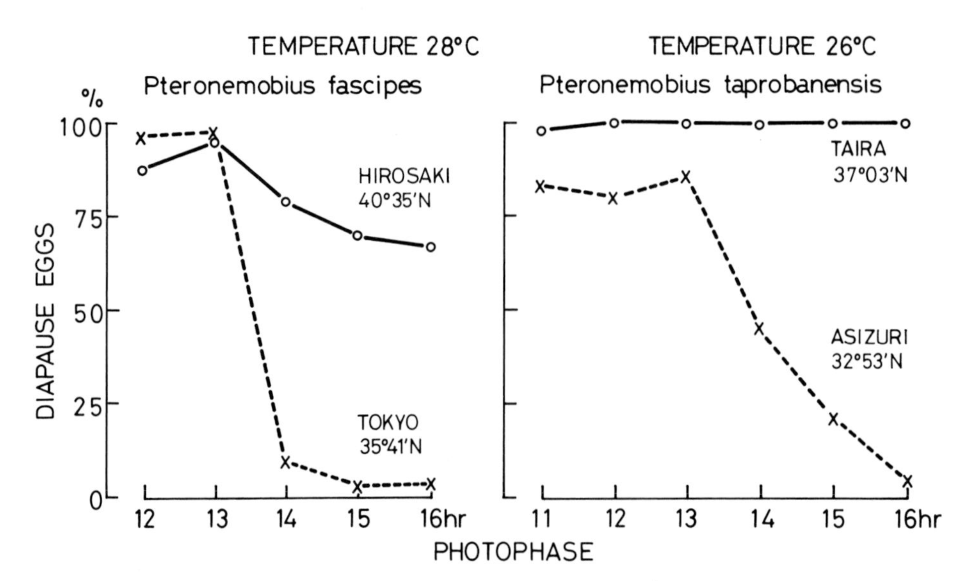

Fig. 6. The influence of parental photoperiod on the incidence of egg diapause in northern and southern populations of two species of ground crickets.

the univoltine species. If our prediction is correct, they are variable in voltinism. A quick method to test this prediction is to compare the diapause characters of different populations which differ in adult size.

The results depicted by Figure 6 seem to support the hypothesis. The egg diapause in both species is facultative in the southern small-sized populations, being programed by the parental photoperiod. This may be taken as evidence for a multivoltine cycle. The large-sized northern populations, on the contrary, have a stronger tendency to enter diapause even at long photoperiods, which means that the predominant type of their life cycle is probably univoltine.

An important question to be asked is what occurs in the transitional zone between the bivoltine and univoltine areas represented by the adult size decreasing from north to south (Fig. 5). Probably, the populations in this zone are polymorphic in voltinism, but we know little about the precise genetic and physiological mechanisms to maintain such polymorphism. Since the univoltine and bivoltine cycles require different durations of development, and any intermediate individual cycle connecting them cannot exist, there would be a possibility for disruptive selection. This appears to be more likely in *P. fascipes* of which the univoltine and bivoltine cycles are controlled by different photoperiodic responses, as will be described later in this section. There is, however, no evidence for disruptive selection, and the adult size varies continuously, showing no dimorphic distribution in the transitional area. This problem will be discussed further later on.

5.2. Diapause Intensity and Voltinism

Another characteristic feature of the geographic variation associated with variable voltinism is found in the intensity of egg diapause in *P. csikii* (Fig. 7). This species seems to be bivoltine as inferred from

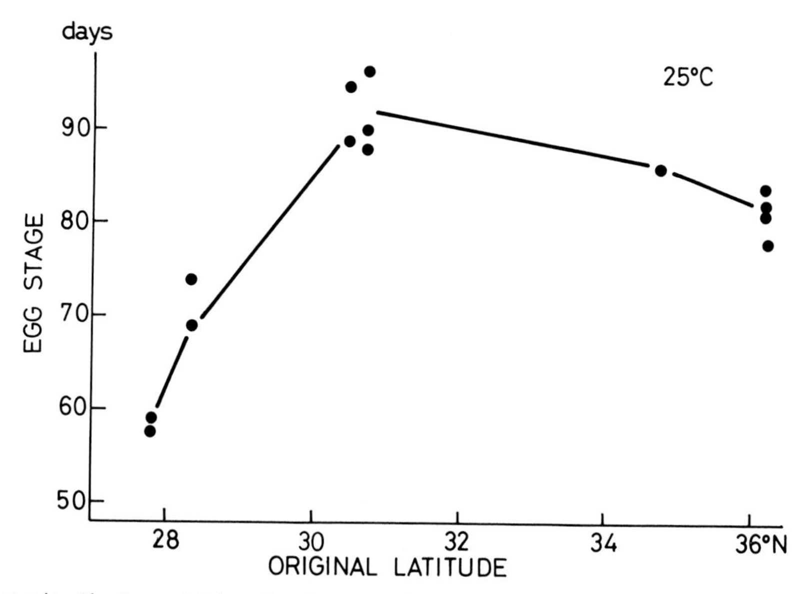

Fig. 7. Latitudinal variation in the duration of diapause egg stage in the beach ground cricket, *Pteronemobius csikii*. Each point represents the mean when the parents were reared at either LD 11:13, 12:12, or 13:11 and 25-27°C.

the photoperiodic programing of their egg diapause and developmental rate (Section 3). Within the bivoltine area, there is a slight southward increase in the intensity of diapause, which might be a result of climatic selection similar to that in the univoltine crickets. The general level of diapause intensity is, however, much lower than in the univoltine *T. emma* (Fig. 2). This is presumably due to the longer time taken for development of two generations than for one. The diapause eggs should be laid later in the bivoltine population than in the univoltine one, and subsequently required to tolerate a smaller amount of heat before winter. This supposition is supported by the fact that in *P. fascipes* a southern bivoltine population undergoes a shorter diapause than does a northern univoltine one (Masaki, 1961).

To the south of 29°N, the intensity of diapause in *P. csikii* is remarkably decreased. This may be related to the very long growing season available for repeating several generations. During field work on the island of Toku (about 28°N) many small juveniles were still found as late as November. This fact, together with the very short critical photoperiod for the induction of egg diapause (Section 3), suggests a very late onset of diapause there. The eggs should therefore lie dormant only a brief period as compared to the northern localities. Moreover, there is no risk of being killed by frost. Under such circumstances, a shorter duration of diapause would be selected for.

There is thus a reversion in the geographic trend of diapause intensity along the linear climatic gradient. This illustrates well how the evolutionary response is influenced by a shift in the internal factor, the adopted type of seasonal strategy. Various physiological traits controlling development and diapause are closely interwoven to form a coadapted system, and therefore any variation in one of them inevitably causes greater or smaller changes in the action of natural selection upon others.

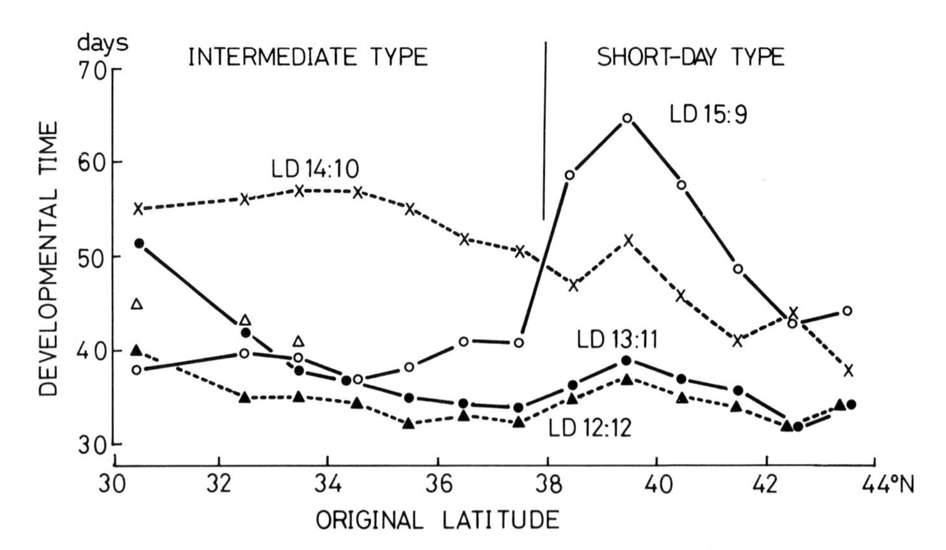

Fig. 8. Latitudinal variation in the nymphal development time of the band-legged ground cricket at 27°C and different photoperiods.

5.3. Photoperiodism in Nymphal Development

It would therefore be expected that the geographic populations of multivoltine crickets show unique variations in still another feature of their seasonal adaptation, the development time of nymphs and its photoperiodic regulation. Figure 8 thus shows the duration of nymphal development in various latitudinal populations of *P. fascipes* reared at different photoperiods (Masaki, 1973). To the north of 38°N, the response is of a short-day type, like the univoltine field crickets, adults emerging earlier in short days than in long days. To the south of 39°N, however, the retarding influence is exerted by an intermediate photophase but not by a long one. In a previous paper (Masaki, 1973), this response has been referred to as a "reversed intermediate type," but a simpler expression "intermediate type" will be used here to denote that it is intermediate between short-day and long-day types. This response apparently fits in a bivoltine cycle. Nymphs that hatch in late spring grow rapidly under the influence of long days, and produce a second brood. The latter are prevented from precocious emergence before autumn by the intermediate daylength during midsummer. The final effect of this particular response is therefore essentially the same as that of the short-day type response of the univoltine forms.

In the transitional zone between the univoltine and bivoltine areas, the two types of photoperiodic response are intermingled: the fast-growing portion of the population shows an intermediate type, and the slow-growing portion a short-day type. Adults thus typically emerge in two separate periods at long photoperiods of LD 15:9 or 16:8, and there is a time lapse of more than a month between them. There is therefore a high possibility that the slow-growing group and progeny of the fast-growing group mature in autumn and interbreed. It may be supposed that the resulting hybrids suffer inferior fitness due to the intermediate developmental rate, which may not match either with a univoltine or with a bivoltine cycle. Such a situation seems, however, to be avoided, if the hybrid response is precisely intermediate between the

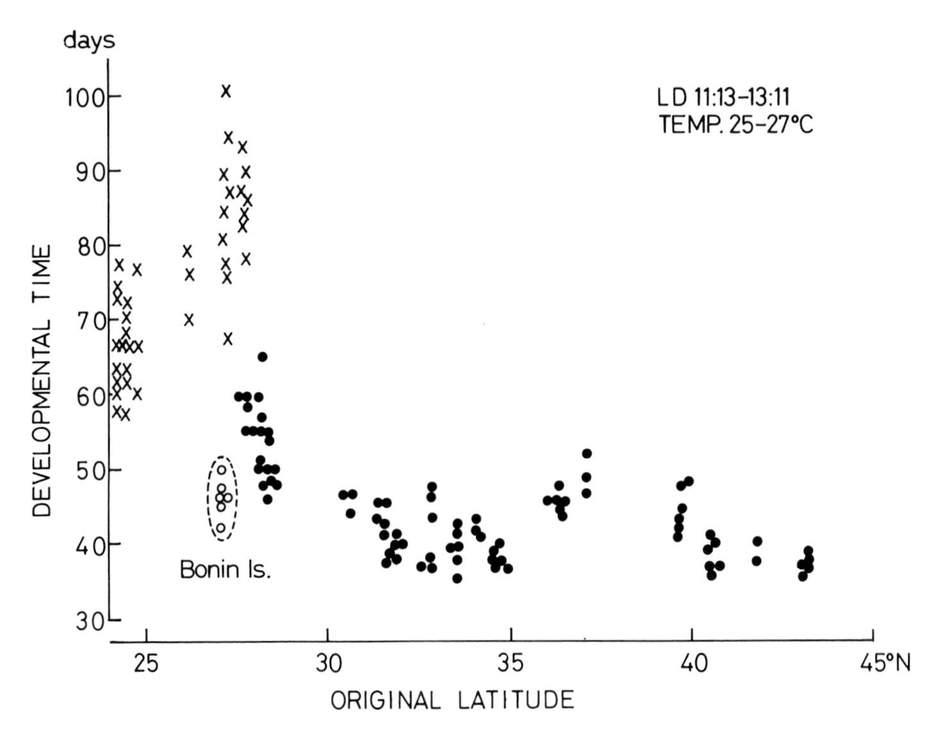

Fig. 9. Latitudinal variation in the nymphal development time of the lawn ground cricket under short-day conditions. Closed circles represent the temperate form and crosses the subtropical form. For the Bonins, see Section 7.

typical short-day and intermediate types. The hybrid would emerge as adults after hibernation much later than do the fast-growing members, but this delay would be compensated for in their offspring by a higher rate of development at an intermediate daylength in midsummer. This may be a possible explanation for the maintenance of species integrity through the transitional zone, but further investigations are required to confirm it.

As we have already seen, *P. taprobanensis* shows a geographic trend of adult size similar to that of *P. fascipes*. This may suggest that the geographic adaptations of their life cycles are also similar. When examined closely, however, a somewhat different picture came into view. Figure 9 gives the geographic profile of variation in the nymphal development time of *P. taprobanensis* under short-day conditions. Populations to the north of 30°N are more or less uniform in the rate of development, those at around 28°N take a considerably longer time to mature, and those at around 27°N (represented by cross) develop still more slowly. Further south at about 24°N, the development time decreases again, but it is yet remarkably longer than in the area to the north of 30°N.

Actually, a shift of the basic type of photoperiodic response is involved in this variation as seen in Figure 10, in which the ratios of the developmental time at long photoperiods to that at LD 12:12 are plotted, in order to give a clear picture of variation in the type of photoperiodic response. Short-day type responses are maintained over a wide range from the northern limit of distribution down to about

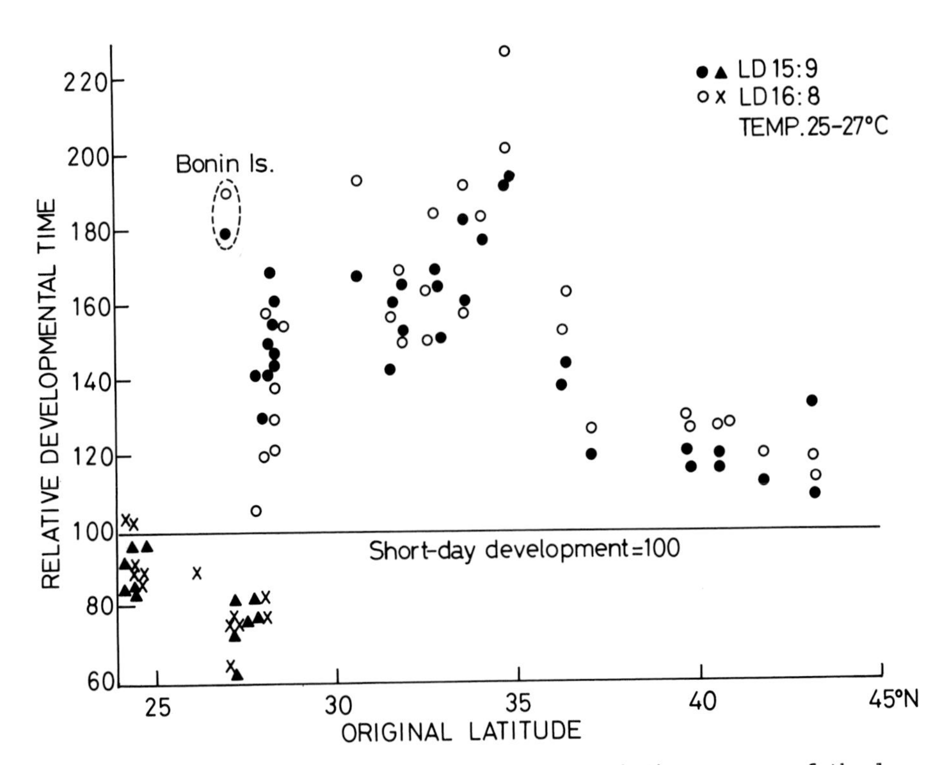

Fig. 10. Latitudinal variation in the nymphal photoperiodic response of the lawn ground cricket. The development time at long photoperiods is converted into the ratio to that at LD 12:12. Closed and open circles represent the temperate form, and crosses and triangles the subtropical forms.

28° N. Within this area, however, there is a considerable variation in the degree of long-day retardation. From the northern limit of distribution at about 44° N, the retardation becomes greater southwards, reaches its maximum at about 35° N, and further south decreases again down to about 28° N.

There is no indication of an intermediate type of response in this species, but the development time at an intermediate photoperiod of LD 14:10 tends to be longer southwards from 43° N to 28° N and, at the same time, the strongest retarding effect becomes exerted with LD 15:9 instead of LD 16:8. As compared with *P. fascipes*, this species therefore shifts its voltinism with a less clear change in the type of photoperiodic response. This difference between the two species is reflected by a difference in their phenologies: the first singing season in the bivoltine area begins earlier for *P. fascipes* than for *P. taprobanensis*, and also the bivoltine area extends more north in the former than in the latter.

Further south beyond 28° N, a drastic change occurs on the island of Toku. The population there shows accelerated, instead of retarded, development at long photoperiods. Their response, therefore, represents a long-day type. This is a very important finding from an evolutionary viewpoint, and deserves to be dealt with in a separate section.

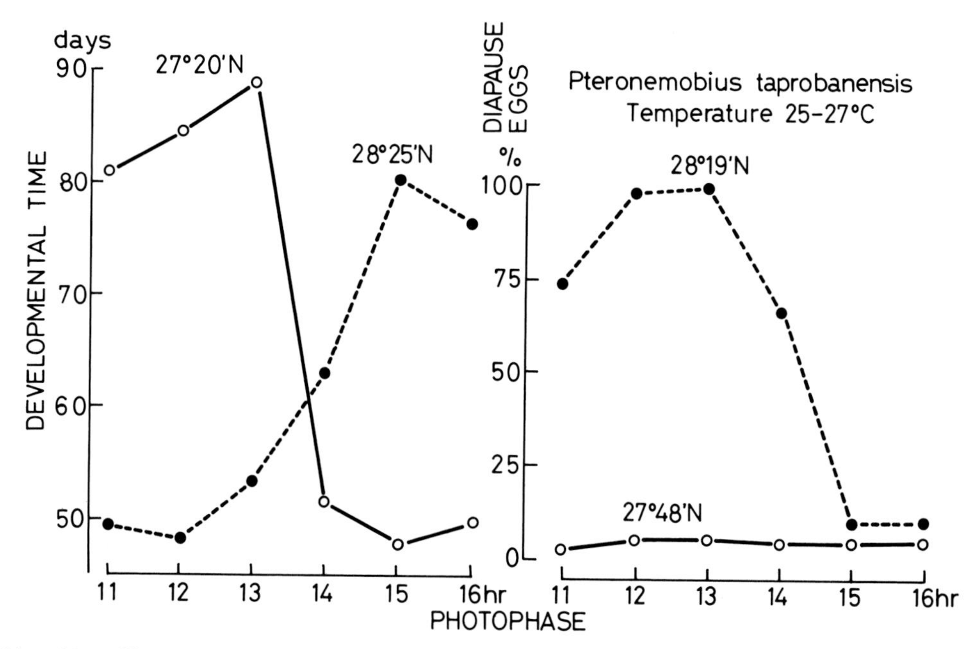

Fig. 11. Effects of photoperiod on the nymphal development (left) and egg diapause in the temperate (28°19' and 28°25'N) and subtropical (27°20' and 27°48'N) forms of the lawn ground cricket.

6. Divergence in Seasonal Life Cycle

6.1. Climatic Forms in *P. taprobanensis*

Apparently, the long-day type regulation of nymphal development matches with hibernation as a nymph, while the short-day type matches with hibernation as an egg. If so, the populations of *P. taprobanensis* to the north and south of 28°N adopt basically different strategies of seasonal adaptations. For convenience, they will be referred to as the temperate and subtropical forms, respectively. Their nymphal photoperiodic responses make a clear contrast, which is associated with a conspicuous difference in their ability to enter an egg diapause (Fig. 11). The eggs of the temperate form enter diapause in response to the short photophase in the parental generation, but those of the subtropical form are almost completely free of diapause at any photophase.

It seems difficult to imagine any graded series of variation interconnecting them. The two different stages for overwintering require contrastingly different types of photoperiodic regulation of nymphal development. The geographic shift from one type to the other is, in fact, abrupt and discontinuous (Fig. 10). The results given in Figure 11 are based on populations coming from near the southern limit of the temperate form (the island of Amami) and the northern limit of the subtropical form (the island of Toku), respectively. They were separated from each other by a shallow and narrow strait of only about 200 m in depth and 50 km in width. Recently, the temperate form was discovered in a restricted northern part of the latter island, where

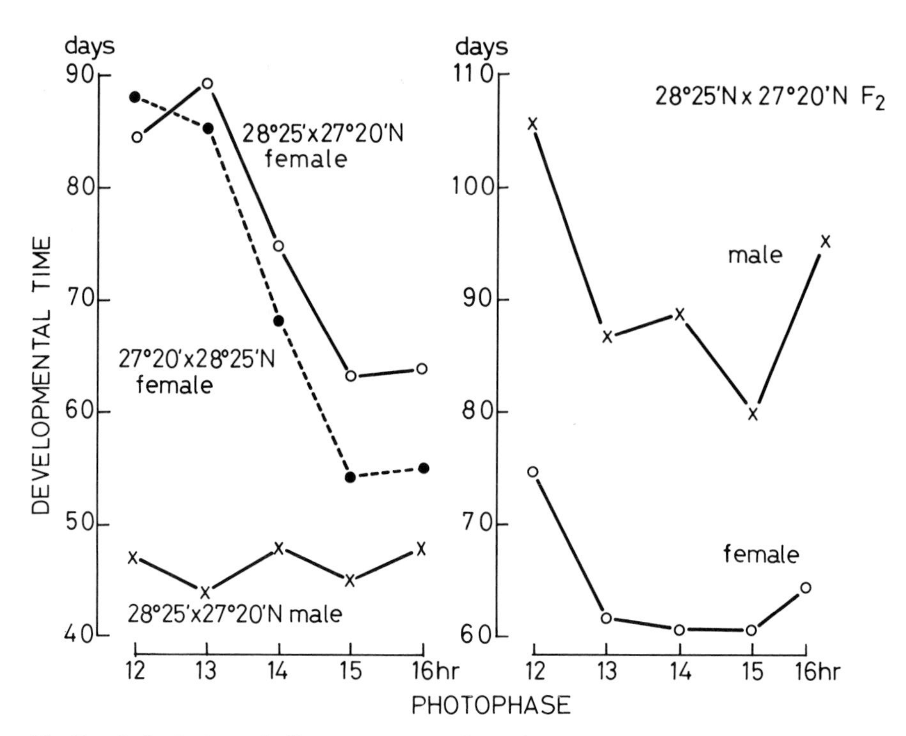

Fig. 12. Nymphal photoperiodic responses of F_1 (left) and F_2 hybrids between the temperate (28° 25'N) and subtropical (27° 20'N) forms of the lawn ground cricket.

the subtropical form elsewhere predominates. The distribution of the two forms is, therefore, parapatric rather than allopatric.

In the laboratory, the two forms easily interbred and produced viable eggs, of which the incubation period was remarkably shortened even when the subtropical form was the male parent (see Section 3). The nondiapause trait was, however, not completely dominant. The hybrid nymph showed very interesting abnormalities (Fig. 12). The F_1 females developed more or less normally, and were highly sensitive to photoperiod. The F_1 males, on the other hand, were quite abnormal. When the mother was the temperate form, they matured in an abnormally short period as compared with the females, and completely lost their photoperiodic response. As a result, there was a wide gap in time of the adult emergence between the two sexes. When the mother was the subtropical form, the F_1 males survived more than 200 days, but in the long run died off without maturing. When obtained, both sexes of F_1 adults were fertile and viable F_2 were obtained. Contrary to the F_1 generation, their development was, on the average, faster in the female than in the male (Fig. 12). There were, however, enormous variations among individuals of both sexes that obscured their photoperiodic responses.

Several other crosses were also made between different temperate and subtropical populations. The photoperiodic response of F_1 females varied with the geographic origin of the parents, but in other respects the results were always consistent. There was a complete breakdown of the photoperiodic response in the F_1 males. There was considerable variation in development time in the F_2 of both sexes. These results may be taken as a manifestation of genetic incompatibility between the

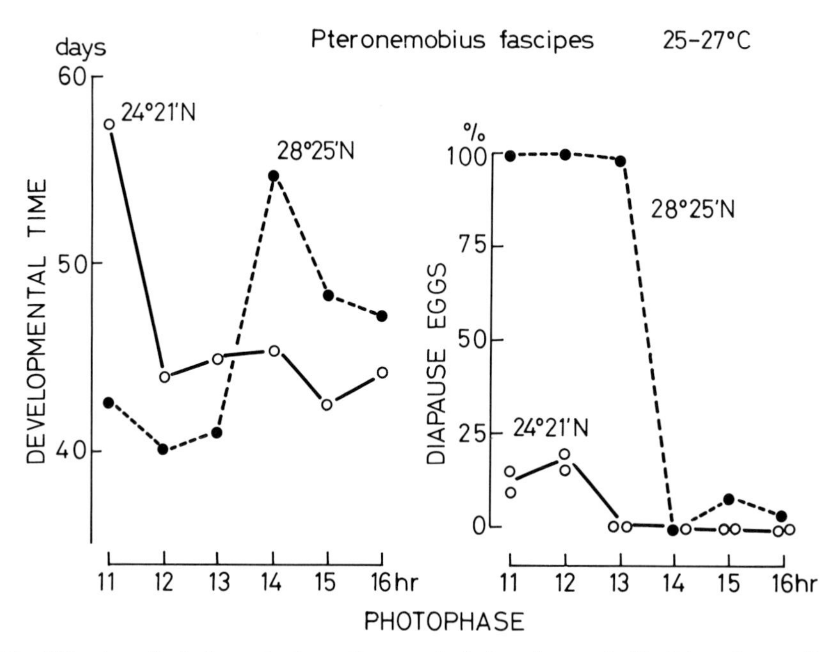

Fig. 13. Effects of photoperiod on the nymphal development (left) and egg diapause in the temperate (28°25'N) and subtropical (24°21'N) forms of the band-legged ground cricket.

temperate and subtropical forms. This fact, together with their distinct seasonal adaptations, suggests that the two forms represent different species.

6.2. Climatic Forms in *P. fascipes*

Whatever the historical background of their present parapatric relationship might be, the divergence in seasonal strategy of the temperate and subtropical forms of *P. taprobanensis* must be the result of natural selection by different climatic conditions: nymph-overwintering is favored in the subtropical climate; egg-overwintering, in the temperate climate. If so, there may be similar divergence in other crickets with similar developmental physiology. This expectation led me to further search for a parallel situation, which was encountered in another ground cricket, *P. fascipes*. At the beginning of my studies, I had suspected that this species did not occur to the south of 28°N, but a few years ago Mr. Matsuura and I found it on the islands of Isigaki, Iriomote, and Yonaguni, located at about 24°N.

The population from 28°N (Amami) showed an intermediate type of photoperiodic response in nymphal development that is prevalent among the bivoltine temperate populations, while the population from 24°N displayed a long-day type response (Fig. 13). In response to a short photoperiod, moreover, the former laid diapause eggs, but such eggs were scarce in the latter, although the photoperiodic effect was still perceived. From these responses, the nymph seemed to be the more common stage of hibernation in the subtropical form.

The cross-breeding tests yielded results which were in important points very similar to those obtained with the previous species (Fig. 14).

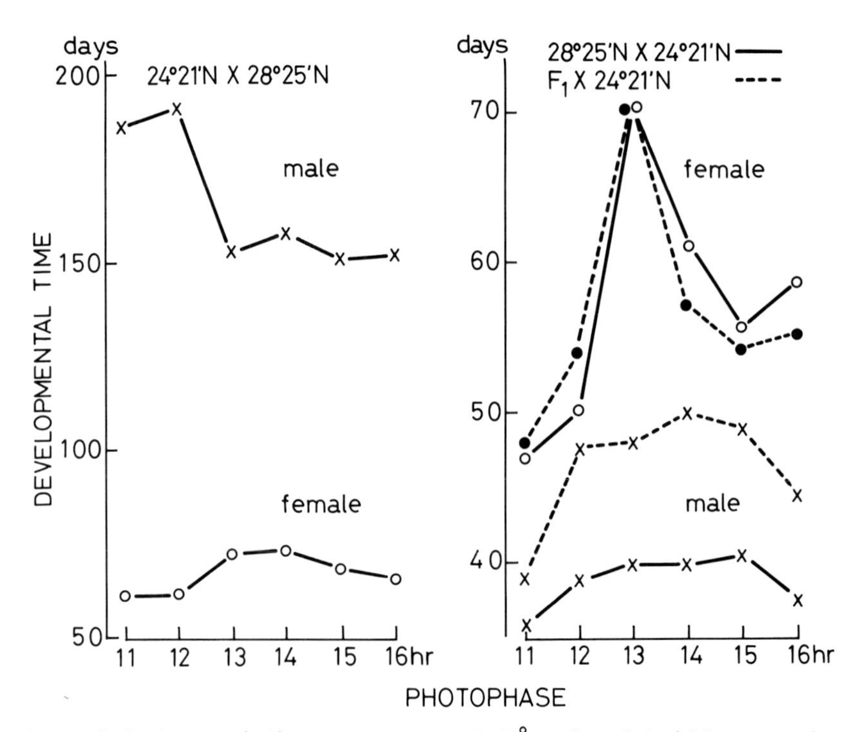

Fig. 14. Nymphal photoperiodic responses at 25-27°C of F_1 hybrid between the temperate (28°25'N) and subtropical (24°21'N) forms of the band-legged ground cricket and its backcross to the latter.

When the mother was the subtropical form the F_1 males were remarkably retarded in development, although they attained the adult stage. In the reverse cross, the photoperiodic response disappeared in the F_1 males, but the F_1 females were highly responsive to photoperiod and their development was retarded by a photoperiod intermediate between the retarding photoperiods for the parents. The F_1 males seemed to be sterile. The F_1 females were fecund and backcross generations were obtained, but again they show abnormal development similar to that of the F_1 generation.

The situation is thus quite similar to that found in *P. taprobanensis*. In both species, the life cycle with an egg diapause in the north is replaced in the south by the one virtually free of egg diapause, this shift being accompanied by a reversal of the photoperiodic regulation of nymphal development. In both species, there is genetic incompatibility between the two "climatic forms," as indicated by similar abnormalities in the development of hybrids.

6.3. Climatic Speciation

In the subtropical climate, a nymph seems to provide a better means of hibernation than an egg. As the winter is not persistently cold, there is always a high probability of untimely development, which is, however, not immediately detrimental because of the mild weather. Yet the winter conditions are not favorable at all for development and reproduction. If insects continue their activities under such conditions, they may suffer a negative *r* value, so that an escape strategy to postpone activities by diapause or similar means until spring is selected for (discussion with K. Ohta). The developmental regulation in this case can be accomplished more efficiently by relying upon the photoperiodic cue. A nymph with its well-developed neurosecretory system may be able to do this, but an egg at an early stage of embryogenesis (Section 2) might not be equipped with any efficient clock mechanism to measure photoperiod. The egg diapause seems, on the other hand, to be better adapted to a longer and persistently cold winter. This supposition comes from the general tendency in Japan for many of the nymph-overwintering crickets to be restricted to the southern half of the country. Although the reason for this is not yet completely clear, "buried" eggs seem to be protected better than are free-living nymphs. Adaptations to temperate and subtropical climates seem to require basically different means of life-cycle regulation, at least in some cases. This splitting of life cycles should have a great impact on the species status whether it occurs in the presence of a topographic barrier or not. Extrinsic isolation alone may favor, but not always result in, speciation in a given restricted time-span. Each species of organism has its own way of adaptation, which can be distinguished from all others. Divergence in adaptation is undoubtedly the most important feature in the process of speciation. It is therefore important to know the factors which accelerate divergence in adaptation. It seems that the climatic gradient on the earth's surface is one of the most important and universal factors, at least for insects which are evolutionarily more or less flexible in their seasonal regulation system.

If a species is split into two distinct entities due to divergence in climatic adaptations as argued above, "climatic speciation" may be an appropriate term to designate such a situation (Masaki, 1972), since a geographic barrier, even if it exists, is not the primary cause of speciation. The possibility that this kind of speciation can occur without any extrinsic isolation at least deserves to be examined more seriously. Since the climatic conditions vary in a linear gradient from north into the temperate region, but also replace the other cycle to the north of the balancing point owing to its superior fitness. The egg-overwintering cycle evolves near the northern margin of a nymph-overwintering population, two different results may occur according to the voltinism. In the case of a univoltine cycle, this divergence will result in a temporal isolation. In the case of a multivoltine cycle, however, there may be overlapping of breeding activities of the different cycles. The egg-overwintering one will not only extend further north into the temperate region, but also replace the other cycle to the north of the balancing point owing to its superior fitness. The situation thus becomes very close to a parapatric relationship. It seems likely that disruptive selection would be effected in the narrow overlapping zone, which may enhance the evolution of isolating mechanisms. This is perhaps a somewhat modified version of one of the speciation models proposed by Alexander (1968).

This model postulates the occurrence of an egg-overwintering cycle near the northern margin of a nymph-overwintering population. Such a situation is, in fact, indicated by both ground crickets mentioned above. In *P. taprobanensis*, populations of the subtropical form from the Ryukyu arc showed hatching peaks in the third week of incubation at 25°C, but a very few eggs took as long as 40-60 days or even more. The percentages (sample size) of such eggs were 2.1 (10355), 0.3 (7186), 0.16 (8241), 0.09 (26947), and 0.08 (9074) in populations from the islands of Toku, Okinawa, Miyako, Isigaki, and Yonaguni, respectively, in the order from northeast to southwest. The nature of these eggs, i.e., whether or not this delay really represents diapause, is very difficult to examine because of their extremely small numbers. In *P. fascipes*, however, the egg diapause of the subtropical form is clearer, and even its photoperiodic control could be perceived at least in the populations derived from Isigaki and Iriomote (Fig. 13). On these islands, therefore, the species may be double-cycled, but the very low incidence of egg diapause suggests the nymph-overwintering cycle being predominant. This occurrence of a few diapause eggs cannot be ascribed to gene flow from the temperate region, because there is no sign of morphological introgression.

In both species, the two climatic forms are morphologically distinct from one another. The subtropical adults are smaller in size with a relatively shorter ovipositor, and lay smaller-sized eggs than the temperate adults. In *P. taprobanensis*, in addition to the above distinction, the latter have a smaller number of file teeth and produce a longer buzzing sound, a trill with a very high pulse rate, than the former. These characters have not been closely examined as yet in *P. fascipes*. In both species, the subtropical forms are closer to the tropical population than to the temperate ones in size, conformation and general color of the body.

7. A Century of Natural Selection

Evidence for evolution of seasonal adaptations is usually circumstantial. A rare chance for witnessing such a historical process may be provided only by recent immigrants establishing themselves in areas climatically different from their homeland. It seems likely that their seasonal adaptations may be modified by selection. This kind of "evolutionary experiment in nature" seems to be going on in an isolated group of small islands located at about 27°N, about 900 km south of the main island of Japan and about 1300 km east of the Ryukyu arc. It is called the Bonins, the name derived from a Japanese word meaning uninhabited, because the islands were not inhabited by man when discovered in 1593. Its isolated location has made this island group full of unique forms of life. More than one-third of the recorded insects have been identified as endemic species or subspecies. The climate is oceanic and subtropical, with an annual mean temperature of 22.6°C and only a 10°C difference between the highest and lowest monthly means. The temperature does not fall persistently below the developmental threshold for most insects even in winter.

P. taprobanensis occurs on at least three islands of this group. This is the only outdoor cricket common to the main archipelago and the Bonins, and therefore suspected to be a relatively recent immigrant in the latter. Since there are two distinct climatic forms in this nominal species, we can inquire which part of the Japanese archipelago is the original place of the Bonin population by analysis of its seasonal adaptation

As shown in Figure 15, most eggs of the Bonin population hatched within three weeks at 25°C, even when their parents were reared under short-day conditions. There were eggs, though very few in number, persisting for a period longer than 100 days. The hatching curve was very similar to that in the Ryukyu populations which are comparable to the Bonin in both the climate and latitude of the habitat. The virtual absence of egg diapause is what might be expected from the mild climate of the islands. The percentage diapause was persistently low at any photoperiod within the range from LD 11:13 to 16:8, being clearly different from that of the temperate form. So far as this finding suggests, the subtropical origin of the Bonin population seems to be the more plausible hypothesis.

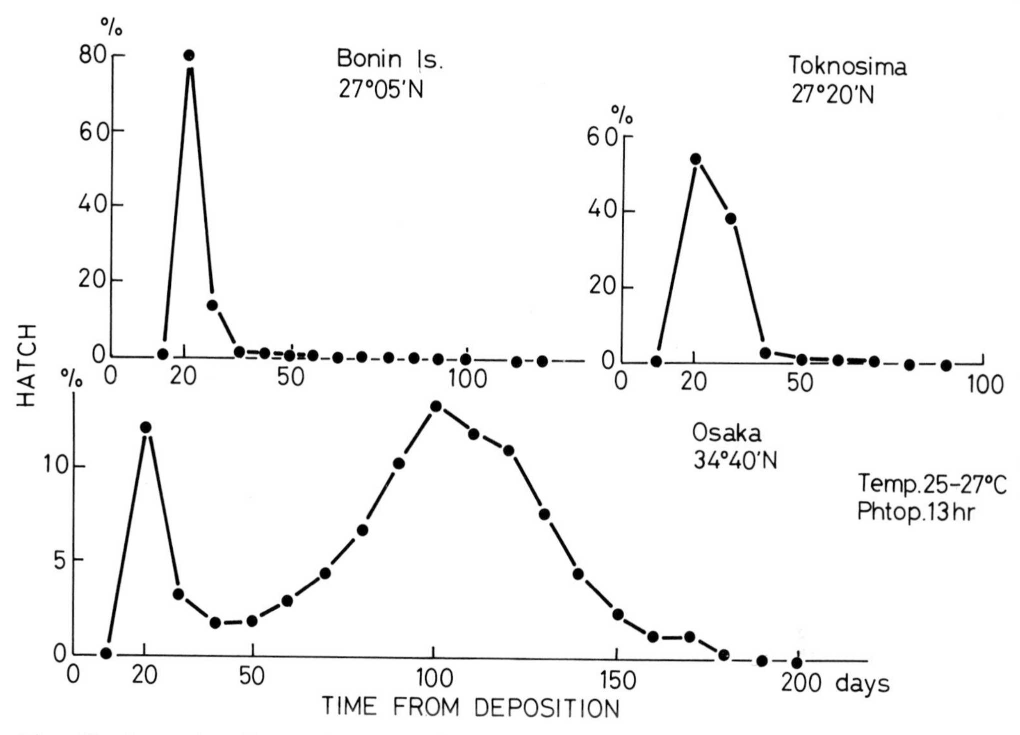

Fig. 15. Comparing the Bonin population of the lawn ground cricket with the temperate (34°40'N) and subtropical (27°20'N) forms in respect of hatching time.

It was therefore surprising to find that Bonin nymphs revealed a conspicuous photoperiodic response of the short-day type very similar to that in the temperate form at about 35°N (Figs. 9, 10, 16). The Bonin population thus represents a disharmonized combination of the northern and southern characteristics. In the Bonins, the cricket was heard singing throughout the year (Arai, personal communication). Owing to the low latitude of the islands, the longest day does not reach 15 hr so that there would not be any discernible retarding effect of photoperiod. At any rate, the nymphal photoperiodic response clearly suggests the northern origin. We are thus in the dilemma of the two opposing hypotheses.

A series of crossing experiments was then carried out to test the genetic compatibility of the Bonins with the temperate and subtropical forms. When the Bonin cricket was crossed to a subtropical population, the hybrid showed a response which was in important points quite similar to that of the hybrid between the temperate and subtropical forms (Fig. 17). The difference between the two results was only quantitative. In both cases, there were rapid development and complete breakdown of photoperiodic response in the F_1 males, and a clear photoperiodic response in the F_1 females (see Section 6.2). Likewise, in the F_2 generation, the females emerged faster than the males, and there were enormous variations among individuals. When the Bonin cricket was crossed to the temperate form derived from the main island (35°N), the nymphal development was normal in both sexes of either F_1 or F_2. The implication is obvious. The Bonin population is genetically closer to the temperate form than to the subtropical form. A cross between sub-

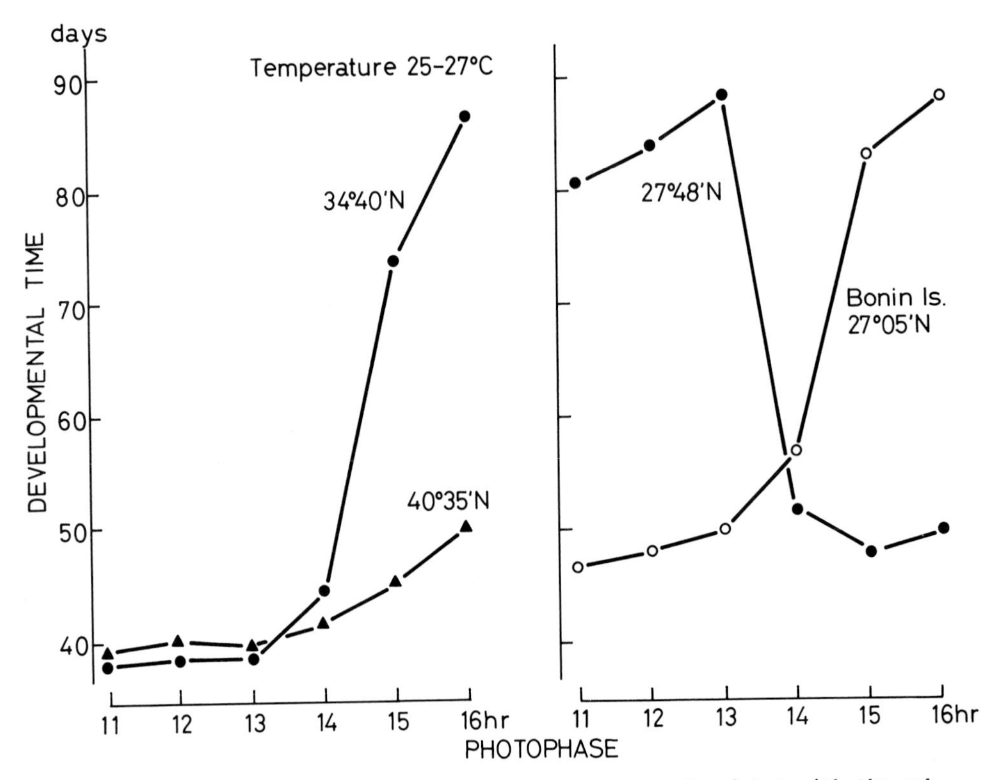

Fig. 16. Comparing the Bonin population of the lawn ground cricket with the subtropical (27° 48'N) and temperate (34° 40' and 40° 35'N) forms in respect of the nymphal photoperiodic response.

tropical populations from distant islands yielded hybrids quite normal in every respect (Fig. 18).

Further evidence in support of the northern origin of the Bonin population came from the geographic trend of variation in a morphometrical trait (Fig. 19). In this species, the length of ovipositor shows a very interesting and regular geographic variation. The populations to the north of 40° N are characterized by a long ovipositor. The ovipositor becomes shorter southwards from 40 to about 33° N, and remains more or less constant in length from 33 to 28° N, the southern limit of the temperate form. Further south in the range of the subtropical form, the ovipositor is shortened abruptly, and a distinct gap occurs. Although the Bonin is latitudinally within the range of the subtropical form, it is without doubt comparable to the temperate form to the south of 35° N in the length of ovipositor. Thus, the morphological trait, genetic compatibility, and nymphal photoperiodic response, all fit in one picture. The Bonin population must have been derived from the north, most probably from the middle part of the main island.

The time of its immigration was probably after the initiation of man's exploitation activities on the islands about 100 years ago. Since this cricket, like many of its relatives, deposits eggs into the soil, there was a high possibility of its eggs being transported with some seedlings from the main island. If such eggs were diapausing, they could have endured the journey quite easily. Since cold temperatures are not

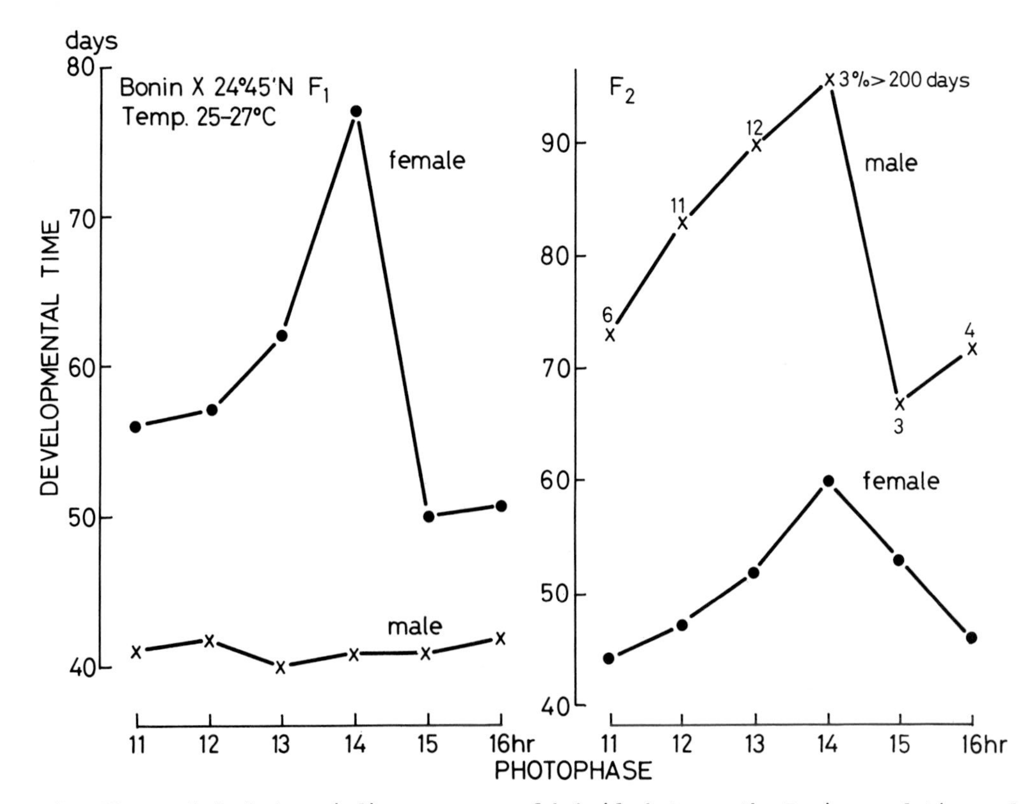

Fig. 17. Nymphal photoperiodic responses of hybrids between the Bonin population and the subtropical form of the lawn ground cricket.

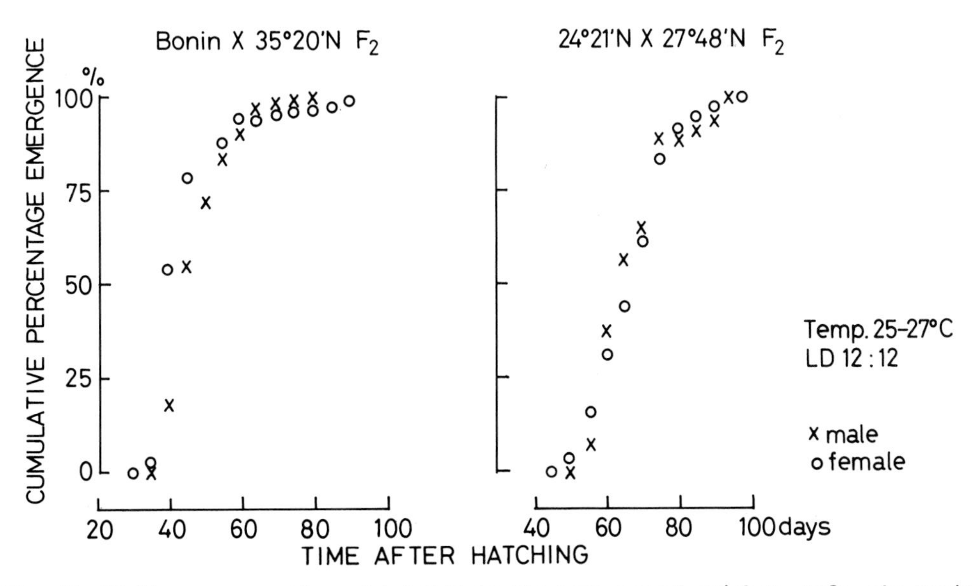

Fig. 18. Adult emergence time of hybrids in the lawn ground cricket. Left, the Bonin population X temperate form. Right, subtropical form X subtropical form.

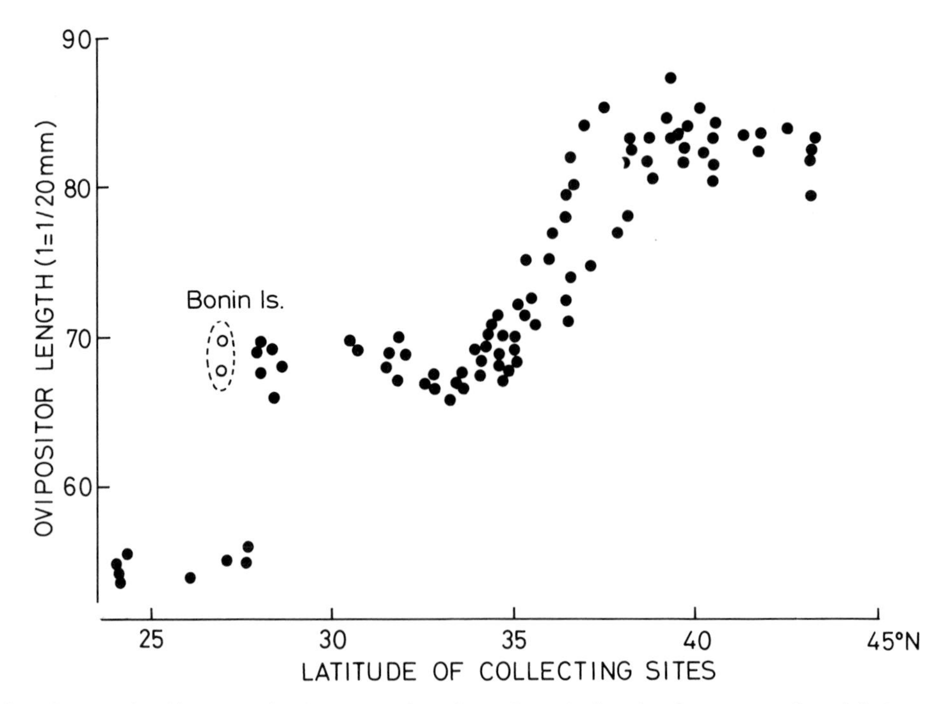

Fig. 19. Latitudinal variation of ovipositor length in the lawn ground cricket.

indispensable for termination of their diapause, many of the introduced eggs might have been able to hatch even in the subtropical climate. Since then, selection by the mild oceanic climate under short-day conditions has been operating against the egg diapause, which has virtually been lost during the last 100 years. The short-day response of the nymph has, on the other hand, persisted as a relic physiological trait, because the delaying action of long days on the nymphal growth has never been effected owing to the low latitude. It has thus not been a target of natural selection. This may be the reason for the disharmonized combination of the virtual absence of egg diapause and short-day type development of the nymph in the Bonin cricket.

As mentioned in Section 3, there are many laboratory experiments in which nondiapause strains have been established from temperate, diapause populations through several generations of selection. The Bonin population of *P. taprobanensis* proves that such a selective response can indeed occur in nature. Similar evolutionary changes might have taken place in other immigrant species not only on the Bonins but also in other parts of the world, particularly remote oceanic islands such as the Hawaiian islands where a number of insects have been introduced from the temperate region. About 20-50 years ago, several immigrant species there were observed to undergo diapause in spite of the equatorial climate (Nishida, 1955). Investigation of what has occured in the last 20-50 years and what will occur in the future in those insects will give important information for understanding the seasonal evolution of insects. Evolutionary changes that might have occurred after the introduction of predatory, parasitic, or pest insects in various countries also deserve intensive studies, particularly when their original places and time of shipment are precisely known. Analysis of their seasonal adaptation may also be instrumental for deducing the original place of invading species, as demonstrated by the Bonin population of *P. taprobanensis*.

8. Concluding Remarks

Diapause has evolved in different species of crickets recurrently and independently of one another at either the egg or nymphal stage, presumably due to the potential capability at these particular stages to evolve cold hardiness. Similar seasonal life cycles found among different species in the temperate regions represent convergent evolution through similar selection pressures on developmental cycles with similar adaptive potentiality.

The genetic system controlling the seasonal development is, in an evolutionary sense, more or less versatile. Cricket populations are thus able to respond to differential selection in the laboratory by changing their developmental programs. This genetic variability enables them to adapt to a wide variety of climatic conditions without deteriorating their harmony with the particular seasonal cycle in each habitat.

The selective response of the development cycle is well illustrated by geographic variations in crickets, which suggest that the outcome of selection is determined by the interplay of external (climatic conditions) and the internal (the type of life cycle) factors. The simplest pattern of variation is found in univoltine field crickets with an egg diapause, which show latitudinal as well as altitudinal clines in the intensity of diapause, the duration of nymphal development, and the degree of long-day retardation of nymphal development, all increasing from north to south in parallel to the climatic gradient. Since the adult size varies as a function of the developmental time, this climatic adaptation is accompanied by a size cline.

Geographic variations due to climatic selection take more complicated patterns, if the seasonal strategy involves a variable voltinism as in certain species of ground crickets. Their adult size increases southwards within each of the univoltine and bivoltine areas, but the trend is reversed in the transitional zone between the two areas. This complication reflects a shift in selection pressure on the development time in accordance with a shift in voltinism, which is effected by variations in the photoperiodic regulations of nymphal development and diapause induction. The intensity of diapause also varies with voltinism, and tends to be weaker with an increasing number of generations per season. The different phases of developmental cycles are thus tightly coadapted to form an integrated system of seasonal strategy.

All these variations are more or less continuous, even when they are complicated by a variable voltinism. This is possibly due to the automatic compensation of hybrid responses to photoperiod, and there is no evidence of disruptive selection in the transitional area where univoltine and bivoltine cycles coexist. So far as the variation is continuous, the species integrity will be maintained.

If, however, a shift in the basic type of life cycle, i.e., from a nymph-overwintering to an egg-overwintering one, occurs, disruptive selection would be effected. In certain ground crickets, it is suggested that the two types of life cycle are adapted better to different climatic zones, the nymph-overwintering cycle to the subtropical climate and the egg-overwintering one to the temperate climate. The geographic replacement of one type by the other would therefore occur, and the border of the two types would be determined by the balance of their fitnesses.

This divergence in the overwintering stage is associated with divergence in the photoperiodic regulation of nymphal growth. Such an overall reorganization of the coadapted system would result in a breakdown of the species integrity. In the two species of ground crickets, at least, the temperate and subtropical forms show hybrid abnormalities and morphological differences. In theory, parapatric speciation due to divergent evolution of this sort seems to be possible without any topographic barrier. The evolution of escape strategy from the seasonal adversity of environment seems to deserve intensive studies not only for its own ecological importance, but also for its significance in speciation phenomena.

A long-term investigation of immigrant populations may shed light on the evolution of seasonal strategy, and an example is afforded by the selective response of a ground cricket introduced from a temperate to a subtropical locality. Presumably in less than a century, this species has virtually lost its ability to enter an egg diapause.

Acknowledgments I am grateful to Dr. H. Dingle for inviting me to this symposium and also for his comments and criticism on the manuscript. Thanks are also extended to Drs. M.J. and C.A. Tauber for their critical reading of the manuscript, and to Messrs. T. Arai and A. Yamamoto for collecting materials from the Bonin islands. This work was supported by grants 61298, 61334, 91256, 92715, 811205, and 056014 from the Ministry of Education of Japan.

References

Alexander, R.D.: Life cycle origins, speciation, and related phenomena in crickets. Quart. Rev. Biol. *43*, 1-41 (1968).

Alexander, R.D., Bigelow, R.S.: Allochronic speciation in field crickets, and a new species, *Acheta veletis*. Evolution *14*, 334-346 (1960).

Alexander, R.D., Meral, G.: Seasonal and daily chirping cycles in the northern spring and fall field crickets, *Gryllus veletis* and *G. pennsylvanicus*. Ohio J. Sci. *67*, 200-209 (1967).

Ando, Y., Miya, K.: Diapause character in the false melon beetle, *Atrachya menetriesi* Faldermann, produced by crossing between diapause and nondiapause strains. Bull. Fac. Agric. Iwate Univ. *9*, 87-96 (1968).

Ankersmit, G.W.: Voltinism and its determination in some beetles of cruciferous crops. Meded. Landbouwhogeschool *64*, 1-60 (1964).

Beck, S.D.: Insect Photoperiodism. New York: Academic Press 1968, viii + 288 pp.

Bigelow, R.S.: Evolution in the field cricket, *Acheta assimilis* Fab. Can. J. Zool. *36*, 139-151 (1958).

Bigelow, R.S.: Interspecific hybrids and speciation in the genus *Acheta* (Orthoptera, Gryllidae). Can. J. Zool. *38*, 509-524 (1960a).

Bigelow, R.S.: Developmental rates and diapause in *Acheta pennsylvanicus* (Burmeister) and *Acheta veletis* Alexander and Bigelow (Orthoptera: Gryllidae). Can. J. Zool. *38*, 973-988 (1960b).

Bigelow, R.S.: Factors affecting developmental rates and diapause in field crickets. Evolution *16*, 396-406 (1962).

Bigelow, R.S., Cochaux, S.A.: Intersterility and diapause differences between geographical populations of *Teleogryllus commodus* (Walker) (Orthoptera: Gryllidae). Aust. J. Zool. *10*, 360-366 (1962).

Brookes, H.M.: The morphological development of the embryo of *Gryllulus commodus* Walker (Orthoptera: Gryllidae). Trans. Roy. Soc. S. Aust. *75*, 150-159 (1952).

Browning, T.O.: The influence of temperature on the completion of diapause in the eggs of *Gryllulus commodus* Walker. Aust. J. Sci. Biol. Sci. *5*, 112-127 (1952a).

Browning, T.O.: On the rate of completion of diapause development at constant temperatures in the eggs of *Gryllulus commodus* Walker. Aust. J. Sci. Res. *35*, 344-353 (1952b).

Bush, G.L.: Sympatric host race formation and speciation in frugivorous flies of the genus *Rhagoletis* (Diptera, Tephritidae). Evolution *23*, 237-251 (1969).

Chopard, L.: Orthopterorum Catalogus. Pars 10, Gryllides. Gravenhage: Uitg. W. Junk N.V. 1967, 211 pp.

Chopard, L.: Fauna of India. Orthoptera, Vol. 2, Grylloidea. Calcutta: Zool. Surv. India 1969, xviii + 421 pp.

Cousin, G.: Essai d'analyse de la speciation chez quelques gryllides du continent America. Bull. Biol. France et Belg. *95*, 155-174 (1961).

Danilevskii, A.S.: Photoperiodism and Seasonal Development of Insects. (In Russian.) Leningrad: Leningrad Univ. Press 1961, 244 pp.

Danilevsky, A.S., Goryshin, N.I., Tyshchenko, V.P.: Biological rhythms in terrestrial arthropods. Ann. Rev. Entomol. *15*, 201-244 (1970).

Fontana, P.G., Hogan, T.W.: Cytogenetics and hybridization studies of geographic populations of *Teleogryllus commodus* (Walker) and *T. oceanicus* (Le Guillou) (Orthoptera: Gryllidae). Aust. J. Zool. *17*, 13-35 (1969).

Fuzeau-Braesch, S.: Etude de la diapause de *Gryllus campestris* (Orthoptera). J. Insect Physiol. *12*, 449-455 (1966).

Gabbutt, P.D.: The bionomics of the wood cricket, *Nemobius sylvestris* (Orthoptera: Gryllidae). J. Anim. Ecol. *28*, 15-42 (1959).

Hogan, T.W.: The onset and duration of diapause in eggs of *Acheta commodus* (Walk.) (Orthoptera). Aust. J. Biol. Sci. *13*, 14-29 (1960a).

Hogan, T.W.: The effects of subzero temperatures on the embryonic diapause of *Acheta commodus* (Walk.) (Orthoptera). Aust. J. Biol. Sci. *13*, 527-540 (1960b).

Hogan, T.W.: Some diapause characteristics and interfertility of three geographic populations of *Teleogryllus commodus* (Walk.) (Orthoptera: Gryllidae). Aust. J. Zool. *13*, 455-459 (1965).

Hogan, T.W.: Physiological differences between races of *Teleogryllus commodus* (Walker) (Orthoptera: Gryllidae) related to a proposed genetic approach to control. Aust. J. Zool. *14*, 245-251 (1966).

Ismail, S., Fuzeau-Braesch, S.: Programmation de la diapause chez *Gryllus campestris*. J. Insect Physiol. *22*, 133-139 (1976).

Lees, A.D.: The Physiology of Diapause in Arthropods. Cambridge: Univ. Press 1955, x + 151 pp.

Levins, R.: Evolution in Changing Environments. Princeton: Princeton Univ. Press 1968, vii + 120 pp.

Lloyd, J.E., Oace, A.E.: Seasonality in northern field crickets. Fla. Entomol. *58*, 31-32 (1975).

MacFarlane, J.R., Drumond, F.H.: Embryonic diapause in a hybrid between two Australian species of field crickets, *Teleogryllus* (Orthoptera: Gryllidae). Aust. J. Zool. *18*, 265-272 (1970).

Masaki, S.: Thermal relations of diapause in the eggs of certain crickets (Orthoptera: Gryllidae). Bull. Fac. Agric. Hirosaki Univ. *6*, 5-20 (1960).

Masaki, S.: Geographic variation of diapause in insects. Bull. Fac. Agric. Hirosaki Univ. *7*, 66-98 (1961).

Masaki, S.: The influence of temperature on the intensity of diapause in the eggs of the Emma field cricket (Orthoptera: Gryllidae). Kontyu *30*, 9-16 (1962).

Masaki, S.: Adaptation to local climatic conditions in the Emma field cricket (Orthoptera: Gryllidae). Kontyu *31*, 249-260 (1963).

Masaki, S.: Geographic variation in the intrinsic incubation period: a physiological cline in the Emma field cricket (Orthoptera: Gryllidae: *Teleogryllus*). Bull. Fac. Agric. Hirosaki Univ. *11*, 59-90 (1965).

Masaki, S.: Photoperiodism and geographic variation in the nymphal growth of *Teleogryllus yezoemma* (Ohmachi et Matsuura). Kontyu *34*, 277-288 (1966).

Masaki, S.: Geographic variation and climatic adaptation in a field cricket (Orthoptera: Gryllidae). Evolution *21*, 725-741 (1967).

Masaki, S.: Photoperiodism and seasonal life cycle of crickets. (In Russian.) In: Problems of Photoperiodism and Diapause in insects. Lenengrad: Leningrad Univ. Press 1972, pp. 25-50.

Masaki, S.: Climatic adaptation and photoperiodic response in the band-legged ground cricket. Evolution *26*, 587-600 (1973).

Masaki, S., Ohmachi, F.: Divergence of photoperiodic response and hybrid development in *Teleogryllus* (Orthoptera: Gryllidae). Kontyu *35*, 83-105 (1967).

Masaki, S., Oyama, N.: Photoperiodic control of growth and wing-form in *Nemobius yezoensis* Shiraki (Orthoptera: Gryllidae). Kontyu *31*, 16-26 (1963).
Nagatomo, T.: Genetic studies on the voltinism in the silkworm. (In Japanese.) Bull. Fac. Agric. Kagoshima Univ. *2*, 1-70 (1953).
Nishida, T.: The phenomenon of arrested development in the Hawaiian Islands. Proc. Hawai. Entomol. Soc. *15*, 575-582 (1955).
Ohmachi, F., Masaki, S.: Intercrossing and development of hybrids between the Japanese species of *Teleogryllus* (Orthoptera: Gryllidae). Evolution *18*, 405-416 (1964).
Ohmachi, F., Matsuura. I.: Observations and experiments on four types in the life history of the Gryllodea. (In Japanese.) J. Appl. Zool. *16*, 104-110 (1951).
Rakshpal, R.: Morphological development of the embryo in diapause and post-diapause eggs of *Gryllus pennsylvanicus* Burmeister (Orthoptera: Gryllidae) and a comparison with non-diapause species of the genus *Gryllus*. Zool. Anz. *168*, 46-53 (1962a).
Rakshpal, R.: Diapause in the eggs of *Gryllus pennsylvanicus* Burmeister (Orthoptera: Gryllidae). Can. J. Zool. *40*, 179-194 (1962b).
Rakshpal, R.: The effect of cold on pre- and post-diapause eggs of *Gryllus pennsylvanicus* Burmeister (Orthoptera: Gryllidae). Proc. R. Entomol. Soc. Lond. (A) *37*, 117-120 (1962c).
Rakshpal, R.: Diapause in the eggs of *Nemobius allardi* Alexander and Thomas (Orthoptera: Gryllidae: Nemobiinae). Zool. Anz. *173*, 282-288 (1964).
Saeki, H.: The effect of the day-length on the occurrence of the macropterous form in a cricket, *Scapsipedus aspersus* Walker (Orthoptera: Gryllidae). Japan. J. Ecol. *16*, 49-52 (1966).
Saunders, D.S.: Insect Clocks. New York: Pergamon Press 1976, viii + 280 pp.
Shiraki, T.: Orthoptera of the Japanese Empire. Part 1. (Gryllotalpidae and Gryllidae). Ins. Mats. *4*, 181-252 (1930).
Tamura, H., Koseki, K.: Population study on a terrestrial amphipod, *Orchestia platensis japonica* (Tattersall) in a temperate forest. Japan. J. Ecol. *24*, 123-139 (1974).
Tanaka, S., Matsuka, M., Sakai, T.: Effect of change in photoperiod on wing form in *Pteronemobius taprobanensis* Walker (Orthoptera: Gryllidae). Appl. Ent. Zool. *11*, 27-32 (1976).
Tauber, M.J., Tauber, C.A.: Insect seasonality: Diapause maintenance, termination, and postdiapause development. Ann. Rev. Entomol. *21*, 81-107 (1976a).
Tauber, M.J., Tauber, C.A.: Environmental control of univoltinism and its evolution in an insect species. Can. J. Zool. *54*, 260-265 (1976b).
Umeya, Y.: Embryonic hibernation and diapause in insects from the view point of the hibernating eggs of the silkworm. (In Japanese.) Bull. Seric. Exp. Sta. *12*, 393-480 (1946).
Walker, T.J.: Experimental demonstration of a cat locating Orthoptera prey by the prey's calling song. Fla. Entomol. *47*, 163-165 (1964a).
Walker, T.J.: Cryptic species among sound-producing Ensiferan Orthoptera (Gryllidae and Tettigonidae). Quart. Rev. Biol. *39*, 345-355 (1964b).
Walker, T.J.: *Gryllus ovisopsis* n. sp.: A taciturn cricket with a life cycle suggesting allochronic speciation. Fla. Entomol. *57*, 13-22 (1974).

Variability in Diapause Attributes of Insects and Mites: Some Evolutionary and Practical Implications

MARJORIE A. HOY

Introduction

Photoperiodic responses of insects and mites have been studied by physiologists, biochemists, ecologists, morphologists, geneticists, and evolutionary biologists. Data obtained using the methodologies of each of these disciplines has led to an understanding of parts of the "diapause phenomenon." This symposium has stressed the evolutionary and ecological aspects of diapause and migration as a means of escape in space and time. This chapter will stress some of the practical implications of the observed intra- and interpopulation variability in diapause responses in insects and mites.

Interpopulation Variability in Photoperiodic Responses

Variability in the photoperiodic responses of geographic populations is known for many species of insects and mites. Danilevskii (1965), Beck (1968), Dingle (1972), Masaki (1965, 1961) and Tauber and Tauber (1973, 1976b) particularly have stressed the adaptiveness of differentiated responses to local climatic conditions. These differentiations are reflected in several aspects of the species' adjustment to local climatic conditions and include the threshold for facultative induction as it relates to temperature and daylength. The temperature limits and optima for diapause termination may also vary, as may the intensity or duration of diapause. Furthermore, species may vary such that certain populations may have an obligatory diapause, while other populations may have a facultative diapause and still other populations may lack a diapause altogether.

Intrapopulation Variability in Photoperiodic Responses

While interpopulation variability and its ecological significance in the adaptation of species to their environment is well recognized (see also Masaki, this volume; Tauber and Tauber, this volume), rather less recognition has been accorded to the variation in diapause attributes *within* populations. This intrapopulation variability has been utilized and recognized primarily when scientists have purposefully subjected insects or mites to genetic selection for diapause and/or nondiapause attributes, or for greater or lesser diapause duration. The role that intrapopulation variability plays in the establishment or nonestablishment of new immigrants or in the distribution of species has been little investigated or documented. The impact of intrapopulation variability on a species' population dynamics is also little understood.

This paper will place considerable emphasis upon the intrapopulation variability in diapause attributes exhibited by many species, and particularly stress the impact of this variability upon applied entomological programs.

Practical Implications of Variability in Diapause Responses

Classical Biological Control Programs

Intra- and interpopulation variability have profound implications for biological control programs. Experiences with a gypsy moth [*Lymantria dispar* (L.)] parasite, *Apanteles melanoscelus* Ratzeburg, will illustrate several points (Hoy, 1975a, b).

A. melanoscelus (Hymenoptera: Braconidae) is a solitary larval endoparasite of the gypsy moth. It was introduced into North America from Sicily in 1911. It established quickly, but its efficacy is limited due in part to heavy hyperparasitism from more than 35 native North American hyperparasites. Adult *A. melanoscelus* emerge in the spring from their overwintering cocoons about the time the gypsy moth eggs hatch and parasitize small host larvae. The new parasite adults produced from the small larvae emerge in June and attack larger gypsy moth larvae. While developing in the larger hosts in June, the parasite enters a photoperiodically induced diapause in the last larval instar (Weseloh, 1973) and then overwinters within a cocoon attached to trees.

While investigating the possibility that hybridization of different geographic strains of *A. melanoscelus* might yield a parasite with increased vigor for inundative releases (Hoy, 1975a), I evaluated the effect of hybridization upon the critical photophase of the hybrid parasite strain (Hoy, 1975b). Photoresponses of colonies originating from Connecticut (CT), France (FR), Yugoslavia (YU), a hybrid between the FR and YU colonies (AB), and a triple hybrid among the CT, FR, and YU colonies (ABC) were evaluated at 24° C under 2-, 8-, 12-, 13-, 15-, 16-, 17-, 18-, 19-, 20-, and 24-hr daylengths.

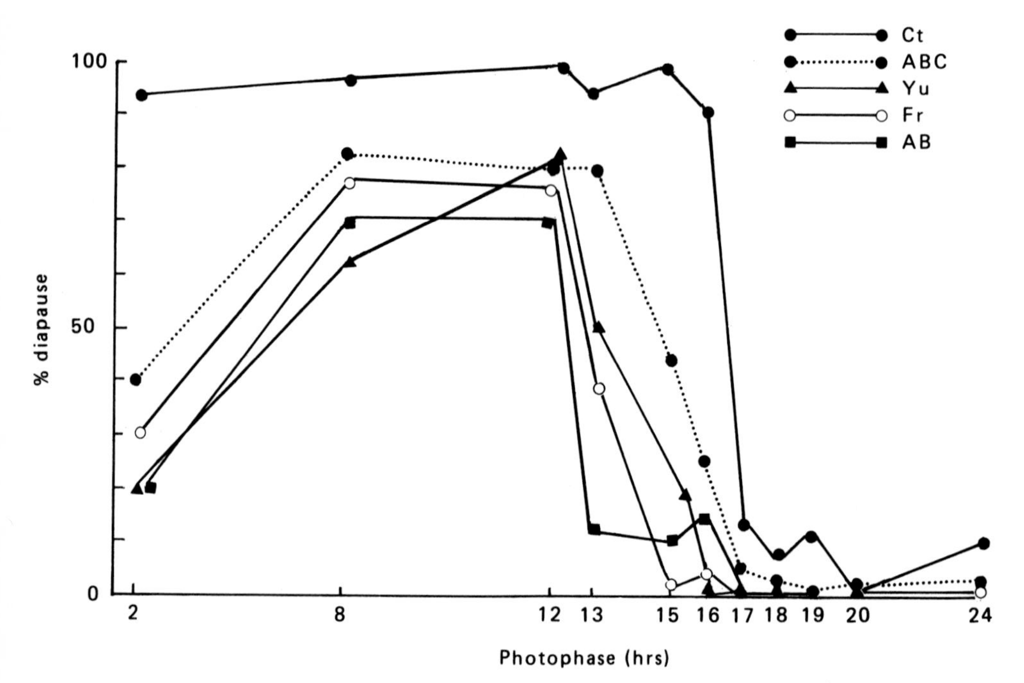

Fig. 1. Photoresponse curves of *A. melanoscelus* colonies reared at 24° C.

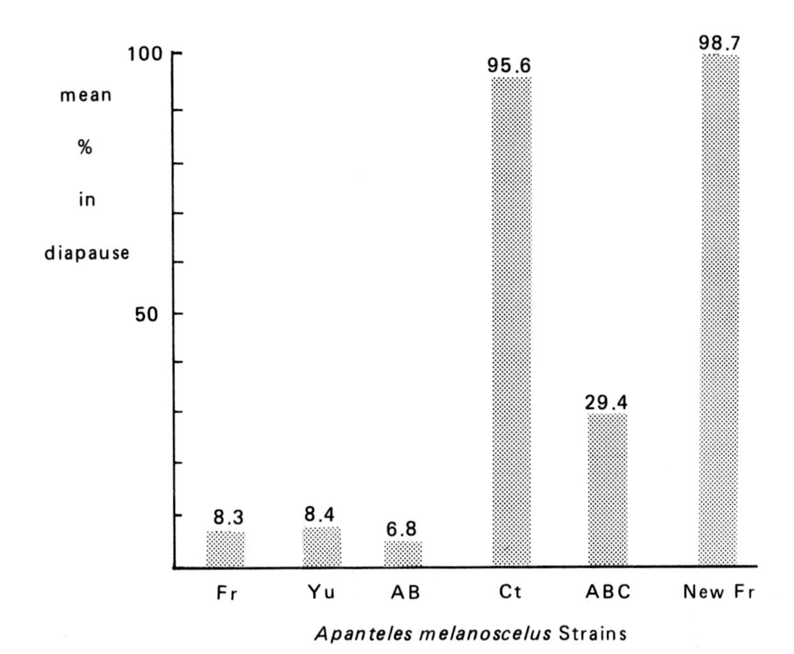

Fig. 2. Mean diapause incidence of *A. melanoscelus* reared outdoors May-July 1974.

The CT colony (Fig. 1) responded quite similarly to the CT colony tested previously by Weseloh (1973). The CT colony's critical photophase for induction is approximately 17 hr and less, but the FR, YU, and AB colonies' critical photophases are approximately 13 hr and less. The ABC hybrid (Fig. 1) has a critical photophase of 15 hr, and thus falls halfway between the responses of the CT and AB hybrid colonies. This intermediate response is consistent with a multigenic model for critical photophase.

When these colonies were reared outdoors from May to July in field cages under natural daylengths and temperatures, the laboratory-determined critical photophases were predictive of the field responses of these colonies. About 7-8% of the FR and YU colonies and of their hybrid (AB) entered diapause (Fig. 2), while 96% of the CT colony did so under these same conditions. The triple hybrid (ABC) again was intermediate in its responses, and about 30% entered diapause. This time interval (June) is the "normal" time when this species enters diapause.

Diapause responses of a freshly collected colony from France (New FR) were comparable to that of the CT colony (Fig. 2). This high level of diapause induction in the New FR colony indicates that the original FR colony might also have been capable of a higher level of diapause induction. Thus, the low response of the FR, of the YU, and of the AB hybrid colonies could have been due to inadvertent genetic selection for lack of a diapause during insectary rearing. Thus, individuals that could develop continuously under short or unregulated daylengths were possibly used to maintain the colony. All colonies were reared under long days (more than 18 hr) to prevent diapause induction after I received them. The CT colony had been colonized a year under these conditions and had maintained its high response to short daylengths. However, the FR and the YU colonies had been reared for 8 and 70 generations, respectively, under uncontrolled daylengths in another laboratory. I therefore suggest that in the interval between field collection and these tests, much of the capacity to diapause under appropriate daylengths had been inadvertently removed. Even if the differences in critical photophase were based upon interpopulation variability, however, the result is the same - a diapause response inappropriate for the area in which the colony was released for permanent establishment. Furthermore, even if the FR or YU parasites interbred with *A. melanoscelus* already present in the area through natural extension of its range, the result is still a parasite not well adapted to this area. The field-produced hybrid's diapause response would probably be similar to that of the ABC hybrid (Fig. 2).

It is also noteworthy that the hybridization of the FR and YU colonies yielded a strain (AB) with a response similar to that of the parent FR and YU colonies (Fig. 2). Thus, hybridization using these two strains did not result in an improved photoperiodic response. It also suggests that the FR and YU colonies are genetically similar and have been subjected to the same selection regime or they would have yielded an increased diapause response upon hybridization due to the masking of several different, but hypothetically recessive genes. Thus, these data suggest that selection had acted similarly and/or that the critical photophase response in this species is inherited in a relatively simple nonquantitative manner.

Inadvertent laboratory selection of other insects, including natural enemies, has been reported relatively often. House (1967) found the parasite, *Pseudosarcophaga affinis* auct. nec Fallen, had inadvertently lost its diapause during insectary production while it was being reared for release against the spruce budworm, *Choristoneura fumiferana* (Clemens). Diapause in this species was considered to be "obligatory" and yet rare individuals capable of development without a diapause apparently provided the intrapopulational variability upon which laboratory selection operated. Glass (1970) found that the oriental fruit moth, *Grapholitha molesta* (Busck), inadvertently lost its ability to diapause over 60 generations of insectary rearing. He later purposefully selected for a nondiapause colony, and diapause declined from 96% to less than 1% by the 24th generation of selection. Barry and Adkisson (1966) found that intrapopulational variability in the pink bollworm, *Pectinophora gossypiella* (Saunders), allowed selection of a strain within 23 generations that was 99.85% free of diapause. During the purposeful selection, the control populations were maintained without intentional selection. Yet, even in this colony, a 25% increase in incidence of nondiapause larvae occurred during the experiment. Lloyd et al. (1967) also discovered the boll weevil, *Anthonomus grandis* Nahrman, had inadvertently lost its ability to diapause during the course of laboratory colonization, and Hodek and Honek (1970) obtained similar results with the pentatomid, *Aelia acuminata* (L.). Finally, Lyon et al. (1972) found that a colony of the western spruce budworm, *Choristoneura occidentalis* Freeman, lost its ability to enter diapause after it had been reared for 20 generations under controlled, nondiapause-inducing environmental conditions.

How many other cases of rapid unintentional selection may also be occuring during colonization of natural enemies for biological control? How many failures of releases for establishment of natural enemies could be due to genetic changes engendered during laboratory rearing rather than due to the inherent climatic nonadaptation (Messenger and van den Bosch, 1971; Messenger, 1971) of the parasites or due to other deficiencies in parasite-host adaptation?

The above examples of inadvertent selection acting upon intrapopulation variability yielded detrimental results. There also exists the possibility that variability in photoperiodic responses can be manipulated in several different ways to yield beneficial results. One method might be to exploit the interpopulational differences in diapause responses to obtain unavailable, but desirable, new strains with a critical photophase of a special, "tailor-made" nature. For example, many species are known to have geographic strains that vary in their photoperiodic responses. Danilevskii (1965) described the photoresponses of Sukhumi and of Leningrad populations of *Apanteles glomeratus* (L.) and of the F_1 and F_2 hybrids between them. These hybrids, like the ABC hybrid of *A. melanoscelus* (Fig. 1), exhibited intermediate photoperiodic response curves. Crosses between different geographic strains of *A. melanoscelus* and other natural enemies could effect a type of genetic improvement by providing natural enemies with desired, but unavailable diapause attributes.

Other potential methods utilizing variability in photoperiodic responses will be discussed later in this paper.

Insectary Rearing Programs

Insects and mites are reared in insectaries for many purposes. Particular rearing programs place different requirements upon the insectaries' product. For example, an insect colony capable of a normal and appropriate diapause upon release into a natural environment may be required for biological control programs, as just discussed. Alternatively, if continuous insectary rearing of hosts for mass production of a nuclear polyhedrosis virus (NPV) or of a beneficial parasite is required, a nondiapausing host strain may be preferable. Thus, it may be desirable to utilize the interpopulation variability and to purposely select for a nondiapausing strain.

This type of genetic manipulation of the gypsy moth, *Lymantria dispar* (L.), has been underway in my laboratory during the past 3 years. The gypsy moth is a polyphagous univoltine lymantriid with a native range extending from western Europe to Korea and Japan (Schedl, 1936). *L. dispar* overwinters as a well-developed larva in diapause within the egg. Egg masses are deposited on trees and other objects in the forest in July and August. Hatch typically occurs the following spring when the leaf buds begin to open on their favored food trees, especially several oak (*Quercus*) species.

The gypsy moth is one of the few species that is still considered to be photoperiodically neutral and thus does not respond to the manipulation of any daylength or temperature combination thus far tested (Leonard, 1968; Masaki, 1956; Hoy, unpublished). Thus, mass rearing of this species for production of beneficial parasites or for nuclear polyhedrosis virus (NPV) is made more difficult because of our inability to manipulate the gypsy moth's diapause. Furthermore, diapause is apparently terminated only after a 90-120-day chill (Masaki, 1956; Schedl, 1936), which makes the generation interval from egg to egg at least six months.

Goldschmidt (1932, 1933) found racial variability in the gypsy moth's diapause duration. He suggested that the interracial differences were polygenically inherited and under a matroclinal influence. The longer diapause duration predominating in the F_1 progeny and the variability exhibited by the F_1 and by the F_2 generations suggested that relatively few genes are involved (Goldschmidt, 1932).

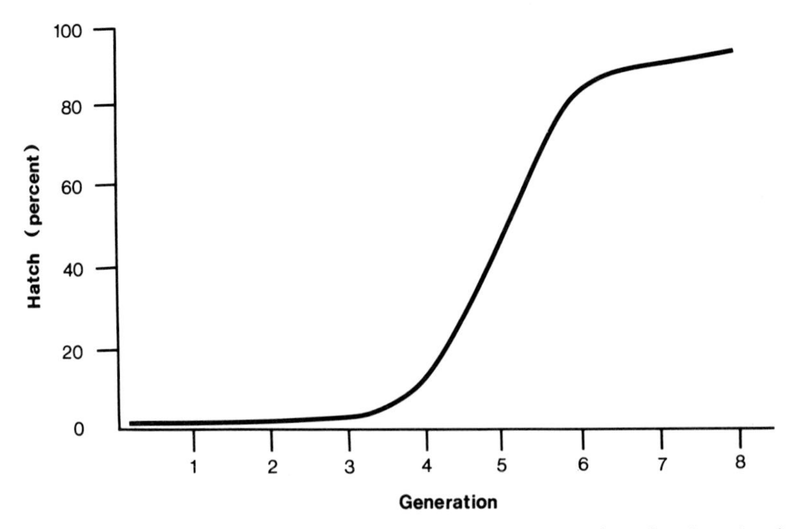

Fig. 3. Response by the gypsy moth, *L. dispar* L., to selection for hatch without chill.

Occasionally, a few gypsy moth larvae hatch in the forest in the fall (Forbush and Fernald, 1896; Schedl, 1936). These rare larvae apparently perish with the onset of cold weather, and I am not aware of any populations of the gypsy moth that are multivoltine. The incidence of the aberrent larvae is always low, but the historical recurrence of these larvae suggested that they might be genetic variants that hatched without a diapause, or that had a diapause of shorter duration. Thus, they might be indicators of intrapopulation variability and be the basis for a selection program for a nondiapausing gypsy moth colony.

The selection program began in October 1973 when I collected 230,000 gypsy moth eggs from northern Connecticut, U.S.A. (Hoy, 1977). These eggs had experienced an unmeasured but undoubtedly small amount of natural chill. After collection, the eggs were surface-disinfected to prevent excessive mortality from the gypsy moth NPV and the number of eggs were estimated volumetrically. The eggs were divided into two lots; half were left under ambient conditions and half were given a "token" chill of two weeks at 5°C, which is much less than that required to terminate diapause in all the eggs (Masaki, 1956). The larvae that hatched were reared at 22°C under an 18-hr daylength on a synthetic diet (Hoy, 1977). During the first four generations, because few larvae hatched, I reared all caterpillars that hatched from both the unchilled as well as the chilled lots of eggs. The brief two-week chill at 5°C doubled the percentage of larvae that hatched from the split egg masses in these early selections.

Response to selection was slow initially (Fig. 3). However, the percentage of larvae that hatched increased from an initial rate of 0.9% to 11% in the F_4 generation. At this point, I increased the selection pressure by selecting egg masses which had a larger proportion hatching, and I also discontinued chilling any eggs. The increased selection pressure yielded a dramatic jump in the F_5 generation and 56% of the eggs hatched. By the F_8 generation, approximately 97% of the eggs hatched without any chill (Fig. 3).

The field-collected eggs used in the initial selection procedure had been deposited in July, but they did not begin hatching in the labora-

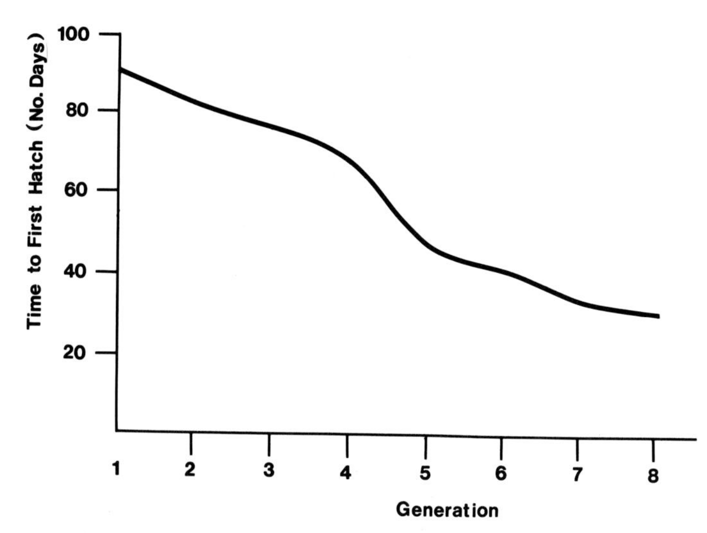

Fig. 4. The mean number of days to initiation of hatch in successive generations of the gypsy moth.

tory until October 30. This was an incubation period of at least 90 days, or close to that which is normal for the gypsy moth (Masaki, 1956). The F_2 egg masses began hatching 82 days after deposition, the F_3 generation required only 76 days on the average to yield the first caterpillars, and this interval from egg mass deposition to initiation of hatch continued to decline during the selection program. By the F_8 generation (Fig. 4), the mean interval from egg mass deposition to first hatch had declined to 31 days. It declined to 28 days in the F_{10} generation, with some egg masses initiating hatch as early as day 25.

Embryonation of gypsy moth eggs at 22°C is estimated to require about 19 days (Schedl, 1936). Thus it appears that the eggs in the selected colony begin hatching shortly after the larva is fully developed. If the unhatched larvae are in diapause between day 19 and day 28 (when the egg masses began to hatch in the F_{10}), it is possible that the selected colony still has a short diapause, and thus selection has effected a change in diapause duration. Also, it is likely that selection in the initial generation was for those variants in the population that could terminate diapause without a chill, or with only a brief two-week chill.

The proportion of eggs hatching without a chill has increased, the interval until initiation of hatch has decreased, and there is a decreased variance in hatching time. During the original F_1 generation, larvae hatched over a 3-4 month interval. By the F_9, the hatch interval shortened to less than 2 weeks, with 50% hatch occurring within 5.15 days, and 90% hatch occurring by 11.9 days (Hoy, unpublished). Thus, the selected colony produces a "burst" of hatch, which would be desirable for insectary rearing programs.

One serendipitous finding has useful practical applications. The selection colony eggs (SEL) have the ability to undergo a long period of reduced development when held at 5°C. A number of F_5 through F_9 gypsy moth egg masses were placed in an unlighted cold room 25 days after the egg masses were deposited, but before hatch began. After one month, at least 10 eggs of each of 10 egg masses were removed monthly to 22°C and an 18-hr daylength to obtain an estimate of the time in which eggs could be chilled without mortality. Clearly, the sixth SEL generation maintained its ability to hatch after 6 months of chill (Table 1). The relationship between this tolerance to chilling at 5°C for up to 6 months and the colony's ability to withstand fluctuating and often much lower temperatures under natural forest conditions has not yet been investigated. However, based on these laboratory data, I suggest that the capacity to diapause and the capacity to withstand cooling without freezing (chilling) may be genetically distinct in the gypsy moth. The practical benefit of this tolerance to chilling is that the colony *need not* be continuously reared unless the investigator desires to do so. The discrimination between cold-hardiness and diapause has been documented rarely in arthropods, although Salt (1961) and Lees (1955) have both previously suggested that diapause and cold-hardiness are separate, albeit concurrently timed, phenomena. It is interesting that the length of time that the SEL eggs can be successfully stored at 5°C compares well with the interval that the wild gypsy moth eggs must endure during normal winter conditions (Kozhantshikov, 1950; Masaki, 1956).

Table 1. Hatch (%) of SEL Gypsy Moth Eggs after Different Chill Intervals (5°C)

Gypsy moth selection generation	Mean % hatch* after chill for									
	1	2	3	4	5	6	7	8	9	10 months
5	--	82	74	64	70	40	42	31	8	1
6	75	80	--	93	89	91	80	37	5	2
7	93	94	78	89	74	62	42	**	**	**
8	92	89	70	59	53	**	**	**	**	**

*Based on minimum of ten eggs from each of ten egg masses.
**Observations not yet completed.

Sullivan and Wallace (1972) found that gypsy moth eggs can withstand supercooling and that the supercooling points of individual gypsy moth eggs averaged -28.8 to -30.4°C, depending upon the acclimation temperature previously experienced. No comparable tests of the SEL gypsy moth colony have yet been made to determine if this degree of cold-hardiness remains, however.

The response to selection for "nondiapause" (or reduced diapause duration?) allows us to conclude that diapause is not obligatory in all members of the gypsy moth population I originally sampled. Selection thus has probably yielded a colony that does not require a chill to terminate diapause and a colony that also has a shorter diapause duration, which is evidenced by a reduced variance in date of hatch of individual egg masses.

The ultimate usefulness of this selected colony has yet to be determined. Evaluations (Hoy, 1978) of a number of biological attributes of the SEL colony have been completed, and additional tests are in progress. Developmental rate, sex ratio, pupal weight, the number of eggs per egg mass, the presence of morphological anomalies in adult males and females, and the sensitivity to the gypsy moth nuclear polyhedrosis virus (NPV) were compared for the F_6 and F_8 generations in contrast to larvae reared from field-collected eggs (Hoy, 1978). There were no obvious morphological anomalies; pupal weight was normal, as was the number of eggs per egg mass, and the viability of eggs. The developmental rate of the SEL_6 generation was slightly slower than that exhibited by the wild-type gypsy moth larvae, however. It is unknown whether this slower developmental rate is due to the inbreeding and inadvertent selection that accompanied the selection program, or whether it is due to the strain differences one might expect when samples are taken from different gypsy moth populations (Leonard, 1972), or to a combination of both factors.

Data for the F_8 and F_9 SEL generations show there has been an increase in the proportion of unembryonated eggs (from 3.8 ± 1.8 to 6.9 ± 1.3 eggs/egg mass) in the ninth generation. Also, there are increasing proportions of embryonated eggs that are not hatching. In the F_8 and F_9 generations, 6.9% and 13.5% of the embryonated eggs did not hatch (Hoy, unpublished). This may also indicate genetic deterioration due to inbreeding. Outcrosses of this SEL colony to the wild type are in progress and may explain the reasons for this possible deterioration.

The SEL colony has been used to rear several gypsy moth parasites including two braconids, *Apanteles melanoscelus* and *Rogas indiscretus* Reardon. *A. melanoscelus* exhibited an apparently normal diapause induction under laboratory conditions when reared in this "nondiapausing" host. Two eupelmids, *Anastatus disparis* Ruschka and *A. kashmirensis* Mathur, and an encyrtid, *Ooencyrtus kuwanai* (Howard), were also reared using these SEL colony eggs as hosts. No apparent quality changes in the parasites were observed.

Genetic selection for other nondiapausing arthropod colonies has been achieved before, with mixed results regarding the usefulness of the resultant colony. Thus, Harvey (1957) successfully selected within six generations a nondiapausing strain of the spruce budworm, *Choristoneura fumiferana* (Clem.), from a few individuals that were able to develop without any cold treatment. Harvey concluded that the nondiapause colony was principally useful for genetic studies where a short generation time facilitates experimentation. The nondiapause stock was eventually discontinued because of the ease with which rearing needs could be met from field stocks that were renewed annually. The nondiapause colony had been started from a very small genetic base and was strongly selected. The high inbreeding engendered decreases in mating success, fecundity, and survival of first instar larvae (Harvey, 1957, and personal communication).

A nondiapausing strain of the western spruce budworm, *C. occidentalis* Freeman, was used by Shon and Shea (1976) to rear two parasites without diapause, and thus shorten the parasites' generation interval from 23 weeks to 7 weeks. However, sex ratios were abnormal, which could have been due to poor mating or to the influences of the genetically altered hosts.

Pickford and Randell (1969) selected a nondiapausing stock of the migratory grasshopper, *Melanoplus sanguinipes* (Fabr.), over more than 75 generations. The resultant colony could be reared continuously in the laboratory and "is an excellent experimental insect." Its viability and vigor were apparently maintained. Pickford and Randell (1969) remarked that "It is easily reared with a minimum of care, and because of its huge reproductive capacity may be obtained in large numbers." A number of papers on the embryology, toxicology, and biochemistry of the colony were published (see Pickford and Randell, 1969, for citations). Thus, it appears possible to produce a useful nondiapause colony using genetic selection procedures.

Slifer and King (1961) report successful selection for both nondiapause and diapause strains of the grasshopper, *Melanoplus differentialis* (Thomas). The true-breeding diapause colony was vigorous until the colony was accidentally lost through technician error. However, one nondiapause colony was very weak by the 12th generation of selection, although two previously-selected nondiapause stocks did not show a drop in vigor, and Slifer and King concluded that the loss in viability was due to the accumulation of (other) deleterious genes during selection rather than to the nondiapause attribute per se.

The most famous example of successful genetic manipulation of diapause is exemplified by the domesticated silkworm, *Bombyx mori* L. Lees (1955) summarizes the genetic studies of this species' diapause by a number of Japanese scientists. Univoltine, bivoltine, and multivoltine strains of this species are available and are used in commercial silk production. Viability and vigor as well as increased silk production are achieved through the hybridization of different strains of the silkworm with known voltinism attributes (Hoy, 1976) and uniform F_1 progeny can be obtained which have a known genetic constitution.

Many authors have selected insects and mites for diapause or nondiapause, although the purpose was not necessarily for production of a laboratory colony; Table 2 gives a summary of the various arthropods in which purposeful or inadvertent selection for diapause or nondiapause has been achieved. The diversity of species and orders represented supports the conclusions that diapause is genetically determined, and that many more arthropods will be amenable to selection. Future selection programs could be designed using additional species and may be able to provide useful and vigorous colonies for multiple uses.

Genetic Control of Pest Insects through Climatic Nonadaptation

Inappropriate adaptations to climate are lethal to arthropods, particularly at certain times of the year. Thus, synchronization of development with the availability of food or with warm periods is critical for survival in a specific habitat. Inter- and intrapopulation variability in arthropods' responses to climate are genetically determined adaptations to specific climates. Intrapopulation variability is required on a long-term basis if evolutionary adaptation to climatic changes is to occur. Yet, excessive variability places a genetic load upon the population through losses of the nonadapted forms which are formed through genetic recombination. Thus tradeoffs in strategy are commonly assumed to occur in the evolutionary strategies of organisms (Bradshaw, 1973; Levins, 1968, 1969).

Several authors have suggested that increasing the genetic load in a population by increasing the species' nonadaptation to climate might lead to suppression or to eradication of the target pest (Klassen et al., 1970a,b; Hogan, 1971, 1974; Foster and Whitten, 1974; LaChance and Knipling, 1962). The conditional lethal traits of potential value in such a genetic manipulation scheme as listed by Klassen et al. (1970a) are (1) inability to diapause, (2) inappropriate duration of diapause, (3) inappropriate choice of hibernal niches, and (4) inability to develop cold-hardiness. Klassen et al. (1970a) suggested each conditional lethal trait might be determined by 1-4 genes and that possibly two or three conditional lethal traits could be simultaneously introduced into the target population. Thus, it was hypothesized that overflooding at a rate of 99:1 with nondiapausing boll weevils, *Anthonomus grandis* Boheman, after population reduction with insecti-

Table 2. Arthropods Selected for Diapause Attributes

Family	Species Order	Selection	Authority
	ACARINA		
Tetranychidae	*Tetranychus urticae* Koch	nondiapause	Helle 1962, 1968
	COLEOPTERA		
Chrysomelidae	*Diabrotica virgifera* LeConte	nondiapause	Branson 1976
	Atrachya menetriesi (Faldermann)	nondiapause	Ando and Miya 1968
Curculionidae	*Anthonomus grandis* Nahrman	nondiapause (inadvertent)	Lloyd et al. 1967
Dermestidae	*Trogoderma granarium* Everts	diapause and nondiapause	Nair and Desai 1973
	DIPTERA		
Calliphoridae	*Lucilia caesar* L.	diapause and nondiapause	Ring 1971
Sarcophagidae	*Pseudosarcophaga affinis* auct. nec Fallen	nondiapause (inadvertent)	House 1967
Tephritidae	*Rhagoletis pomonella* (Walsh)	nondiapause	Baerwald and Bousch 1967
	HETEROPTERA		
Lygaeiidae	*Oncopeltus fasciatus* (Dallas)	nondiapause	Dingle 1974
Pentatomidae	*Aelia acuminata* (L.) *A. rostrata* Boh.	nondiapause	Honek 1972 Hodek and Honek 1970

Table 2. Arthropods Selected for Diapause Attributes *(Continued)*

Family	Species Order	Selection	Authority
	HYMENOPTERA		
Diprionidae	*Diprion hercyniae* (Htg.)	nondiapause	Preble 1941 (ex Harvey)
Megachilidae	*Megachile pacifica* Panzer	for univoltine strain	Hobbs and Richards 1976
	ORTHOPTERA		
Acrididae	*Locusta migratoria* Remaudiere	nondiapause	LeBerre 1953
	Melanoplus sanguinipes (Fabricius)	nondiapause	Pickford and Randell 1969
	M. differentialis (Thomas)	diapause and nondiapause	Slifer and King 1961
Gryllidae	*Gryllus campestris*	nondiapause	Ismail and Fuzeau-Braesch 1976
	LEPIDOPTERA		
Bombycidae	*Bombyx mori* (L.)	univoltine, bivoltine, multivoltine strains	Lees 1955
Gelechiidae	*Pectinophora gossypiella* (Saunders)	nondiapause, unintentional in controls	Barry and Atkinson 1966
	P. gossypiella	greater/lesser diapause duration	Langston and Watson 1975
Lymantriidae	*Lymantria dispar* (L.)	nondiapause	Hoy 1977
Noctuidae	*Heliothis zea* (Boddie)	nondiapause	Herzog and Phillips 1974
	H. zea	diapause	Herzog and Phillips 1976

	LEPIDOPTERA (Cont'd.)		
	H. zea	diapause duration	Holtzer et al. 1976
Olethreutidae	*Grapholitha molesta* (Busck)	nondiapause	Glass 1970
Phycitidae	*Ephestia elutella* Hb.	nondiapause	Waloff 1949
Pyralidae	*Pionea forficalis* (L.)	diapause and nondiapause	King 1974
	Ostrinia nubilalis (Hubner)	univoltinism	Arbuthnot 1944
Saturniidae	*Hyalaphora cecropia* (L.)	early and late termination	Waldbauer and Sternburg 1973
	Antheraea pernyi Guer.	univoltinism	Chetverikov 1940; Tanaka 1951
Tortricidae	*Choristoneura fumiferana* (Clem.)	nondiapause	Harvey 1957
	C. occidentalis	nondiapause	Lyon et al. 1972

cides might yield such a reduction of the overwintering population (to 0.1% of the original level) that eradication might be achieved. Klassen et al. (1970a) also suggested an alternative model in which a nondiapause strain is partially sterilized and released. This combination might allow pest suppression during the cotton growing season and provide additional lethality during the winter.

Foster and Whitten (1974) suggest that the genetic control of the Australian sheep blowfly, *Lucilia cuprina* (Wiedemann), might be achieved through the decreased likelihood of survival of this species over the winter. These authors are screening mutagen-treated flies for dominant genetic factors using stocks containing several genetic markers. The use of marker genes to determine more precisely the genetic basis of "nondiapause" is very desirable. The most precise genetic analysis of "diapause" has been the study of voltinism in the silkworm (Lees, 1955).

Hogan (1971, 1974) summarized extensive research on the possibility of using genetic control techniques to control the field cricket, *Teleogryllus commodus* (Walker), in Australia. Hybrids between *I. oceanicus* (Le Guillou) and a sibling species *T. commodus*, are viable but nearly completely sterile. Interspecific hybridization may thus provide the opportunity to transfer the "nondiapause" attribute from one species to another. Hogan (1974) reviewed all of the problems inherent in the project, which include a need for mechanisms to transport the nondiapause trait into the target species, and for a lack of variability in the lethality of the trait under conditions where a low intensity of diapause induction would exist during different years. Assortative mating is a problem in the field, and hybrid sterility is not complete. Thus, Hogan (1974) reported that the use of nondiapausing fertile hybrids has been suspended as an eradicative approach, but may have promise as a population suppression method.

Genetic suppression or eradication programs using diapause attributes as conditional lethal traits have the constraints of standard lethal systems which are induced through the use of chemosterilants or through radiation. Thus, a thorough knowledge is required of the target species' life cycle, phenology, natural population density, reproductive potential, and the presence or absence of density-dependent factors operating within the population dynamics of the species. In addition, adequate methods for mass production of viable, competitive insects are also required. Thus, the implementation of this method will be expensive, quite species-specific, and will require a considerable degree of technical and biological sophistication.

Furthermore, while entomologists were generally surprised at the ease with which insects developed resistance to insecticides, it should not be a surprise to entomologists if resistance to one or more of the various genetic control schemes develops. Whitten and Pal (1974) suggested, "If resistance develops, the form it will take is unlikely to be known until these types of genetic control have been implemented on a large scale." I believe that their comments may apply to the use of climatic nonadaption as a genetic control method, as well, should it ever be attempted.

Changes in Voltinism in Pest and Beneficial Arthropods

Intra- and interpopulation variability has allowed a rapid evolutionary change in voltinism in several species. This rapid change has generally been documented because of the pest status of the species concerned, and thus entomologists were monitoring the pests' biology over a long period.

For example, the codling moth, *Laspeyesia pomonella* (L), has apple, walnut, and plum races in California (Phillips and Barnes, 1975). It entered California by 1873, where it was found on pears and apples for the first 40 years and later on walnuts. By 1963, it had become a pest of plums in Fresno and Tulare Counties. The development of these "host races" is of considerable economic importance. Diapause initiation, diapause termination, and developmental heat requirements during diapause development are different in these races. Phillips and Barnes (1975) suggested that the apple race gave rise to the walnut race, which in turn gave rise to the plum race. One may thus speculate that chronological isolation based on differences in diapause, as well as differential ovipositional preferences, and a relative lack of dispersal out of the different host plant sites, may ultimately lead to sufficient reproductive isolation that a subspecies status may be reached.

The European corn borer, *Ostrinia nubialis* (Hubner), is another prominent example of a pest with a number of biotypes in North America, each having differing responses to diapause in relation to climate and to agricultural practices. Thus, voltinism varies from the North to the South in the United States. The northern populations are univoltine (Beck and Apple, 1961), but third and fourth generations are able to develop in southern Georgia, U.S.A. (Sparks and Showers, 1975). Changes in the voltinism of pest insects have been documented most often, but other exotic species also may have altered their voltinism subsequent to their establishment in new habitats. Furthermore, additional changes in voltinism in pest insects may yet occur.

Therefore, discussion of the potential for a change in voltinism in the gypsy moth is of interest. It is clear that while the gypsy moth has an "obligatory" diapause that is not yet amenable to manipulation via environmental cues, there is considerable genetic variability in diapause attributes present in this species. The selection of the "nondiapause" colony was done with the intent that it would be used only in the laboratory or in the insectary (Hoy, 1978). Clearly, if laboratory adaptation has progressed to a point that the colony is precluded from establishment in the field, there is no reason for concern. However, if this stock were purposefully or inadvertently released, there are conditions under which we might speculate that a multivoltine strain of the gypsy moth could be developed under forest conditions.

The scenarios that follow are hypotheses.

(a) If small numbers of SEL gypsy moths were released, the survival of the pure line would be unlikely, due to interbreeding with the more numerous wild moths, to climatic nonadaptation in the Northern United States where winters average below 5°C for long periods, and to a lack of food for any second-generation larvae. (Tests of tolerance of the SEL colony to temperatures below 5°C have not yet been completed.) If the SEL colony interbred with the wild type in the field, the few additional "nondiapause" genes injected into the native population would probably be selected against.

(b) Should large numbers of SEL gypsy moths be released, the question of the impact of these "nondiapausing" genes arises. Part of the uncertainty is based on a lack of knowledge of the genetic basis of diapause or nondiapause in the gypsy moth, although a genetic analysis of this SEL colony is underway (Lynch and Hoy, in progress). However, based on a preliminary analysis of the genetic crosses, we believe that it is unlikely that the nondiapause trait is determined by *many* pairs of genes, i.e., an "indefinitely large number of loci affecting the character" (Lerner, 1958), and that the impact of allelic substitution may not be trivial in comparison to the total amount of observed variation in the character. (A discussion of the genetic basis of diapause follows in the next section.)

If large numbers of SEL males were released into a wild population of the gypsy moth, it is likely that the hybrid F_1 progeny would have a "diapause." [The F_1 progeny hatch, in the laboratory, at very low rates without a chill (Lynch and Hoy, in progress)]. These releases therefore would not have an immediate lethal impact on the population, but conceivably could cause a delayed lethal effect and thus a type of genetic control of the gypsy moth. However, the feasibility of this procedure would depend upon the extensive development of the technical, biological, and ecological knowledge that is required for genetic control generally (Pal and Whitten, 1975).

(c) Another hypothetical possibility also exists. The gypsy moth is extending its range in the United States from New England into the South and the West. The gypsy moth advanced into southern New Jersey, Maryland, and West Virginia within the past few years, and it has recently been discovered in San Jose, California, where it has been present at least since 1974. If large numbers of the SEL colony were released into areas with mild winters where foliage is often available year around the nondiapause trait might survive in the endemic population, because the F_1 progeny might be sufficiently vigorous and hardy to allow development of a multivoltine strain, particularly since some of these populations are relatively isolated from other normal gypsy moth populations and thus would not be swamped with "normal" gypsy moths. The development of a multivoltine strain would be a serious problem for economic entomologists. The gypsy moth is quite polyphagous, with more than 200 potential host plants, including cotton (Forbush and Fernald, 1896). Furthermore, survival of young larvae might be facilitated as the larvae could develop on refoliating vegetation.

The question of why the gypsy moth is not already multivoltine naturally arises. The gypsy moth has an extensive natural range but *apparently* does not complete a second generation successfully anywhere (Schedl, 1936; Forbush and Fernald, 1896). Univoltinism has been hypothesized as a mechanism whereby a species does not overexploit its food resources. As such, it has obvious adaptiveness, since host trees defoliated a second time in the same year are often killed. However, the inherent intraspecific variability within the gypsy moth, plus the new circumstances provided by man in which lush vegetation is present on a year-round basis especially in warmer climates, because of irrigation, fertilization and other crop management practices, might allow the development of a new evolutionary opportunity. The likelihood of a multivoltine gypsy moth has been increased through the laboratory selection procedure since presumably rare alleles were sequestered and preserved under very protected laboratory conditions in which all natural mortality agents were removed or greatly reduced during laboratory selection. Thus, mortality due to weather, parasites, predators, or pathogens, as well as to pesticides was excluded. Laboratory selection procedures thus significantly increased the likelihood that the rare variants would be preserved, but it does not seem completely unlikely that a multivoltine strain could develop even under field conditions. The genetic variability necessary is clearly present in the natural population.

An intentional change in voltinism of a beneficial arthropod has also been achieved. The beneficial bee, *Megachile pacifica* Panzer, was genetically selected for univoltinism, and the strain was shown to be "effective" under field conditions. Hobbs and Richards (1976) developed the univoltine strain of the alfalfa leafcutter bee after it was introduced into western Canada from the northwestern United States. Because of the shorter season in western Canada, the second-generation bees emerged too late to reproduce or to pollinate. After 8 years of natural selection under field conditions, the proportion of bees having a second generation had declined from 25% to 4.6%. The elimination of the "nondiapausing" bees continued during four seasons of directed selection during which time the proportion of bees producing a second generation dropped to 0.4%. Whether this "improved" strain persists in the environment remains to be seen. However, the apparent success of this program is encouraging. One factor that may have contributed to its apparent success may have been that the selection program was carried out under semicontrolled field conditions rather than the laboratory, where genetic adaptation to laboratory conditions often occurs.

The beneficial neuropterans, *Chrysopa downesi* Banks and *C. carnea* Stephens, may represent examples of speciation that has occurred through differentiation of photoperiodic responses (Tauber and Tauber, 1976b). The major difference in their photoresponse patterns lies in the requirement by *C. downesi* for a short-day:long-day regime for diapause prevention and termination. The two species can hybridize and produce viable F_2 progeny in the laboratory. In their natural environment they are asynchronous and, although sympatric, are effectively isolated.

Examples no doubt will be found that demonstrate that reproductive isolation, through the asynchrony which may arise as a result of photoperiodic adaptations, may eventually lead to the accumulation of sufficient genetic differences to allow a population to reach the level of species identity. Photoperiodic adaptations thus may serve as a mechanism involved in the process of divergence and speciation. Furthermore, the rate of such changes may be rapid, particularly in contrast to a geological time scale.

Genetic Analyses of Diapause

Studies of the genetic basis of diapause often have been fragmentary, and sometimes have yielded conflicting results. Difficulties in diapause genetic analysis are due to the nature of diapause; i.e., it is a threshhold character in which temperature and photoperiodic cues trigger "all-or-none" individual responses. Therefore we must view diapause as a population phenomenon.

Genetic analyses of diapause have usually proceeded using one of three methods: analysis of crosses between diapausing and artificially selected nondiapausing colonies; analysis of crosses between different geographic races; and analysis of diapause response of inbred and hybrid lines. Many genetic analyses of diapause were apparently conducted as an afterthought. The differing approaches used have yielded varying and sometimes conflicting conclusions regarding the inheritance of diapause. However, it would be surprising if diapause were not inherited differently in different species, since diapause has no doubt evolved independently in different arthropod groups.

The following sections review primarily those papers published since the review of diapause inheritance by Beck (1968), Danilevskii (1965), and Lees (1955).

Genetic Analysis through Crosses of Geographic Strains

The genetic basis of diapause has been most completely analyzed in the silkworm, *Bombyx mori*, and Lees (1955) has summarized the studies of Toyama, Tanaka, and Nagatomo. Three sex-linked alleles and three autosomal dominant genes determine the inheritance of voltinism in the silkworm. (The Lepidoptera have the reverse pattern of sex linkage, with the female being the heterogametic sex.) The sex-linked alleles are epistatic, while the autosomal dominant genes are cumulative. With this knowledge of the mode of inheritance of the silkworm's diapause, stocks with specific "hibernation values" can be constructed through the hybridization of univoltine and bivoltine races. Diapause is dominant over continuous development, and F_1 females produced from crosses of two races have an intermediate type of diapause, although the reciprocal crosses are not alike. It is also possible to construct two strains with equal "hibernation values," but having differing genotypes.

Adult reproductive diapause has been found and analysed in several species of *Drosophila* (Lakovaara et al., 1972; Lumme et al., 1975). In *D. littoralis*, maternal factors were implicated. Furthermore, Lumme et al. (1975) concluded that the basis of diapause in *D. littoralis* was determined by a limited number of genes more potent than the "additive" genes of orthodox quantitative inheritance (see also Lumme, this volume). In contrast, Lakovaara et al. (1972) believed that diapause was polygenically controlled in *D. ovivororum*.

Rabb (1969) crossed a North Carolina and a St. Croix strain of the tobacco hornworm, *Manducca sexta* (Johannson), to analyse pupal diapause of the reciprocal hybrids. Results were inconclusive regarding the mode of inheritance of the incidence and intensity of pupal diapause.

Genetic Analysis through Crosses of Genetically Selected Strains

Most attempts to analyse the genetic basis of diapause have been made using strains that were purposely or inadvertently selected for lack of a diapause. McCoy et al. (1968) crossed an inadvertently selected laboratory (nondiapause) colony and a wild strain of boll weevils (*Anthonomus grandis* Boheman) and observed the diapause responses in the F_1, F_2 and backcross progeny. The F_1 progeny resembled the "low-diapausing parent, which fact suggests that the nondiapause trait may be dominant." No difference existed in the F_1 progeny of reciprocal crosses, thus ruling out sex linkage, but the responses of the F_2 progenies were different, which suggested that a maternal influence was present. The backcross data suggested that ability to diapause is "controlled by a simply inherited recessive gene."

Hodek and Honek (1970) crossed high diapause response (wild) strains of the pentatomid *Aelia acuminata* (L.) with low diapause response strains which had experienced 16-17 generations of selection. The F_1 progeny were intermediate in their responses. Later, Honek (1972) selected *A. acuminata* for three generations using two methods. The lower level of selection preserved a high diapause response, and Honek concluded that diapause inheritance is determined by several different genes.

The bollworm, *Heliothis zea* (Boddie), responded rapidly to selection for nondiapause (Herzog and Phillips, 1974). Within three generations of selection, the proportion entering diapause dropped from 64.2% to 5.6%. The authors concluded that the rapid and early response for nondiapause "suggests that only a few pairs of genes are involved in maintaining diapause at the usually high level of about 80% under these conditions." The plateau in selection response reached after nine generations suggested that "multigenic factors" were also involved.

Ismail and Fuzeau-Braesch (1976) analyzed diapause in the univoltine cricket, *Gryllus campestris*, through crosses of a diapause and a genetically selected nondiapausing strain. *G. campestris* has two types of development. One consists of rapid growth followed by a diapause in the next to the last larval stage, the other of slow growth without diapause. Selection eliminated diapause of the crickets reared at 30°C as well as prevented normal survival when the crickets were reared at 20°C. Diapause was dominant in short-day conditions and recessive in long-day conditions at 30°C in the F_1 progeny. Ismail and Fuzeau-Braesch (1976) therefore propose that the two types of development, diapause and slow growth, are controlled by the same "unitary system."

King (1974) selected a British, bivoltine strain of the pyralid, *Pionea forficalis* (L), for three generations for tendency to diapause and for lack of tendency to diapause, using late-emerging adults and first adults to emerge, respectively. The correlation of the date of emergence of the F_1 progeny with the date of emergence of the families of origin was computed. The correlation was 0.509 for the male parent and 0.033 for the female parent. Thus, emergence response is inherited through the male parent, whose family emergence characteristics accounted for about 50% of the variance among the progeny. The apparent transmission through the male parent appears to be unique. Polygenic control of response to photoperiod was also suggested.

Nair and Desai (1973) concluded that diapause is a polygenic character in the dermestid, *Trogoderma granarium* Everts, based on the progress of selection for diapause or nondiapause. Density-dependent diapause and density-independent diapause occur in this species and the presence of density-dependent diapause in populations with density-independent diapause was totally eliminated by selection, thereby demonstrating that more than one set of genes are involved in the determination of "diapause." Nair and Desai concluded that density-dependent diapause is determined by a major recessive factor and that density-independent diapause is determined by a dominant factor. The proportion of density-dependent diapause became high after one generation of selection for, while the proportion of density-independent diapause reached a low plateau in only one generation of selection against.

Ando and Miya (1968) crossed a selected nondiapause strain of the false melon beetle, *Atrachya menetriesi* Faldermann, with a "normal" strain. Whether the F_1 eggs entered diapause was determined by the genetic constitution of the female parent, but not that of the male parent. However, the male parental influence appeared in the F_2. Normal diapause eggs and diapause eggs obtained by crossing nondiapause selected males to diapause females were different genetically. The latter had a weaker diapause intensity, as determined by the time required for a chill to terminate diapause. Ando and Miya (1968) were unable to analyze the inheritance type from the ratio of diapause and nondiapause eggs of the F_2 and F_3 and concluded that the diapause trait is "genetically complex." They also found that their nondiapause strain had become so badly inbred that egg viability and fecundity decreased, and the colony was lost after the nineteenth generation. This was attributed to inbreeding per se, however, and not to the effects of the nondiapause genes.

Waldbauer and Sternberg (1973) determined that the bimodality of emergence from diapause in the field of the moth *Hyalophora cecropia* (L.) is genetically determined. Selection for group I (early) emergence for two or three generations yielded a jump from 8 to 75% early emergence in this group, but additional genetic analyses were not done.

Morris and Fulton (1970) found that pupal diapause intensity in the fall webworm, *Hyphantria cunea* Drury, is sex-linked. Pupal weight and larval and pupal survival were significantly correlated with midparent heat requirement for diapause termination, which suggested that the genes controlling heat requirements "also influence pupal weight and larval and pupal survival, or are linked to genes controlling these characters."

The pink bollworm, *P. gossypiella*, has been selected for nondiapause (Barry and Adkisson, 1966) and for longer and shorter diapause duration (Langston and Watson, 1975). Genetic analysis of progeny of a nonselected colony and of a nondiapause colony (Barry and Adkisson, 1966) suggested that diapause is "a multigenic character" and that sex linkage was not involved, although nondiapause tended to be dominant to diapause. Selection after 11 generations on part of the colony was relaxed and the subcolony was reared for ten additional generations. The incidence of nondiapause decreased gradually from 95.6% to 82%. The authors believed that the rapidity of response to selection in the early generations was due to "a few major" genetic factors determining diapause, and regression during the relaxed selection pressure suggested that several minor genetic factors were operating as well.

Langston and Watson (1975) selected for early and late diapause termination in *P. gossypiella* and performed reciprocal crosses between the two lines. The results suggested that termination time, either early or late, was not inherited as a dominant but that sex-linkage or maternal effects were present.

Complete genetic analysis of the nondiapause strain of the spruce budworm, *C. fumiferana*, was not done, but Harvey (1957) suggested that diapause is not sex-linked and is apparently determined by multiple genes. However, he also concluded that the rapidity of response to selection suggested that there are relatively few genes involved in diapause determination.

Dingle (1974) selected the milkweed bug, *Oncopeltus fasciatus* (Dallas), for nondiapause for eight generations. The selection was shown to be shifting the critical photophase, because the bugs reproduced early at 12 hr of light after selection but at 11 hr continued to diapause. Dingle estimated the realized heritability based on the selection response and showed that the mean heritability from all generations was 1.18. (A value of 1.00 means that the phenotypic variance for the trait is completely the result of additive genetic variance.)

Ring (1971) crossed two strains of the blowfly, *Lucilia caesar* (L.), that had low and high incidence of diapause under normally diapause-preventing conditions. The F_1 progeny had diapause attributes intermediate between the parents and the reciprocal crosses were significantly different. Little segregation occurred in the F_1 and F_2 generations and the author concluded that there was a multiple-gene mode of inheritance for diapause in this fly, with the male entering diapause more readily than the female. Ring (1971) concluded that "several genes, both sex-linked and autosomal, are involved" in this species' diapause.

Genetic variation in time required for diapause termination of the bollworm, *Heliothis zea* (Boddie), was demonstrated to be highly heritable through two-way directional selection for early and late emergence (Holtzer et al., 1976). Herzog and Phillips (1974, 1976) selected for diapause and for nondiapause in *H. zea*. Mode of inheritance studies were not conducted, but the authors concluded that "only a few pairs of genes" are involved in maintaining diapause.

The photoperiodic responses of the aphid, *Myzus persicae* (Sulz.) (Blackman, 1971), have been shown to be heritable. Thus, clones established from random samples of summer populations in southern England were stable and of four types: (1) holocyclic, (2) intermediate, (3) androcyclic, and (4) truly anholocyclic. Progeny of a cross between oviparae of a holocyclic clone and males of an androcyclic clone were studied. The 14 F_1 progeny clones segregated into holocyclic and androcyclic genotypes and the results thus suggested a "fairly simple inheritance mechanism" was determining the inheritance of the life cycles of this aphid.

Bigelow (1962) did not do detailed analyses of the genetics of diapause or of developmental rate in *Gryllus* species. However he suggests that both developmental rate and the temperature at which diapause is triggered are polygenically determined in all species he tested and the differences between species might be due to differences in frequency of alleles. Data from *inter*specific hybrids showed that genes affecting the developmental rate are located on the X-chromosome of some cricket species.

Genetic Analysis through the Use of Inbred Lines

Helle (1962) successfully selected for nondiapause in the two-spotted spider mite (*Tetranychus urticae* Koch) within six generations and found that nondiapause was genetically independent from resistance to organophosphorus compounds, i.e., was nonpleitropic to insecticide resistance.

Helle later (1968) analyzed diapause in this arrhenotokous species by using nonselected, but inbred lines. After seven generations of inbreeding, the photoresponses of the lines were analyzed and a diversity of photoresponse curves was obtained. Low response (LR) lines were crossed with high response (HR) lines reciprocally. The F_1 progeny exhibited HR, indicating that diapause was dominant to nondiapause. Backcrosses to LR (haploid) males demonstrated the transmission of the diapause trait was different in the two types of F_1 hybrids. Helle suggested two possible reasons for the differences; a cytoplasmic determinant or a difference in inheritance related to a hypothetical haplo-diploid sex-determination system. Genetic analysis using inbred but unselected lines was carried out because Helle reasoned that the selected nondiapause strains that are commonly used for analysis give possibly misleading data since the selection procedure itself pools genes and thus possibly masks genes with *major* effects. The inbreeding without selection, in contrast, is more likely to allow individual major genes to become homozygous (in at least a few lines), and therefore detectable, although the influence of minor "modifying" genes may be lost due to the inbreeding. Of 19 inbred lines, three had a markedly lower diapause response over the range of photoperiods tested (0-14 hr), although the critical range was not affected. One line appeared to have a different threshold. In five lines, the "normal" response appeared more intense compared to the initial colony's response. Thus, genetic variability was documented for the initial colony. Crosses within and between the LR lines showed similar responses -- thus, there was no increase in diapause response due to the hybridization of different LR lines, which might occur if recessive nondiapause genes were being masked through heterozygosity.

The rapidity of response to selection for nondiapause females allowed Helle to conclude that the genetic basis of LR is not very complex and this is substantiated by the fact that in three of 19 lines, inbreeding led to the fixation of the LR trait. These three lines could be homozygous for a recessive gene which determines LR. This is further supported by the fact that the three LR lines yield progenies when crossed that also have a LR, thus indicating a common and simple basis for the inheritance of the LR in these three lines.

Genetic Dissection of "Diapause"

It seems clear from the previous discussion of the genetic analyses of diapause that there are several "parts" of diapause that are genetically determined. The critical photophase, the duration of diapause, the day degrees required for diapause termination and the degree of cold-hardiness exhibited during diapause all respond to selection, as demonstrated by several species.

Diapause probably evolved independently in different arthropodan groups. It would thus be surprising if the genetic determinants were identical in the different groups, and one may speculate that egg diapause is qualitatively different from larval, pupal, or adult reproductive diapause.

No clear conclusion can be drawn about the genetic basis of diapause in the broad sense, in part because of the inadequacies of the genetic analyses conducted to date. I suspect that diapause is difficult to analyze genetically because it has two aspects that make analysis difficult: it is a threshold character, and it probably falls into a "no man's land" with respect to analysis of the numbers of genes involved. The only complete genetic analysis of diapause that is available is for the silkworm, *Bombyx mori*, in which three autosomal genes and three sex-linked alleles determine voltinism. This means that in any one silkworm there are four pairs of genes determining diapause. Four pairs of genes are far fewer than the numbers of genes commonly associated with orthodox quantitative genetics. However, they yield more complex ratios than mono- or digenic crosses. In fact, four pairs of genes yield ratios that begin to be indistinguishable from the "many" genes of polygenic inheritance. Only if analyses can be done using genetic markers will it be possible to pinpoint the numbers of genes involved. The genetic variability exhibited by most arthropods, as just discussed in the preceding section, is impressive. This genetic variability may provide a tool for the genetic and physiological analysis of "diapause" in species in addition to *B. mori*. One could take at least two different approaches to the genetic "dissection" of the components of diapause.

One technique would involve the genetic selection of several strains, each differing in diapause attributes. For example, if several lines of selected "nondiapausing" individuals were crossed, one might learn in how many different ways "nondiapause" is achieved. Crosses between the nondiapause lines and the normal wild strain would allow mode of inheritance tests for the types of "nondiapause." Other lines could be selected for other diapause attributes and analyzed similarly. If selection for greater diapause duration can be completed without altering the temperature requirements for diapause termination or without altering the responses to extreme chill, then we could conclude that genes controlling diapause duration are different from those controlling these other aspects of diapause.

Another dissection strategy might be to treat a population with mutagens, such as EMS (ethylmethane sulfonate), which cause point mutations of the DNA. Screening for variants in diapause responses could then be done. This would be a time-consuming and tedious process, but might allow the most detailed analysis of the genetic basis of diapause, especially if such a program were carried out with a species for which genetic markers are available.

Such genetic dissections would allow a more precise understanding of the genetic basis of diapause, and more practically, could allow a comparison of the physiological and biochemical attributes of the selected and "normal" strains. In this way we could develop an understanding of the detailed physiological components of diapause induction, maintenance, and termination.

Acknowledgments Part of the work reported herein was funded by a U.S.D.A.-sponsored program entitled: The Expanded Gypsy Moth Research and Applications Program.

I thank Drs. Ronald Weseloh and K. Hagen for their helpful comments on the first draft of this paper, and G. S. Walton for statistical assistance.

References

Ando, Y., Miya, K.: Diapause character of the false melon beetle, *Atrachya menetriesi* (Faldermann), produced by crossing diapause and non-diapause strains. J. Fac. Agr. Iwate Univ. *9*, 87-95 (Transl. from Japanese) (1968).

Arbuthnot, K.D.: Strains of the European corn borer in the United States. Techn. Bull. U.S. Dept. Agr. *869*, 1-20 (1944).

Baerwald, R.J., Boush, G.M.: Selection of a nondiapausing race of apple maggot. J. Econ. Entomol. *60*, 682-684 (1967).

Barry, B.D., Adkisson, P.L.: Certain aspects of the genetic factors involved with the control of the larval diapause of the pink bollworm. Ann. Entomol. Soc. Am. *59*, 122-125 (1966).

Beck, S.D.: Insect Photoperiodism. New York: Academic Press 1968, pp. 288.

Beck, S.D., Apple, J.W.: Effects of temperature and photoperiod on voltinism of geographical populations of the European corn borer, *Pyrausta nubilalis*. J. Econ. Entomol. *54*, 550-558 (1961).

Bigelow, R.S.: Factors affecting developmental rates and diapause in field crickets. Evolution *16*, 396-406 (1962).

Blackman, R.L.: Variation in the photoperiodic response within natural populations of *Myzus persicae* (Sulz.). Bull. Entomol. Res. *60*, 533-546 (1971).

Bradshaw, W.E.: Homeostasis and polymorphism in vernal development of *Chaoborus americanus*. Ecology *54*, 1247-1259 (1973).

Branson, T.F.: The selection of a non-diapause strain of *Diabrotica virgifera* (Coleoptera: Chrysomelidae). Ent. Exp. & Appl. *19*, 148-154 (1976).

Chetverikov, S.S.: The selection of the Chinese oak silkworm (*Antheraea pernyi* Guer.) for univoltinism. In: Selection and Acclimatization of Oak Silkworms. Sel'khozgiz, M. (ed.). 1940, pp. 16-22.

Danilevskii, A.S.: Photoperiodism and Seasonal Development of Insects. Edinburgh: Oliver & Boyd, 1965, pp. 282.

Dingle, H.: Migration strategies of insects. Science *175*, 1327-1335 (1972).

Dingle, H.: The experimental analysis of migration and life-history strategies in insects. In: Experimental Analysis of Insect Behavior. Browne, L. B. (ed.). New York: Springer 1974, pp. 329-342.

Forbush, E.H., Fernald, C.H.: The gypsy moth, *Porthetria dispar* (Linn.). A report of the work of destroying the insect in the commonwealth of Massachusetts, together with an account of its history and habits both in Massachusetts and Europe. Boston: Wright & Potter Printing, St. Printers, St. Bd. of Agric., Mass. 1896.

Foster, G.G., Whitten, M.J.: The development of genetic methods of controlling the Australian sheep blowfly, *Lucilia cuprina*. In: The Use of Genetics in Insect Control. Pal, R., Whitten, M. J. (eds.). No. Holland: Elsevier 1974, pp. 19-43.

Glass, E.H.: Changes in diapause response to photoperiod in laboratory strains of Oriental fruit moth. Ann. Entomol. Soc. Am. *63*, 74-76 (1970).

Goldschmidt, R.: Untersuchungen zur Genetik der geographischen variation. V. Arch. Entwickl. Mech. Org. *126*, 674-768 (1932).

Goldschmidt, R.: Untersuchungen zur Genetik der geographischen variation. VII. Arch. Entwickl. Mech. Org. *130*, 562-615 (1933).

Harvey, G.T.: The occurrence and nature of diapause-free development in the spruce budworm *Choristoneura fumiferana* (Clem.) (Lepidoptera: Tortricidae). Can. J. Zool. *35*, 549-572 (1957).

Helle, W.: Genetics of resistance to organophosphorus compounds and its relation to diapause in *Tetranychus urticae* Koch (Acari). H. Veerman & Zonen N.V. Wageningen 1962, pp. 1-41.

Helle, W.: Genetic variability of photoperiodic response in an arrhenotokous mite (*Tetranychus urticae*). Ent. Exp. & Appl. *11*, 101-113 (1968).

Herzog, G.A., Phillips, J.R.: Selection for a non-diapause strain of the bollworm, *Heliothis zea* (Lepidoptera: Noctuidae). Environ. Entomol. *3*, 525-527 (1974).

Herzog, G.A.: Selection for a diapause strain of the bollworm, *Heliothis zea*. J. Heredity *67*, 173-175 (1976).

Hobbs, G.A., Richards, K.W.: Selection for a univoltine strain of *Megachile* (Eutrichaeaea) *pacifica* (Hymenoptera: Megachilidae). Can. Entomol. *108*, 165-167 (1976).

Hodek, I., Honek, A.: Incidence of diapause in *Aelia acuminata* (L.) populations from southwest Slovakia (Heteroptera). Acta Soc. Zool. Bohemoslov. *34*, 197-183 (1970).

Hogan, T.W.: An evaluation of a genetic method for population suppression of *Teleogryllus commodus* (Wlk.) (Orth., Gryllidae), in Victoria. Bull. Entomol. Res. *60*, 383-390 (1971).

Hogan, T.W.: A genetic approach to the population suppression of the common field cricket *Teleogryllus commodus*. In: The Use of Genetics in Insect Control. Pal, R., Whitten, M. J. (eds.). No. Holland: Elsevier 1974, pp. 57-70.

Holtzer, T.O., Bradley, J.R., Jr., Rabb, R.L.: Geographic and genetic variation in time required for emergence of diapausing *Heliothis zea* (Lep., Noctuidae). Ann. Entomol. Soc. Am. *69*, 261-265 (1976).

Honěk, A.: Selection for non-diapause in *Aelia acuminata* and *A. rostrata* (Heteroptera, Pentatomidae) under various selective pressures. Acta Ent. Bohemoslov. *69*, 73-77 (1972).

House, H.L.: The decreasing occurrence of diapause in the fly *Pseudosarcophaga affinis* through laboratory-reared generations. Can. J. Zool. *45*, 149-153 (1967).

Hoy, M.A.: Forest and laboratory evaluations of hybridized *Apanteles melanoscelus* (Hymen.: Braconidae), a parasitoid of *Porthetria dispar* (Lep.: Lymantriidae). Entomophaga *20*, 261-268 (1975a).

Hoy, M.A.: Hybridization of strains of the gypsy moth parasitoid, *Apanteles melanoscelus*, and its influence upon diapause. Ann. Entomol. Soc. Am. *68*, 261-264 (1975b).

Hoy, M.A.: Genetic improvement of insects; fact or fantasy? Environ. Entomol. *5*, 833-839 (1976).

Hoy, M.A.: Rapid response to selection for a non-diapausing gypsy moth. *Science* *196*, 1462-1463 (1977).

Hoy, M.A.: Selection for a non-diapausing gypsy moth: some biological attributes of a new laboratory strain. Ann. Entomol. Soc. Am. (in press) (1978).

Ismail, S., Fuzeau-Braesch, S.: Programmation de la diapause chez *Gryllus campestris*. J. Insect Physiol. *22*, 133-139 (1976).

King, A.B.S.: Photoperiodic induction and inheritance of diapause in *Pionea forficalis* (Lepidoptera: Pyralidae). Ent. Exp. & Appl. *17*, 397-409 (1974).

Klassen, W., Creech, J.F., Bell, R.A.: The potential for genetic suppression of insect populations by their adaptations to climate. Ag. Res. Serv. USDA, Misc. Pub. 1178, 1970.

Klassen, W., Knipling, E.F., McGuire, J.U.: The potential for insect-population suppression by dominant conditional lethal traits. Ann. Entomol. Soc. Am. *63*, 238-255 (1970).

Kozhantshikov, I.W.: Features of the hibernation and diapause of the gypsy moth (*Ocneria dispar* L.). C. R. Acad. Sci. USSR (N.S.) *73*, 605-607 (1950).

LaChance, L.E., Knipling, E.F.: Control of insect populations through genetic manipulations. Ann. Entomol. Soc. Am. *55*, 515-520 (1962).

Lakovaara, S., Saura, A., Koref-Santibanez, S., Ehrman, L.: Aspects of diapause and its genetics in northern drosophilids. Hereditas *70*, 89-96 (1972).

Langston, D.T., Watson, T.F.: Influence of genetic selection on diapause termination of the pink bollworm. Ann. Entomol. Soc. Am. *68*, 1102-1106 (1975).

Leberre, J.R.: Contribution a l'etude biologique du criquet migrateur des Landes (*Locusta migratoria gallica* Remaudiere). Bull. Biol. *87*, 227-273 (1953).

Lees, A.D.: The Physiology of Diapause in Arthropods, Cambridge: Cambridge Univ. Press 1955, pp. 150.

Legner, E.F.: Observations on hybridization and heterosis in parasitoids of synanthropic flies. Ann. Entomol. Soc. Am. *65*, 254-263 (1972).

Leonard, D.E.: Diapause in the gypsy moth. J. Econ. Entomol. *61*, 596-598 (1968).

Leonard, D.E.: Recent developments in ecology and control of the gypsy moth. Ann. Rev. Entomol. *19*, 197-229 (1974).

Lerner, I.M.: The Genetic Basis of Selection. New York: John Wiley & Sons 1958, pp. 291.

Levins, R.: Evolution in Changing Environments. Some Theoretical Explorations. Monographs in Population Biology, Princeton: Princeton Univ. Press 1968, pp. 120.

Levins, R.: Dormancy as an adaptive strategy. In: Dormancy and Survival, Symp. Exp. Biol. XXIII. New York: Academic Press 1969, pp. 1-10.

Lloyd, E.P., Tingle, F.C., Gast, R.T.: Environmental stimuli inducing diapause in the boll weevil. J. Econ. Entomol. *60*, 99-102 (1967).

Lumme, J., Lakovaara, S., Oikarinen, A., Lokki, J.: Genetics of the photoperiodic diapause in *Drosophila littoralis* (Dipt., Drosophilidae). Hereditas, *79*, 143-148 (1975).

Lyon, R.L., Richmond, C.E., Robertson, J.L., Lucas, B.A.: Rearing diapause and diapause-free western spruce budworm (*Choristoneura occidentalis*) (Lepidoptera: Tortricidae) on an artificial diet. Can. Entomol. *104*, 417-426 (1972).

Masaki, S.: The effect of temperature on the termination of diapause in the egg of *Lymantria dispar* Linne (Lepidoptera: Lymantriidae). Japanese J. Appl. Zool. *21*, 148-157 (1956).

Masaki, S.: Geographic variation of diapause in insects. Bull. Fac. Agric. Hirosaki Univ. *7*, 66-98 (1961).

Masaki, S.: Geographic variation in the intrinsic incubation period; a physiological cline in the Emma field cricket (Orthoptera: Gryllidae: *Teleogryllus*). Bull. Fac. Agric. Hirosaki Univ. *11*, 59-90 (1965).

McCoy, J. R., Lloyd, E. P., Bartlett, A. C.: Diapause in crosses of a laboratory and a wild strain of boll weevils. J. Econ. Entomol. *61*, 163-166 (1968).

Messenger, P. S.: Climatic limitations to biological controls. Proc., Tall Timbers Conf. Ecol. Anim. Cont. by Habitat Mgmt. *3*, 97-114 (1971).

Messenger, P. S., van den Bosch, R.: The adaptability of introduced biological agents. In: Biological Control. Huffaker, C. B. (ed.). New York: Plenum Press 1971, pp. 68-92.

Morris, R. F., Fulton, W. C.: Heritability of diapause intensity in *Hyphantria cunea* and correlated fitness responses. Can. Entomol. *102*, 927-938 (1970).

Nair, K. S. S., Desai, A. K.: Studies on the isolation of diapause and non-diapause strains of *Trogoderma granarium* Everts (Coleoptera, Dermestidae). J. Stored Prod. Res. *9*, 181-188 (1973).

Phillips, P. A., Barnes, M. M.: Host race formation among sympatric apple, walnut, and plum populations of the codling moth, *Laspeyresia pomonella*. Ann. Entomol. Soc. Am. *68*, 1053-1060 (1975).

Pickford, R., Randell, R. L.: A non-diapause strain of the migratory grasshopper *Melanoplus sanguinipes* (Orthoptera: Acrididae). Can. Entomol. *101*, 894-896 (1969).

Prebble, M. L.: The diapause and related phenomena in *Gilpinia polytoma* (Hartig). Can. J. Res. *19(D)*, 295-322 (1941).

Rabb, R. L.: Diapause characteristics of two geographical strains of the tobacco hornworm and their reciprocal crosses. Ann. Entomol. Soc. Am. *62*, 1252-1256 (1969).

Ring, R. A.: Variations in the photoperiodic reaction controlling diapause induction in *Lucilia caesar* L. (Diptera: Calliphoridae). Can. J. Zool. *49*, 137-142 (1971).
Salt, R. W.: Principles of insect cold-hardiness. Ann. Rev. Entomol. *6*, 55-74 (1961).
Schedl, E.: Der Schwammspinner (*Porthetria dispar* L.) in Euroasien, Afrika und Neuengland. Monog. Zur Ange. Entomol. Band XXII 1936, pp. 238.
Shon, F. L., Shea, P. J.: Increased rearing efficiency of two hymenopterous parasites using a non-diapausing host species, *Choristoneura occidentalis* (Lep., Torticidae). Environ. Entomol. *5*, 277-278 (1976).
Slifer, E. H., King, R. L.: The inheritance of diapause in grasshopper eggs. J. Heredity *52*, 39-44 (1961).
Sparks, A. N., Showers, W. B.: Current status of European corn borer in South Georgia. J. Georgia Entomol. Soc. *10*, 347-352 (1975).
Sullivan, C. R., Wallace, D. R.: Variations in the photoperiodic response of *Neodiprion sertifer*. Can. J. Zool. 46, 1082-1083 (1968).
Tanaka, Y.: Studies on hibernation with special reference to photoperiodicity and breeding of the Chinese tussar silkworm. V. Nippon Sanshigaku Zasshi (J. Sericult, Sci. Japan) *20*, 132 (1951).
Tauber, M. J., Tauber, C. A.: Insect Phenology: Criteria for analyzing dormancy and for forecasting postdiapause development and reproduction in the field. Search (Agricult.) Cornell Univ. Agr. Expt. Sta., Ithaca, New York *3*(12), 1-16 (1973).
Tauber, M. J., Tauber, C. A.: Environmental control of univoltinism and its evolution in an insect species. Can. J. Zool. *54*, 260-265 (1976a).
Tauber, M. J., Tauber, C. A.: Insect seasonality: diapause maintenance, termination, and postdiapause development. Ann. Rev. Entomol. *21*, 81-107 (1976b).
Waldbauer, G. P., Sternburg, J. G.: Polymorphic termination of diapause by *Cecropia:* Genetic and geographical aspects. Biol. Bull. *145*, 627-641 (1973).
Waloff, N.: Observations on larvae of *Ephestia elutella* Hubner (Lep., Phycitidae) during diapause. Trans. R. Entomol. Soc. Lond. *100*, 147-159 (1949).
Weseloh, R. M.: Termination and induction of diapause in the gypsy moth parasitoid, *Apanteles melanoscelus*. J. Insect Physiol. *19*, 2025-2033 (1973).
Whitten, M. J., Pal, R.: Introduction. In: The Use of Genetics in Insect Control., Pal, R., Whitten, M. J. (eds.). No. Holland: Elsevier 1974, pp. 1-16.

Phenological Adaptation and the Polymodal Emergence Patterns of Insects

GILBERT P. WALDBAUER

This paper is concerned with the phenological implications of emergence, i.e., the resumption of development and activity, by temperate zone insects which undergo a hibernal diapause. More specifically, it describes and considers the adaptive significance of the polymodal emergence curves which my co-worker, Dr. James G. Sternburg, and I have found to be typical of *Hyalophora cecropia* and of some populations of *Callosamia promethea*. It also discusses some of the accounts in the literature of the polymodal emergences of other temperate zone insects, and suggests a classification of polymodal emergence patterns.

Escape in time implies the volution of adaptive phenological relationships, in other words, the adoption by an organism of a seasonal occurrence that coincides with the availability of its resources and that avoids unfavorable physical or biotic factors. Among temperate zone insects, diapause is the most important of the mechanisms which synchronize development and activity with the seasons. Danilevskii (1965) said, "... the eco-physiological features of diapause form the basis of the entire life cycle of insects," and, more recently, Tauber and Tauber (1976) stressed the central importance of diapause in the determination of insect seasonality.

It has long been recognized that the diapause-inducing mechanism causes insects to enter diapause in time to survive the cold of winter, and that the diapause-terminating mechanism assures the timely resumption of development in spring, while preventing its premature resumption in autumn (Shelford, 1929; Andrewartha, 1952). Photoperiod is usually the most important factor in the induction of facultative diapause, but the termination of hibernal diapause often requires exposure of varying durations to low, winterlike temperatures, although photoperiod and other stimuli may also be involved (e.g., Danilevskii, 1965; Beck, 1968; Mansingh, 1971; Mansingh and Smallman, 1971; Wigglesworth, 1972; Tauber and Tauber, 1976).

The processes (or process) which terminate diapause have been designated by various terms, including diapause development (Andrewartha, 1952), diapause-ending processes (Schneiderman and Horwitz, 1958), the reactivation process (Danilevskii, 1965), and the activation process (Mansingh, 1971) [see Sheldon and MacLeod's (1974) cogent discussion of this terminological problem]. I agree with Sheldon and MacLeod that "diapause-ending processes" is the only suitable term which has so far been proposed. This term, however, is awkward to use because it has neither a verbal nor an adjectival form. In this communication I will simply refer to the diapause-ending processes as conditioning (in the dictionary sense of putting into a proper state). This word is understandable in so many different biological contexts that it has obviously maintained its neutrality, and probably carries a minimum of misleading connotations.

The generalization has often been made that conditioning is completed and diapause terminated before the end of winter, and that morphogenesis will begin just as soon as it is warm enough (Beck, 1968; Danilevskii et al., 1970). For example, Danilevskii (1965) said, "Field observations show that, because of individual variation in the degree of diapause, the proportion of reactivated insects increases more or less gradually throughout the winter. But as the temperature then is considerably below the developmental threshold, reactivated individuals remain in an obligitory state of rest....Ultimately, by spring, the whole wintering stock appears to be potentially capable of resuming development immediately, and this begins uniformly as soon as the temperature rises to the necessary level." Similarly, the work of Williams (1956) has led to the supposition that all of the pupae of cecropia begin to develop as soon as it is warm enough, synchronization of individuals with each other occurring because all develop at nearly the same temperature and, therefore, at nearly the same rate.

However, as shown below, cecropia and many other insects are exceptions to this generalization, and it will probably eventually be found that most insects are exceptions. Among temperate zone insects there is a great diversity of phenological strategies for the resumption of activity after the winter has passed; different species are likely to differ markedly as to the seasonal occurrence and the degree of synchronization of their emergence, as illustrated by the following examples. Unless otherwise specified, the examples below pertain only to insects under field conditions.

Various Diapause-Terminating Strategies

In many species emergence is delayed without producing a polymodal emergence curve. The seasonal history of a variety of species suggests that emergence is frequently delayed beyond the earliest possible date predicted by either the species' threshold temperature for development or the effect of temperature on the rate of development or both. In some species, the emergence dates of individuals vary greatly, resulting in long, drawn-out emergence periods for the population. For example, the fall webworm moth (*Hyphantria cunea*) in New Brunswick and Nova Scotia has an exceptionally long emergence period that commonly extends over two to three months (Morris and Fulton, 1970; Morris, 1971). In other species all individuals are delayed more or less equally, resulting in a tightly synchronized unimodal emergence curve, as for example, in the univoltine strain of *Callosamia promethea* (see below). According to Corbet (1963), dragonflies which emerge as adults in the spring have a tightly synchronized "explosive emergence," while dragonflies whose emergence occurs in summer have a much longer and less synchronized emergence. Corbet suggested that these differences in synchronization and season of emergence are associated with longevity and interspecific competition. Willey (1974) found that the emergence of the subalpine dragonfly *Somatochlora semicircularis* is earlier and more tightly synchronized than the emergence of low altitude species, a difference which she attributed to the brevity of the season in the subalpine environment. Powell (1974) reported that entire populations of some yucca moths may delay emergence for more than a year. Kreasky (1960) found that diapausing eggs of the high-altitude grasshopper *Melanoplus borealis* do not hatch until after they have experienced two winters. All viable eggs hatched as a tightly synchronized group when he moved them to an incubator after they had spent two winters in the field. A similar experiment showed that almost all of the diapausing eggs of *Melanoplus alpinus* must pass through three winters before they will hatch (Kreasky, 1960).

In other insect species varying degrees of delay by portions of the emerging population result in polymodal emergence curves. The pattern of emergence may be bimodal, trimodal, or even more complex. The emergence of an age cohort may extend over two or more years, or it may occur polymodally within one season. The literature on polymodal emergence patterns is widely scattered, and I have not made an exhaustive search for cases of this phenomenon. A scrutiny of some recent reviews (deWilde, 1962; Danilevskii, 1965; Masaki, 1961; Beck, 1968; Danilevskii et al., 1970; Tauber and Tauber, 1976; Chippendale, 1977) reveals few cases, suggesting that polymodal emergence patterns either seldom occur or that they tend to go unrecognized. I suspect that they are more common than they seem to be. There are three distinct types of polymodal emergence patterns which, for the sake of convenience, I will designate Types A, B, and C.

Type of Polymodality

In species with type A polymodality, diapause does not extend over more than one winter. All individuals which enter diapause in the same year terminate diapause the following growing season, exhibiting a bimodal or, in at least one species, a trimodal emergence curve. As discussed in detail below, this sort of emergence strategy occurs in at least two species of saturniid moths which overwinter as pupae (Sternburg and Waldbauer, 1969; Waldbauer and Sternburg, 1973; Sternburg and Waldbauer, unpublished). The zebra swallowtail butterfly (*Eurytides marcellus*) also overwinters as a pupa and has a bimodal emergence curve. However, in this species the dimorphism involves color and form as well as the emergence date (Scudder, 1889, pp. 1273-1278). Danilevskii (1965, p. 203) cited a report that the emergence curve of cotton bollworm moths (*Chloridea obsoleta*) from the overwintering pupae is bimodal in the Caucasus. The emergence curve of tobacco hornworm moths (*Manduca sexta*) from the overwintering pupae is bimodal in North Carolina (Rabb, 1966). The phantom midge (*Chaoborus americanus*) overwinters as a last instar larva and has a bimodal emergence curve in Michigan. The early emerging morph is larger and yellower than the pale late-emerging morph (Bradshaw, 1973).

Type B Polymodality

Some insects exhibit a polymodal emergence pattern which is completed within one season, but which differs fundamentally from Type A polymodality in its origin. In these insects the modes of the emergence curve represent individuals which belong to different year classes. The bimodal emergence curve of adults of the European dragonfly *Anax imperator* is an example (Corbet, 1963). The first mode appears in spring, is well synchronized, and may include over 90% of the year's emergence. It represents the metamorphosis to the adult stage of nymphs that are in their third summer, and that entered diapause as full-grown nymphs the previous summer. The second mode appears about 25 days later and is less well-synchronized. It represents the metamorphosis to the adult stage of individuals that are in their second summer. They overwinter in the penultimate instar, and in the second summer pass through the last instar without diapausing. Corbet (1963) listed three other European dragonflies which have a bimodal adult emergence curve which may have a similar origin: *Aeshna affinis*, *Calopteryx virgo*, and *Pyrrhosoma nymphula*. Kormondy (1959) found that the North American *Tetragoneuria cyanosura* also has a bimodal emergence curve.

Type C Polymodality

Many insect species are known in which the emergence of a year class (individuals which enter diapause in the same season) extends over two or more years. Literature reviewed by Powell (1974) indicates that, in addition to the groups mentioned below, portions of a year class may delay development for a year or more in certain species of katydids, chrysomelid beetles, gall midges, anthophorid bees, and moths of the families Geometridae, Incurvariidae, Lasiocampidae, Saturniidae and Tortricidae. Powell (1974) reported that the emergence of some moths of the family Ethmiidae may extend over two or more years. Kreasky (1960) found that only some of the eggs of two high-altitude grasshoppers, *Melanoplus bruneri* and *Chorthippus longicornis*, would hatch if they were moved to an incubator after spending two winters in the field. The remaining eggs hatched after they were refrigerated for four months and then returned to the incubator, indicating that they required exposure to three winters. The wheat blossom midges *Contarinia tritici* and *Sitodiplosis mosellana* (Cecidomyidae) have a similarly prolonged larval diapause (Barnes, 1952). A few *C. tritici* complete development without diapausing, most emerge as adults after one winter, and a few emerge after two winters. *S. mosellana* always enter diapause and usually most emerge as adults after the first winter although the completion of development may be delayed as long as 12 years.

The spread of emergence over two or more years seems to be especially common among sawflies. The resumption of development by diapausing prepupae of the European pine sawfly (*Neodiprion sertifer*) may be delayed for one or more years in Europe (Kangas, 1941) and in Canada (Griffiths, 1959). *N. sertifer* is univoltine and may overwinter as either an egg or as a prepupa in a cocoon. The portion of the larval population that develops during the long days of early summer enters an aestival diapause that is likely to be prolonged to a hibernal diapause if the diapausing larvae are exposed to cool temperatures (Sullivan and Wallace, 1967). Hence, a portion of the population may emerge and oviposit later the same summer, while the rest may remain in diapause for from one to three years, emerging as adults in concert with the ovipositing population of the current summer (Griffith, 1959; Sullivan and Wallace, 1967). On the Gaspé Peninsula of Quebec, the European spruce sawfly (*Gilpinia polytoma*) is univoltine, and its diapause commonly extends over four years, and occasionally over six years. A group of over 2000 larvae that entered diapause in the same summer and that were held outdoors emerged as follows: 19.5% the following summer, 2.3% the second summer, 40.1% the third summer, 37.6% the fourth summer, and 0.5% the fifth summer. Data from many years show a similar pattern, indicating that the pattern is not the result of variations in weather. In New Brunswick *G. polytoma* is bivoltine, and its emergence extends over only three years, with the usual pattern of emergence being: 64.5% the first year, 24.1% the second year, and 11.4% the third year (Prebble, 1941).

The apple maggot (*Rhagoletis pomonella*), which overwinters as a pupa in the soil, is unusual because it appears to exhibit all of the three types of polymodality described above. First, Oatman (1964) reported the occurrence of Type A bimodality. He found that under field conditions in Wisconsin, most individuals from cultivated apples emerged as adults after the first winter, and that in three out of five years their emergence curves were bimodal, with one peak in June and another in July. I wonder if in this case the bimodal curve represents not the emergence of a single population, but rather the emergence of a mixed population of the apple and hawthorn races of *R. pomonella*. This assumes the repeated invasion of apple by the hawthorn race. The two races are reported to have distinct adult emergence peaks -- separated by about four or five weeks -- that seem to correspond to the modes reported by Oatman (Bush, 1969, 1975). Oatman (1964) also reported Type C bimodality. Some flies reared in two of the five years did not emerge until after a second winter, 5% in one year and 15% in the other. Allen and

Table 1. Yearly distribution between the two emergence groups of adult cecropia which emerged from cocoons that were collected from the wild population in the vicinity of Urbana, Illinois, and held in a screened, outdoor insectary

		Emerging with Group I		Emerging with Group II	
Year of emergence	Total no. emerged	Mean %	Median emergence date	Mean %	Median emergence date
1966a	321	10.0	May 30	90.0	June 25
1967a	167	9.0	May 26	91.0	June 20
1968a	557	4.7	May 24	95.3	June 20
1969b	409	7.3	May 24	92.7	June 24
1970b	165	9.7	May 19	90.3	June 15
1971b	270	15.6	May 21	84.4	June 17
1973c	60	20.0	May 22	80.0	June 22
1975c	202	4.5	May 24	95.5	June 23
1976c	135	24.4	May 25	75.6	June 26

a Data from Sternburg and Waldbauer (1969).
b Data from Waldbauer and Sternburg (1973).
c Unpublished.

Fluke (1933) also found that the emergence of a year class may extend over two seasons in Wisconsin, with a maximum of 37% emerging the second year. Finally, Allen and Fluke (1933) reported Type B bimodality; flies from two-year-old pupae emerging from 8 to 12 days later than flies from one-year-old pupae. This is similar to the situation described above for *Anax imperator*, in which two emergence modes that occur in the same season represent individuals that belong to different year classes.

The emergence of the closely related blueberry maggot (*R. mendax*) may extend over as many as four years, with about 80% emerging after one year, 16% the second year, 4% the third year, and 0.4% the fourth year. Carryover flies emerged four to six days later than flies which emerged after the first winter (Lathrop and Nickels, 1933).

Diapause Termination by Some Saturniid Moths

Observations and experiments which Dr. J.G. Sternburg and I have conducted since 1965 show that the emergence of adult *Hyalophora cecropia* from overwintered pupae consistently presents a Type A bimodal pattern (Sternburg and Waldbauer, 1969; Waldbauer and Sternburg, 1973). At Urbana, Illinois, the emergence of the early mode (Group I) is usually confined to the second half of May, with the median emergence date varying from May 19 to May 30; the emergence of the late mode (Group II) extends from about mid-June to early July, with the median emergence date varying from June 15 to June 26 (Table 1). Table 1 also shows that the proportional distribution of the population between the two groups varies from year to year.

The moths included in Table 1 and in the upper portion of Figure 1 emerged from pupae in cocoons that were held in an outdoor, screened insectary after having been collected from the wild overwintering popu-

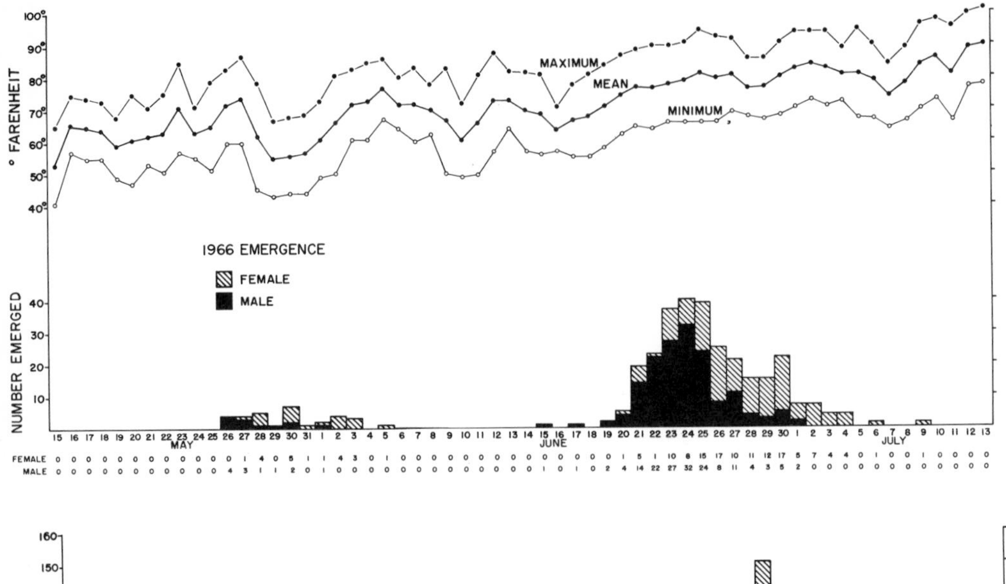

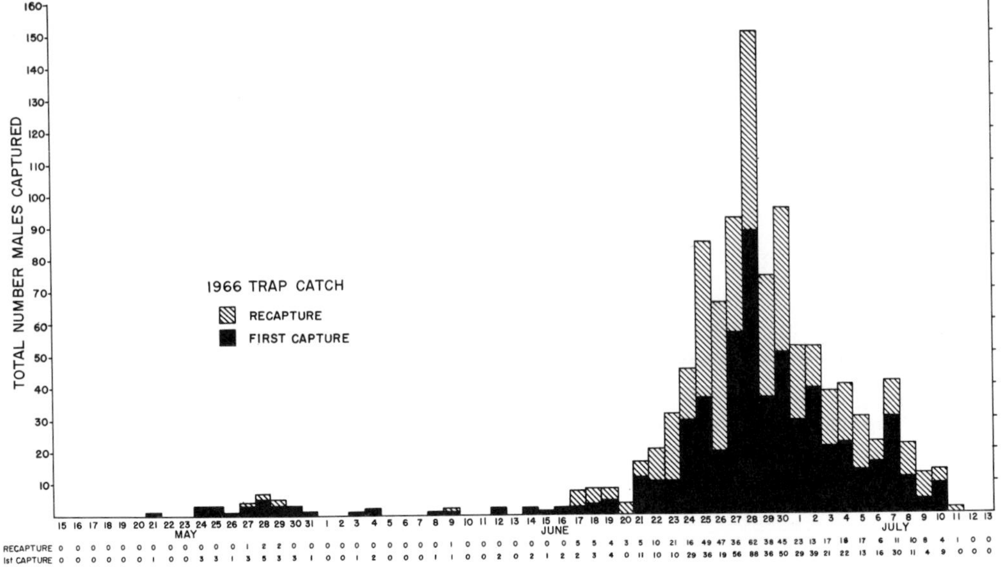

Fig. 1. Trap catch of male cecropia and emergence from locally collected cocoons held in a screened, outdoor insectary. Urbana, Illinois, 1966 (from Sternburg and Waldbauer, 1969).

lation in the vicinity of Urbana. However, the lower portion of Figure 1 represents wild, adult, male cecropia caught at or near Urbana in traps baited with pheromone-releasing, virgin females. A comparison of the two curves in Figure 1 [see Sternburg and Waldbauer (1969) for corresponding curves for 1967 and 1968] shows that the emergence pattern of the population left undisturbed in the wild is essentially the same as the emergence pattern of a sample taken from the same wild population and held in the outdoor insectary where all were exposed to identical conditions. This shows that our handling of the cocoons did not affect the emergence pattern, and that the bimodal emergence pattern does not

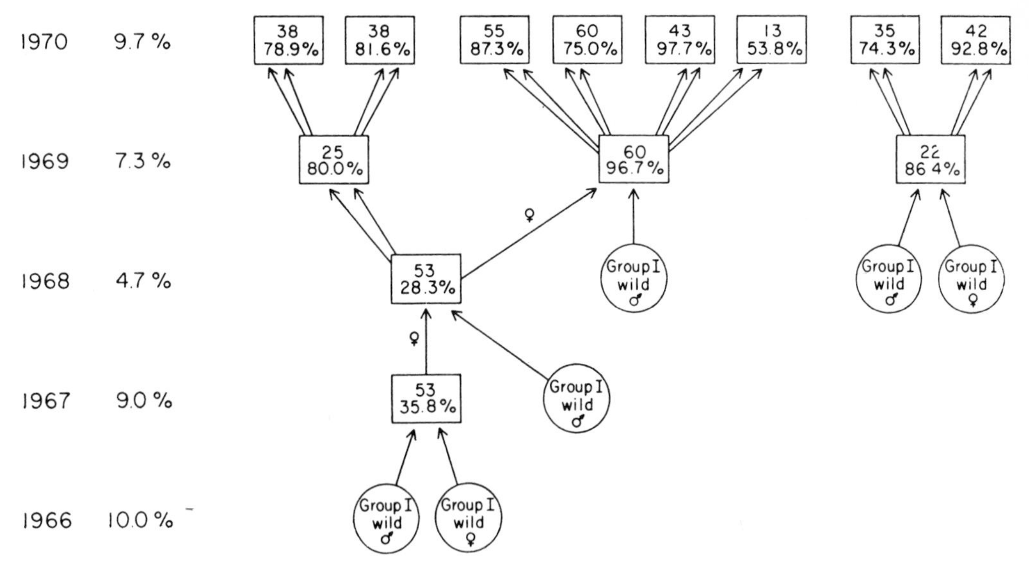

Fig. 2. The descent of the eight groups of cecropia siblings, line selected from the Urbana population for emergence with Group I, that emerged as adults in 1970. On the left are the year of emergence and the percentage of the wild population that emerged with Group I in each year. A circle indicates a wild parent, a box the progeny resulting from a controlled mating. The upper figure in each box shows total progeny reared; the lower shows the percentage that emerged with Group I. Arrows show parentage; sex is indicated only where necessary (from Waldbauer and Sternburg, 1973).

result from the effects of different microenvironments on the pupae in the field during the winter. The slight shift to the right of the trap catch is to be expected because the traps sampled an accumulating population of males, and because the traps became more attractive as the population of virgin, wild females decreased. Virgin females release the sex-attractant pheromone but mated females do not.

The males of each of the emergence groups emerge earlier than the females (Sternburg and Waldbauer, 1969; Waldbauer and Sternburg, 1973). Therefore, emergence Groups I and II are distinguished not only by an interval of no emergence, but also by the alternating emergence of males and females (Fig. 1).

The bimodal emergence curve of cecropia is not just a local phenomenon confined to the vicinity of Urbana. Cecropia's emergence pattern is bimodal in Chicago (Marsh, 1941; Waldbauer and Sternburg, 1973) as well as at several other Illinois localities ranging from Chicago to the southern tip of the state (see below). Rau and Rau (1912, 1914) found its emergence curve to be bimodal at St. Louis, Missouri, and C.B. Worth (personal communication, see Waldbauer and Sternburg, 1973) found its emergence curve to be bimodal in Cape May County, New Jersey.

The correspondence between the trap catch and the emergence in the insectary suggests that cecropia's bimodal emergence pattern is under genetic control. Line selection for a strain of cecropia which emerges predominantly with Group I rather than with Group II showed that the emergence pattern is indeed genetically determined (Waldbauer and Sternburg, 1973). Figure 2 shows that, after four years or less, we had selected from the Urbana population strains that averaged over 80% emergence with Group I, as compared with the average of 9.7% and 7.3% for

Table 2. Comparison of the initiation of development and emergence of the adult from cecropia pupae held in a screened, outdoor insectary (from Willis et al., 1974)

	No.	Median date on which development first noted	Median emergence date	Mean no. days to complete development
Observation group				
Group I	11[a]	April 1	May 31	59
Group II	58	June 3	June 26	25
Undisturbed group				
Group I	26	--	May 24	--
Group II	531	--	June 20	--

a Data include estimates of emergence dates.

the local wild population in the same years. It is important to note that both Group I and Group II parents produced progeny which emerged with either group. In fact, wild parents of both groups usually produced a preponderance of progeny which emerged with Group II.

Under field conditions the selected strains are essentially in phase with the corresponding groups of the wild population. There is no tendency for selection to produce intermediates; rather, the interval between groups increases because the Group I emergers tend to cluster more tightly about the median (Waldbauer and Sternburg, 1973). Dr. Sternburg and I are still maintaining these strains, and they still emerge predominantly with Group I. Our findings leave no doubt that we are dealing with a population which is polymorphic for emergence date.

The difference between the two emergence morphs of cecropia could be due to a difference in the time of diapause termination or, perhaps, to different rates of development, in individuals which terminate diapause at the same time. To settle this question Willis et al. (1974) followed the development of cecropia pupae held in an outdoor insectary, examining them at least once a week from the end of March until all of the moths had emerged. They looked for the earliest recognizable sign of development, i.e., apolysis of the leg epidermis, and then followed the pupae through the later stages of development as described in the timetable of Schneiderman and Williams (1954).

The Group I individuals began to develop in late March, but the median date for the initiation of development by Group II was June 3, 63 days later than the median initiation date of Group I, and 3 days later than even the median emergence date of Group I. Group I developed at less than half the rate of Group II (Table 2) because of differences in mean ambient temperatures, 13°C while Group I developed and 23°C while Group II developed (Willis et al., 1974).

A study of the latitudinal adaptations of cecropia's diapause-terminating mechanism gave revealing results (Waldbauer and Sternburg, 1973). One of the experimental techniques was to transfer samples of cecropia cocoons from the Chicago (41° 51' north latitude) population to the outdoor, screened insectary at Urbana (40°6') late in the autumn. We also made reciprocal transfers, so that samples of both Urbana and Chicago cocoons were held in screened, outdoor insectaries at Urbana and at the Morton Arboretum near Chicago.

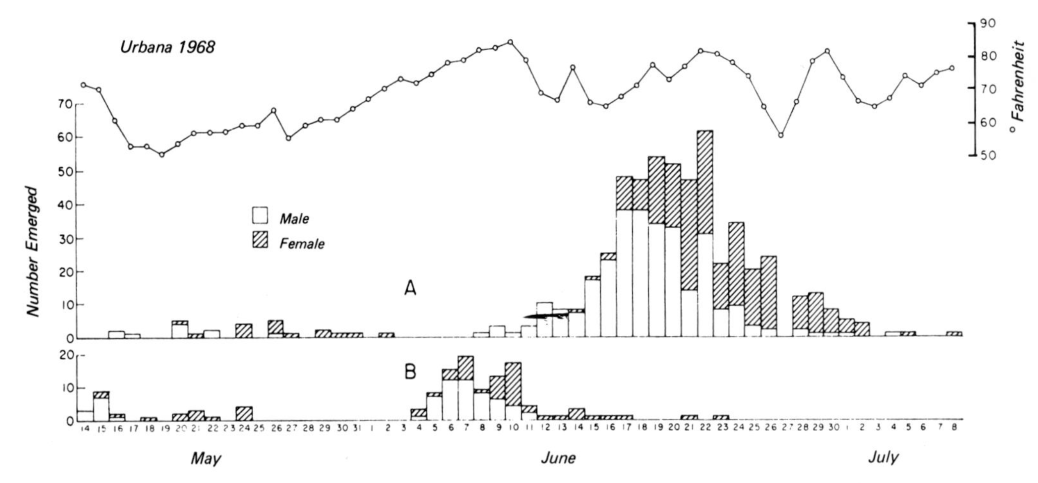

Fig. 3. Emergence of adult cecropia at Urbana, Illinois, in 1968 from wild cocoons collected in the field and moved into a screened, outdoor insectary by late fall; (A) cocoons collected in the vicinity of Urbana; (B) cocoons collected in Chicago. The daily mean temperatures are plotted above the histograms (from Waldbauer and Sternburg, 1973).

At either location the transferred Group I cecropia emerged more or less in phase with the local population. However, at either location the Group II Chicago moths always emerged significantly earlier than the Urbana moths (Figs. 3 and 4). Rearing cecropia from the Chicago population at Urbana demonstrated that the difference in emergence times between the two populations has a genetic basis: Chicago cecropia reared in Urbana remained on "Chicago time" in the F_1 generation and in the F_2 generation (Fig. 4).

In any case -- whether a transfer is involved or not -- the emergence of Group II comes sooner after the emergence of Group I in the Chicago population. With Chicago moths the elapsed time between the median emergence dates of the two groups varied from 17 to 23 days ($\bar{x}$ = 19.2), and with Urbana moths it varied from 26 to 31 days ($\bar{x}$ = 27.9), (Waldbauer and Sternburg, 1973). This raises the logical possibility that the factor required to initiate the development of the Group II moths is simply the passage of time. However, other evidence (see below) indicates that the duration of the delay in the initiation of development is temperature-dependent, although it is not simply a matter of the temperature exceeding a threshold for development. Photoperiod is probably not involved since a transfer of Urbana cocoons to the north, where the days lengthen more rapidly, delays rather than accelerates emergence.

The results of additional experiments (Waldbauer and Sternburg, unpublished) confirm and extend these findings on the latitudinal adaptations of cecropia's diapause-terminating mechanism. Cocoons were collected from sites in Illinois ranging the length of the state from north to south: Chicago (40°51'), Kankakee (41°7'), Urbana (40°6'), Charleston (39°28'), and the University of Illinois' Dixon Springs Experiment Station near Vienna (37°25'). This transect includes a wide climatic range; tamarack bogs occur just north of Chicago, and cypress swamps are common in the vicinity of Dixon Springs. The progeny of these moths were raised outdoors on netted trees at Urbana and the cocoons were overwintered in the screened, outdoor insectary at Urbana. We assumed

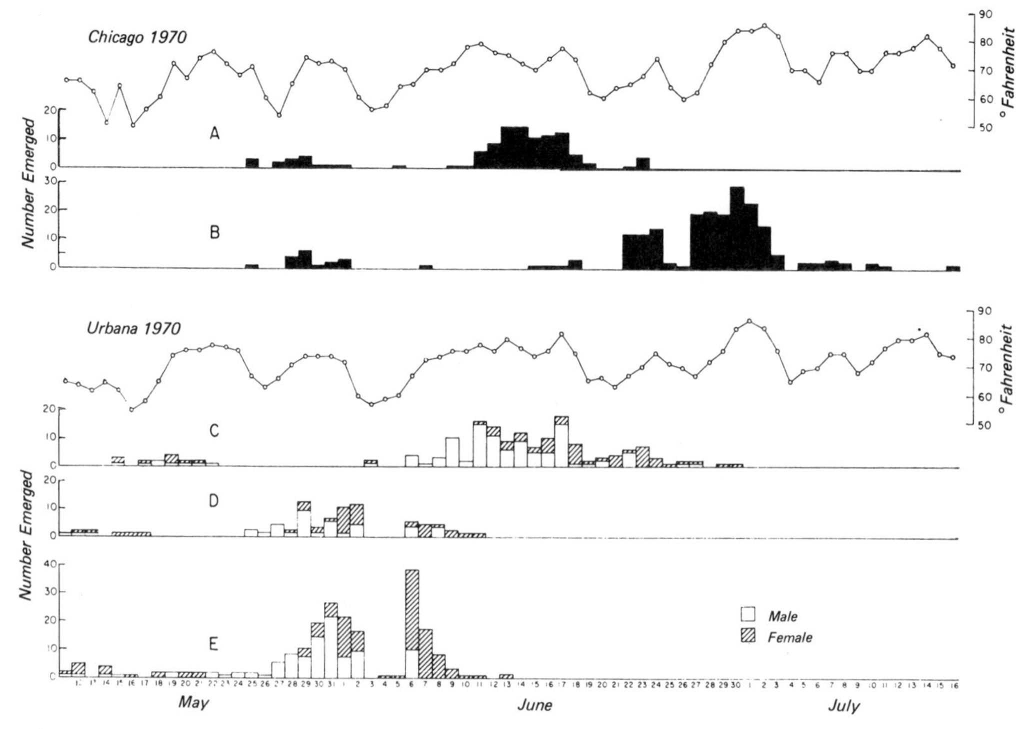

Fig. 4. Emergence of adult cecropia in 1970 from cocoons moved into a screened, outdoor insectary at either the Morton Arboretum near Chicago (A and B) or at Urbana (C, D, and E) by late fall. The origin of the cocoons is as follows: (A) wild, collected in Chicago, (B and C) wild, collected in Urbana, (D) wild, collected in Chicago, and (E) F_2 progeny reared in Urbana from wild parents collected in Chicago in 1967-68. The daily mean temperatures are plotted above the histogram (from Walbauer and Sternburg, 1973).

that the differences in the phenological responses of the populations from various latitudes are genetically fixed, as we had demonstrated earlier for the Urbana and Chicago populations (Waldbauer and Sternburg, 1973).

Again all of the Group I progeny of the transferred moths emerged more nearly in phase with the local Urbana population than did their Group II progeny. Both the Group I and Group II progeny of parents from north of Urbana again emerged earlier than the Urbana moths, and, to our surprise, the moths whose parents had originated south of Urbana also emerged earlier than their Urbana counterparts. In explanation we offer the hypothesis that the populations from both north and south of Urbana have been selected for early emergence by the relative brevity of the favorable growing season in their areas. In the north the favorable season is likely to be cut short by cold weather, while in the south it is likely to be cut short by the hot, dry weather of mid-summer when the leaves of the various food trees and shrubs of cecropia are likely to become nutritionally less suitable (Feeny, 1970).

Reciprocal transfers of cocoons between Urbana and Dixon Springs gave largely the expected results. At Dixon Springs emergence began in mid-April rather than in mid-May as at Urbana. At both sites the Group I moths began to emerge at the same time, but the emergence of the Urbana population continued for a longer time. At both Dixon Springs and Urbana, the Dixon Springs Group II moths emerged earlier than the Urbana Group II moths.

Our observations of latitudinal effects on the emergence pattern of cecropia indicate that the Group I moths tend to be opportunists which more or less conform with Danilevskii's (1965) generalization that diapause is terminated before the end of winter, and that development begins as soon as the temperature rises to the necessary level in the spring. On the other hand, the Group II pupae from all areas are capable of delaying the initiation of development for two months, until after many of the Group I moths have already emerged (Table 2).

In order to determine the progress of conditioning in overwintering cecropia pupae, we noted the emergence of adults from pupae that we transferred at intervals during the winter to laboratory conditions that favor development. The overwintering pupae were held in the screened outdoor insectary. At 40-day intervals from the first of December to the end of March, equal numbers of cocoons were transferred from the insectary to an incubator with a constant temperature of 25°C and a 17-hr photophase and a 7-hr scotophase (Waldbauer and Sternburg, unpublished).

We found that the effects of conditioning accumulated gradually. As winter exposure increased, the incubator time required for half of the moths to emerge decreased, and the bimodality of the emergence curve became more pronounced. By the end of March winter conditioning was complete. At that time the Group I moths emerged after 18 to 25 days (median = 21) in the incubator. According to Schneiderman and Williams (1954) cecropia's adult development requires 22 days at 25°C. Thus, it appears that some of the Group I moths had already started to develop in the insectary, while the rest began to develop immediately or almost immediately after being transferred to the incubator. The Group II moths emerged from the 39th to the 60th day (median = 46) after transfer. They obviously delayed the initiation of development for an average of 25 days. A similar experiment, done four years later, gave almost identical results; the median emergence of the Group I and II moths occurred on the 20th and 46th days, respectively, an average delay of 26 days in the initiation of development by the Group II moths. Under field conditions the Group II moths did not initiate development until 63 days after the Group I moths (Table 2), but the mean temperature during the interim was only 13.5°C. These results suggest that the initiation of development by Group II is triggered by the accumulation of a sufficient number of day-degrees above some as yet undetermined threshold.

Callosamia promethea, another saturniid moth, presents a more complicated emergence pattern than does cecropia (Sternburg and Waldbauer, unpublished). While cecropia is apparently univoltine throughout its range, promethea is univoltine in the north, but bivoltine in the south. The dividing line between the univoltine and bivoltine populations of promethea is just a few miles north of Urbana.

When transferred to the outdoor insectary at Urbana, samples of a univoltine population -- from about 70 miles to the north -- show an essentially unimodal emergence curve. One out of 200-300 moths sometimes emerges at the end of May, but the rest emerge in one tightly synchronized peak with a median date of about July 5. On the other hand, samples of a bivoltine population -- from about 50 miles to the south -- show a trimodal emergence whose peaks have median dates of about May 17, June 17, and June 30 at the Urbana insectary. One pair of parents may produce progeny which emerge with all three peaks. The median emergence date of the partial second generation is about August 12. Progeny of moths of any of the three emergence peaks may go into diapause rather than reproduce immediately. However, an increasing proportion of individuals enters diapause with each succeeding emergence peak. The progeny of first-peak moths rarely diapause, but the progeny of third-peak moths almost all enter diapause.

Discussion

The examples cited above show that there is indeed a variety of phenological strategies for the resumption of development in the spring, and that some of these strategies involve polymodal emergence patterns which may have two or more peaks within one growing season (Types A and B), or two or more peaks which are separated by intervals of a year (Type C). The existence of these phenomena raises several general questions: Are these patterns genetically controlled and do the emergence modes, therefore, represent distinct morphs? What is the adaptive significance of polymodal emergence patterns, and how did such patterns evolve? And finally, what are the underlying physiological mechanisms which permit some individuals to delay emergence for weeks or even years, while other individuals complete development under the same conditions? The answers to these questions are still fragmentary.

The results of line selections leave no doubt that the Type A polymodality of cecropia is genetically determined; the Group I and Group II emergers are dintinct, genetically mediated morphs (Waldbauer and Sternburg, 1973). Bradshaw (1973) presented indirect evidence that the two emergence modes of *Chaoborus americanus* represent genetically determined morphs. I do not know of similar evidence for the other species cited above which exhibit either Type A or Type B polymodality, however, there is no reason to doubt that their emergence patterns are also genetically determined. Powell (1974) stated that Type C polymodality seems to be an expression of genetic heterogeneity, but that evidence to support this view is lacking. However, Prebble (1941) found that in the European spruce sawfly a pattern of reduced emergence in the second year, followed by increased emergence in the third and fourth years repeated itself for a number of years and was independent of climatic variations in different seasons. This strongly suggests genetic determination of the pattern. Similarly, Sullivan and Wallace (1967) believed that the European pine sawfly is a species composed of several types of individuals with different diapause potentials.

The phenotypic expression of genetically determined polymodal emergence curves of Type A, B, or C may be significantly affected by environmental factors. For example, Allen and Fluke (1933) suggested that a large brood of apple maggots that emerged after a second year of diapause came from pupae that had remained in the soil during the abnormally hot and dry preceding summer. Sullivan and Wallace (1967) found that if European pine sawfly cocoons from short-day laboratory rearings (diapause-inducing) were exposed to 10°C instead of 21°C, then they underwent a prolonged diapause with emergence peaks at about 100, 200, 400, and 740 days rather than the usual three- to four-week diapause. European corn borer larvae *Ostrinia nubilalis* grown in the laboratory at daylengths near the critical daylength for diapause induction enter a less intense diapause than do corn borers grown at shorter daylengths (McLeod and Beck, 1963). Tauber and Tauber (1970, 1972) found a similar relationship between the diapause-inducing daylength and the intensity of the reproductive diapause of *Chrysopa carnea*. To my knowledge, neither the European corn borer nor *Chrysopa carnea* exhibit a polymodal emergence pattern. However, the modes of a polymodal emergence curve can be viewed as representing groups of individuals with different intensities of diapause (see also Masaki, 1961, especially pp. 86-88).

Waldbauer and Sternburg (1973) said that the adaptive strategy involved in cecropia's dimorphic, Type A termination of diapause is best expressed by the metaphor which warns against putting all of the eggs in one basket. They said that, "Cecropia's dimorphism avoids placing all of the progeny of a pair in one "temporal basket" in the following growing season at the critical period encompassing adult development, emergence, and reproduction. Partitioning the progeny between an early and a late emergence group might allow at least some of the progeny of a pair to excape various detrimental factors which may occur at different times in the growing season, but do not occur in every year or vary in severity from year to year. . . ." Bradshaw (1973) found that the early emerging morph of the aquatic *Chaoborus americanus* did well in a year when spring was early and the weather became continuously warmer. However, in a year in which an early thaw and plankton bloom were followed by a refreezing of the pond, few of the early emergers survived although the late emergers did well. Although Type A and Type B polymodalities have different origins, they result in similar emergence patterns and probably serve the same adaptive ends.

Several authors have pointed out the adaptive value of Type C polymodality. For example, Griffiths (1959) said that the survival of diapausing European pine sawfly larvae through one or more overwintering periods buffers the population against the catastrophic elimination of an entire year class. Sullivan and Wallace (1967) agreed, despite the association of increased mortality and decreased fecundity with prolonged diapause. Powell (1974) suggested that the Type C polymodality of ethmiid moths enables them to cope with drought years.

It is noteworthy that many insects with polymodal emergence curves are associated with marginal areas where conditions tend to vary unpredictably from year to year. Among them are northern forms such as the douglas-fir cone moth *Barbara colfaxiana* which exhibits a Type C polymodality extending over as many as three winters (Hedlin, 1960), the European pine sawfly (Griffiths, 1959), and the European spruce sawfly (Prebble, 1941); high elevation or subalpine species such as the grasshoppers *Melanoplus bruneri* and *Chorthippus longicornis* (Kreasky, 1960); species in arid areas such as the katydids *Idiostatus bechteli* and *I. elegans* whose egg hatch may extend over more than one year (Rentz, 1973), the yucca moths (Incurvariidae), and some species of ethmiid moths (Powell, 1974).

The timing and the degree of synchronization of an overwintering population's reentry into the phenological flow may, of course, have a profound effect on its fitness. The selective forces which determine the timing and pattern of the emergence may be lumped into three major categories: (1) the availability of resources (including the absence of unfavorable factors) for an adult or its progeny, (2) the availability of partners for the purpose of reproduction, and (3) the temporal predictability of the resources. The first group of selective forces affects both the timing and the degree of synchronization of the emergence. Their effect is most obvious in species that have a high degree of host specificity, as illustrated by the close synchronization of *Rhagoletis* emergence patterns with the ripening pattern of their host fruits (Boller and Prokopy, 1976). The second factor, the availability of mates, promotes the synchronization of emerging individuals with each other. There will be a high degree of synchronization among species which are short-lived as adults, particulary if they occur in low population densities. For example, cecropia and other saturniids do not feed as adults and do not live more than a few days in the field (Rau and Rau, 1912, 1914; Ferguson, 1972). As already pointed out above, the members of each of cecropia's or promethea's emergence groups tend to be well-synchronized with each other (Sternburg and Waldbauer, 1969; Waldbauer and Sternburg, 1973; Sternburg and Waldbauer, unpublished). The third factor, the temporal unpredictability of all or some of the resources, exerts an opposing selection pressure which tends to spread the emergence out over time instead of synchronizing it. In other words, if there is no environmental cue which predicts the continued favorability of conditions after emergence, there should be a selection for the population to "hedge its bet" by increasing the duration of the emergence period or by becoming polymorphic for emergence date. Polymorphism seems more likely to occur when factors such as a short adult life demand synchronization of the emergence of individuals with each other as a means of increasing reproductive success.

Bradshaw (1973) suggested that *Chaoborus americanus* faces an environment which is coarse-grained or patchy (Levins and MacArthur, 1966) with respect to the optimum time to terminate diapause. In other words, the spring thaw may be early and followed by another freeze, or the thaw may be early and followed by continued favorable conditions, or finally, the thaw may be late, in which case it is always followed by continued favorable conditions. Bradshaw argued that a cue that will predict whether or not the pond will refreeze is not present when *Chaoborus* faces the developmental option of continuing in diapause or initiating development, and that, therefore, selection favors the existence of a polymorphism: an opportunistic early emerging morph which does well in years when an early thaw is followed by continued favorable conditions, and a conservative, late emerging morph which does not initiate development until continued favorable conditions are a certainty.

Mating will tend to be assortative in species with polymodal emergence curves, i.e., members of one mode are more likely to mate with each other than with members of another mode. In cecropia it is highly unlikely that Group I emergers ever mate with Group II emergers because there is usually a gap of at least a week between the last of the Group I emergers and the first of the Group II emergers. Furthermore, the females mate only once, and are almost invariably mated on the first or second night after they emerge (Sternburg and Waldbauer, unpublished). However, there is obviously gene flow between the two groups because pairs from either group usually produce a preponderance of progeny which emerge with Group II (Fig. 2). Nevertheless, yearly variations in the proportions of Group I and II emergers (Table 1) suggest the operation of differential selection which favors Group I more in some years than in others.

The final question concerns the physiological mechanisms which permit some individuals to delay emergence while other individuals of the same species complete their development. The endocrine mechanism which permits diapause to be prolonged for a year or more is not known (Chippendale, 1977). There are few data which pertain specifically to the developmental delays involved in Type A polymodality. However, statements which are common in the literature would, if we accept them as generalizations, lead us to conclude that developmental delays within a season are not mediated by the same mechanisms which control diapause. For example, Danilevskii (1965) explained delayed development by saying that, although diapause has been terminated, morphogenesis does not begin until a token stimulus is received, this being a temperature significantly exceeding the level necessary for further development.

Tauber and Tauber (1976) suggested that, except for a few species which require a specific stimulus to end diapause, the occurrence of diapause termination within a population is distributed over a broad time span, and that in these species phenological synchronization is achieved largely through factors which control morphogenesis during the post-diapause period. These views suggest that the diapause-terminating mechanism acts as no more than a very "coarse adjustment" which simply prevents the initiation of development before spring.

However, there is good reason to believe that a condition of physiological diapause exists until development actually begins. There is no doubt that at some time during the winter a diapausing individual regains its competency to develop. Tauber and Tauber (1976) interpret this as the end of diapause, but I believe that it is more likely to be a decrease in the intensity of diapause or a shift to a different phase of diapause. The difference is not simply semantic, as shown by evidence, cited by Mansingh (1971), that there is little physiological difference between individuals which are competent to resume development and individuals which have not yet achieved competency. For example, competent individuals are still cold-hardy and still maintain a low metabolic rate.

Danilevskii's (1965) generalization does appear to fit Group I cecropia, the opportunistic early emergers which begin to develop just as soon as it is warm enough in the spring (Table 2). However, his generalization obviously does not fit Group II cecropia, which can delay the initiation of development for 63 days under the relatively cool conditions of the field, and for 21 days even at a constant temperature of 25°C. It is easier to conceive of a physiological mechanism that generates polymodal emergence patterns if we view diapause termination as taking place in more than one step. Perhaps Group II cecropia are in a more intense diapause than Group I cecropia, and, therefore, require an extra step to fully terminate diapause. In cecropia the extra step is probably mediated by an exposure of sufficient duration to temperatures above an undetermined threshold (see above, and Waldbauer and Sternburg, 1973).

This idea is supported by unpublished results (Sternburg, Waldbauer, and George R. Wilson) which indicate that an endocrine mechanism mediates the delay of development by Group II cecropia, and that this mechanism is not inconsistent with either the hormonal failure theory or the juvenile hormone regulation theory for the endocrine control of diapause (Chippendale, 1977). Cecropia pupae were transferred from the outdoor insectary to a constant temperature of 25°C at the end of March, when the Group I individuals were presumably fully conditioned. A control group of these pupae, injected with a saline solution, exhibited the usual bimodal emergence curve, with a small Group I peak and a much larger Group II peak. However, another group of these pupae was injected with 2.5 μg per g of β-ecdysone in saline, and almost all of them emerged in concert with Group I of the controls. In other words, the developmental delay of Group II was prevented by the injection of β-ecdysone.

It may not be realistic to speak of discrete steps; the actual mechanism may be more complex. Nevertheless, at this point I will risk oversimplification in order to state the argument as simply as possible.

A diapause-terminating mechanism which could proceed in either one or two steps would serve as a versatile phenological adjustment, and would explain the phenomena described above. Populations which emerge together early in spring may consist of individuals that do not require the second step, while populations whose emergence is uniformly delayed may consist of individuals that all require the second step. Populations with a polymodal emergence curve may consist of a mixture of individuals that require either one or both steps. The first of these steps, the regaining of competency to develop, serves as a phenological "coarse adjustment," while the final termination of diapause, signaled by the actual resumption of development, serves as a phenological "fine adjustment." The effects of temperature on development are the "ultra-fine adjustment."

References

Allen, T.C., Fluke, C.L., Jr.: Notes on the life history of the apple maggot in Wisconsin. J. Econ. Entomol. *26*, 1108-1112 (1933).

Andrewartha, H.G.: Diapause in relation to the ecology of insects. Biol. Rev. *27*, 50-107 (1952).

Barnes, H.F.: Studies of fluctuations in insect populations. XII. Further evidence of prolonged larval life in the wheat-blossom midges. Ann. Appl. Biol. *39*, 370-373 (1952).

Beck, S.D.: Insect Photoperiodism. New York: Academic Press 1968, 288 pp.

Boller, E.F., Prokopy, R.J.: Bionomics and management of *Rhagoletis*. Ann. Rev. Entomol. *21*, 223-246 (1976).

Bradshaw, W.E.: Homeostasis and polymorphism in vernal development of *Chaoborus americanus*. Ecology *54*, 1247-1259 (1973).

Bush, G.L.: Sympatric host race formation and speciation in frugivorous flies of the genus *Rhagoletis* (Diptera, Tephritidae). Evolution *23*, 237-251 (1969).

Bush, G.L.: Sympatric speciation in phytophagous parasitic insects. In: Evolutionary Strategies of Parasitic Insects. Price, P.W. (ed.). New York: Plenum Press 1975, pp. 187-206.

Chippendale, G.M.: Hormonal regulation of larval diapause. Ann. Rev. Entomol. *22*, 121-138 (1977).

Corbet, P.S.: A Biology of Dragonflies. Chicago: Quadrangle Books 1963, 247 pp.

Danilevskii, A.S.: Photoperiodism and Seasonal Development of Insects. [Translated from Russian by J. Johnston.] Edinburgh: Oliver and Boyd 1965, 283 pp.

Danilevskii, A.S., Goryshin, N.I., Tyshchenko, V.P.: Biological rhythms in terrestrial arthropods. Ann. Rev. Entomol. *15*, 201-244 (1970).

deWilde, J.: Photoperiodism in insects and mites. Ann. Rev. Entomol. *7*, 1-26 (1962).

Feeny, P.: Seasonal changes in oak leaf tannins and nutrients as a cause of spring feeding by winter moth caterpillars. Ecology *51*, 565-581 (1970).

Ferguson, D.C.: Bombycoidea (in part). In: The Moths of America North of Mexico. Dominick, R.B., et al. (eds.). London: E.W. Classey 1972, pp. 155-275.

Griffiths, K.J.: Observations of the European pine sawfly, *Neodiprion sertifer* (Geoff.), and its parasites in southern Ontario. Canad. Entomol. *91*, 501-512 (1959).

Hedlin, A.F.: On the life history of the douglas-fir cone moth, *Barbara colfaxiana* (Kft.) (Lepidoptera: Olethreutidae), and one of its parasites, *Glypta evetriae* Cush. (Hymenoptera: Ichneumonidae). Canad. Entomol. *92*, 826-834 (1960).

Kangas, E.: Beitrag zur Biologie und Gradation von *Diprion sertifer* Geoffr. (Hym., Tenthredinidae). Ann. Entomol. Fennici 7, 1-31 (1941).

Kormondy, E.J.: The systematics of *Tetragoneuria*, based on ecological, life history, and morphological evidence (Odonata: Corduliidae). Misc. Pub. Mus. Zool. Univ. Michigan *107*, 1-79 (1959).

Kreasky, J.B.: Extended diapause in eggs of high-altitude species of grasshoppers, and a note on food-plant preferences of *Melanoplus bruneri*. Ann. Entomol. Soc. Amer. *53*, 436-438 (1960).

Lathrop, F.H., Nickels, C.B.: The biology and control of the blueberry maggot in Washington County, Maine. U.S.D.A. Tech. Bull. 275, 76 pp. (1933).

Levins, R., MacArthur, R.: Maintenance of genetic polymorphism in a spatially

heterogeneous environment: variation on a theme by Howard Levene. Amer. Nat. *100*, 585-590 (1966).
Mansingh, A.: Physiological classification of dormancies in insects. Canad. Entomol. *103*, 983-1009 (1971).
Mansingh, A., Smallman, B.N.: The influence of temperature on the photoperiodic regulation of diapause in saturniids. J. Insect Physiol. *17*, 1735-1739 (1971).
Marsh, F.L.: A few life-history details of *Samia cecropia* within the southwestern limits of Chicago. Ecology *22*, 331-337 (1941).
Masaki, S.: Geographic variation of diapause in insects. Bull. Fac. Agric. Hirosaki Univ. *7*, 66-98 (1961).
McLeod, D.G.R., Beck, S.D.: Photoperiodic termination of diapause in an insect. Biol. Bull. *124*, 84-96 (1963).
Morris, R.F.: Observed and simulated changes in genetic quality in natural populations of *Hyphantria cunea*. Canad. Entomol. *103*, 983-906 (1971).
Morris, R.F., Fulton, W.C.: Heritability of diapause intensity in *Hyphantria cunea* and correlated fitness responses. Canad. Entomol. *102*, 927-938 (1970).
Oatman, E.A.: Apple maggot emergence and seasonal activity in Wisconsin. J. Econ. Entomol. *57*, 676-679 (1964).
Powell, J.A.: Occurrence of prolonged diapause in ethmiid moths. Pan-Pac. Entomol. *50*, 220-225 (1974).
Prebble, M.L.: The diapause and related phenomena in *Gilpinia polytoma* (Hartig). V. Diapause in relation to epidemiology. Canad. J. Res. *19*, 437-454 (1941).
Rabb, R.L.: Diapause in *Protoparce sexta* (Lepidoptera: Sphingidae). Ann. Entomol. Soc. Amer. *59*, 160-165 (1966).
Rau, P., Rau, N.: Longevity in saturniid moths: an experimental study. J. Exp. Biol. *12*, 179-204 (1912).
Rau, P., Rau, N.: Longevity in saturniid moths and its relation to the function of reproduction. Trans. Acad. Sci. St. Louis *23*, 1-78 (1914).
Rentz, D.C.: The shield-backed katydids of the genus *Idiostatus*. Mem. Amer. Entomol. Soc. 29, 211 pp. (1973).
Schneiderman, H.A., Horwitz, J.: The induction and termination of facultative diapause in the chalcid wasps *Mormoniella vitripennis* (Walker) and *Tritneptis klugii* (Ratzeburg). J. Exp. Biol. *35*, 520-551 (1958).
Schneiderman, H.A., Williams, C.M.: The physiology of insect diapause. IX. The cytochrome oxidase system in relation to the diapause and development of the cecropia silkworm. Biol. Bull. *106*, 238-252 (1954).
Scudder, S.H.: The Butterflies of the Eastern United States and Canada. Vol. II. Cambridge, Mass.: S.H. Scudder 1889, 708 pp.
Sheldon, J.K., MacLeod, E.G.: Studies on the biology of Chrysopidae IV. A field and laboratory study of the seasonal cycle of *Chrysopa carnea* Stephens in central Illinois (Neuroptera: Chrysopidae). Trans. Amer. Entomol. Soc. *100*, 437-512 (1974).
Shelford, V.E.: Laboratory and Field Ecology. Baltimore: William and Wilkins Co. 1929, 608 pp.
Sternburg, J.G., Waldbauer, G.P.: Bimodal emergence of adult cecropia moths under natural conditions. Ann. Entomol. Soc. Amer. *62*, 1422-1429 (1969).
Sullivan, C.R., Wallace, D.R.: Interaction of temperature and photoperiod in the induction of prolonged diapause in *Neodiprion sertifer*. Canad. Entomol. *99*, 834-850 (1967).
Tauber, M.J., Tauber, C.A.: Adult diapause in *Chrysopa carnea:* stages sensitive to photoperiodic induction. J. Insect Physiol. *16*, 2075-2080 (1970).
Tauber, M.J., Tauber, C.A.: Geographic variation in critical photoperiod and in diapause intensity of *Chrysopa carnea* (Neuroptera). J. Insect Physiol. *18*, 25-29 (1972).
Tauber, M.J., Tauber, C.A.: Insect seasonality: diapause maintenance, termination, and postdiapause development. Ann. Rev. Entomol. *21*, 81-107 (1976).
Waldbauer, G.P., Sternburg, J.G.: Polymorphic termination of diapause by cecropia: genetic and geographical aspects. Biol. Bull. *145*, 627-641 (1973).
Wigglesworth, V.B.: The Principles of Insect Physiology. London: Chapman and Hall 1972, 827 pp.

Willey, R.L.: Emergence patterns of the subalpine dragonfly *Somatochlora semicircularis* (Odonata: Corduliidae). Psyche *81*, 121-133 (1974).

Williams, C.M.: Physiology of insect diapause. X. An endocrine mechanism for the influence of temperature on the diapausing pupae of the cecropia silkworm. Biol. Bull. *110*, 201-218 (1956).

Willis, J.H., Waldbauer, G.P., Sternburg, J.G.: The initiation of development by the early and late emerging morphs of *Hyalophora cecropia*. Ent. exp. & appl. *17*, 219-222 (1974).

Phenology and Photoperiodic Diapause in Northern Populations of *Drosophila*

JAAKKO LUMME

Introduction

This paper deals with an important aspect of adaptation to the northern environment, the temporal synchronization of populations with seasons, i.e., phenology of *Drosophila* species. Winter dormancy is the main synchronizer in northern areas. Extensive reviews of this subject do not contain any information on *Drosophila* (e.g., Lees, 1955; Danilevskii, 1965; Beck, 1968; Danilevsky et al., 1970).

The wide use of the species of the genus *Drosophila* as objects of study in many fields of biology is due to the practical advantages they have in everyday experimentation in the laboratory. The great amount of information which has accumulated since 1909 is a secondary, but still strong argument for using *Drosophila* whenever possible. Their primary economic importance can be considered small, which partly explains some of the gaps in knowledge concerning them. In particular the seasonality of insects has been elucidated most effectively by applied entomology in many parts of the world, even at the level of "basic" research. For those concerned with practical problems, *Drosophila* of course has not been the object of primary interest.

In almost all branches of research related to ecology, or in those demanding accurate ecological knowledge as a basis (e.g., population genetics today!), *Drosophila* species are still poor subjects for study. Their ecology has been studied only fragmentarily, but interest is rapidly increasing. No detailed monographs on the ecology of the genus are available, but Throckmorton (1975) gives a good review and a representative list of literature on this topic in connection with a modern presentation on the phylogeny of the family.

The geographical distribution of the species of the genus is strongly uneven: a great majority of *Drosophila* inhabit tropical regions of the world. Because the distribution of *Drosophila* workers is uneven, too, the species from southern temperate areas are overrepresented in the literature. Furthermore, many of the really well-known and widely used species are cosmopolitan and commensals of human culture. The ecology of the domestic species outside the areas of origin is a special case, of limited interest to field ecologists. Some species, however, inhabit the "edges of tundra" (Throckmorton, 1975). Collection data from the northernmost areas colonized by *Drosophila* have been published by Wheeler and Throckmorton (1960) from Alaska, Basden (1956) and Basden and Harnden (1956) from northern Norway, and Lumme et al. (1974a) from Fennoscandia northwards from latitude 65° N. Lakovaara et al. (1974) give a list of species found in Norway, Finland, Sweden, and Denmark which contains 33 species of *Drosophila* (Table 1). Although this figure includes many unestablished immigrants (e.g., *Drosophila melanogaster* Meigen, *D. simulans* Sturtevant, *D. busckii* Coquillet, *D. funebris* Fallen, *D. immigrans* Sturtevant, and *D. hydei* Sturtevant), it is very small in comparison with the number of species in the whole genus. Lists of known species (Wheeler, 1959; Wheeler

Table 1. ***Drosophila*** species recorded from Fennoscandia and Denmark (Lakovaara *et al.*, 1974), their taxonomic position and remarks on the phenology in the north. Species marked with asterisk have been recorded north of latitude 65° N.

Subgenus	species group	species	notes on overwintering and phenology
Scaptodrosophila		*deflexa** Duda	larval diapause (Basden, 1954)
Dorsilopha		*busckii** Coquillett	commensal of human culture, not overwintering outdoors
Lordiphosa		*fenestrarum** Fallén	? (not attracted by baits)
		forcipata Collin	?
		nigricolor de Meijere	?
Sophophora	*melanogaster*	*melanogaster** Meigen	commensal
		*simulans** Sturtevant	commensal
	obscura	*alpina** Burla	pupal diapause (Lumme *et al.*, in press)
		ambigua Pomini	commensal even in England (Shorrocks, 1972)
		*bifasciata** Pomini	photoperiodic adult diapause (Lakovaara *et al.*, 1972)
		*eskoi** Lakovaara & Lankinen	adult diapause (Lumme *et al.*, in press)
		*obscura** Fallén	photoperiodic adult diapause (Begon, 1976)
		subobscura Collin	not synchronized in England (Begon, 1976)
		*subsilvestris** Hardy	does not overwinter as adult (Lumme *et al.*, in press)
		tristis Fallén	adult diapause? (Basden, 1954)
Hirtodrosophila		*cameraria** Haliday	English populations not synchronized? (Charlesworth and Shorrocks, 1976)
		ingrica Hackman	?

		*lundstroemi** Duda	?
		*subarctica** Hackman	photoperiodic adult diapause (Lakovaara *et al.*, 1972)
Drosophila	*funebris*	*funebris** Fallén	commensal
	virilis	*littoralis** Meigen	photoperiodic adult diapause (Lumme *et al.*, 1974b, 1975, in press; Lumme and Oikarinen, in press)
		*ezoana** Takada & Okada	photoperiodic adult diapause (Lakovaara *et al.*, 1973)(1)
		*lummei** Hackman	photoperiodic adult diapause (Lakovaara *et al.*, 1973)
		*montana** Stone, Griffen & Patterson	photoperiodic adult diapause (Lakovaara *et al.*, 1973)(2)
	repleta	*hydei** Sturtevant	commensal
	immigrans	*immigrans* Sturtevant	commensal
	quinaria	*limbata* von Roser	?
		*phalerata** Meigen	photoperiodic adult diapause (Geyspits and Simonenko, 1970; Charlesworth and Shorrocks, 1976; Muona and Lumme, in press)
		*transversa** Fallén	photoperiodic adult diapause (Charlesworth and Shorrocks, 1976; Muona and Lumme, in press)
	testacea	*testacea** von Roser	photoperiodic adult diapause (Lumme *et al.*, in press)
	miscellaneous	*confusa* Staeger	?
		*histrio** Meigen	adult diapause? (Lumme *et al.*, in press)
		picta Zetterstedt	?

(1) a name *D. lakovaarai* was erroneously used

(2) a synonym *D. ovivororum* Lakovaara & Hackman was used

and Hamilton, 1972) already contain 1254 names. The low number of northern species may indicate qualitatively the difficulties which the organisms face in adapting to severe climatic conditions.

While all facets of the ecology and adaptation of *Drosophila* species in the north are poorly known, the phenology of other insect groups has been extensively studied, especially in those species of economic importance (e.g., the review by Tauber and Tauber, 1973). Thus not very much that is novel can be expected from the research on *Drosophila*, at least at the level of the collection of basic observations. Why then coose *Drosophila* at all? We consider it important to study the process of evolution leading to the adaptation of some, even if only a few, species to northern conditions. Which are the relevant differences between successful and nonsuccessful colonizers? When the basic information is collected and combined with the knowledge already available on population genetics, phylogeny, developmental genetics, and physiology of *Drosophila*, we expect fruitful conclusions to arise. The diversity of the genus (exceptionally wide morphological and ecological variation exists within this group contained in a single genus) enables us to find several different mechanisms of adaptation and thus some general trends in the evolution of northern species. Furthermore, some fundamental problems concerning, for example, the physiology of different adaptation mechanisms (photoperiodism!) have so far not been solved in any insect species. The advantages of *Drosophila* in such work are self-evident. However, a great deal of the work at the beginning is necessarily a repetition of research already done in other insect species, in other words an applied entomology of *Drosophila*.

The temporal synchronization of populations to seasons is of course not the only prerequisite for survival in hard climatic conditions, but surely one of great importance. There is no abrupt change from temperate to cold climate, nor any clear-cut specific group of "northern" species. The change is gradual, and the problem of synchronization is general, concerning species in all life zones. In the north, wintertime is the most crucial phase of life. Individuals must overwinter without a continuous supply of energy, they must resist low temperatures and low humidity, and they must be able to continue development *as a population* when the short suitable season begins.

Phenology of *Drosophila* Populations

During mainly collection-oriented work, extensive data have been collected on the seasonal occurrence of *Drosophila* species in different parts of the world. Especially in starting more detailed studies, the following list of some published observations can be useful. Patterson (1943) has studied the *Drosophila* fauna in Texas and Cooper and Dobzhansky (1956) the fauna in California. In Europe, Burla (1951, 1961) and Bächli (e.g., 1972a,b) have worked in Switzerland, Basden (1953, 1954) in Scotland, Herting (1955) in West German, Rocha Pité (1972) in Portugal, and Shorrocks (1975), Begon (1976), and Charlesworth and Shorrocks (1976) in England. Data from Brazil are given by Dobzhansky and Pavan (1950), from Colombia by Hunter (1966), from Lebanon by Pipkin (1952), and from Queensland, Australia, by Mather (1956). Wakahama (1962), Toda (1973), and Kimura (1976) have worked in Japan and Lachaise (1974) in the Ivory Coast of West Africa.

The general impression arising from the seasonal occurrence of different species is all but clear. A general observation is that the flying time of most species is quite long and not restricted to a short period of the summer (or humid season) as in many other insects. In many cases, population densities seem to increase slowly during all of the suitable season (warm, humid), indicating continuous reproduction and the presence of several successive generations. In England, *Drosophila obscura* Fallén has three to four, and *D. subobscura* Collin between four to six generations per year (Begon, 1976); when the life of individual adults is quite long, the generations overlap and the population growth (after limiting winter) can be compared to that of a bacterial population in the exponential phase. In the very north the situation is different, as is seen in the following.

Seasonal Development of Populations at the Latitude 65°N in Finland

The northernmost locality in the list above was Scotland (about 56°N lat.). Since 1973 the seasonal occurrence and life cycles of drosophilids have been studied in the vicinity of the town of Oulu, which is situated on the coast of Bothnian Bay at latitude 65°N, Finland (Lumme et al., 1974b). In this area the length of the thermal summer is on the average 150 days. The July mean temperature is slightly over 16°C.

During the summers of 1974 and 1975 flies were collected at about two-week intervals during all of the warm season by means of attractive baits (malt bait of Lakovaara et al., 1969). The disadvantage of this method is that only adults can be trapped, and the ends of the summer give poor results due to low temperature. These difficulties are present in almost all work on natural populations of *Drosophila*: quantitative collection methods have not been developed and natural breeding sites are often practically unknown. The total number of species of *Drosophila* was 21, a good proportion of those 33 ever recorded in Scandinavia and Finland (Lakovaara et al., 1974). The results are presented in detail elsewhere (Lumme et al., in press; Muona and Lumme, in press).

Because of the frequency data of different species give a very superficial and (due to an unsatisfactory collection method) obscure picture on the phenology, the reproductive status of populations was checked by dissecting the females of all species (for classifications of ovarial stages, see Lumme et al., 1974b; Begor 1976). In addition, some small-scale outdoor culture experiments were done (on the malt medium of Lakovaara, 1969) to elucidate the rate of development of different species at different times of the summer.

As an example, the life cycles of two species are presented here, namely *Drosophila littoralis* Meigen and *D. transversa* Fallen, which illustrate well some of the typical phenomena found. The annual variation of the reproductive condition of their adult populations is presented in Figure 1. The figures indicating the number of females studied also roughly tell the frequency of these species in our samples (but not absolute densities); sex ratio is even in both of these species.

In the first samples obtained in spring, females with completely undeveloped ovaries are found, but their "aged" appearance indicates clearly that they have overwintered as adults. In the spring of 1975, all *D. transversa* females checked at the beginning of May were completely undeveloped, but at the same time the maturation of *D. littoralis* females had already started. None of them were laying eggs, however. This difference between these two species in spring development was consistent in all data: *D. transversa* starts flying and reproducing slightly later than *D. littoralis*.

In both species the populations develop to full reproductive maturity

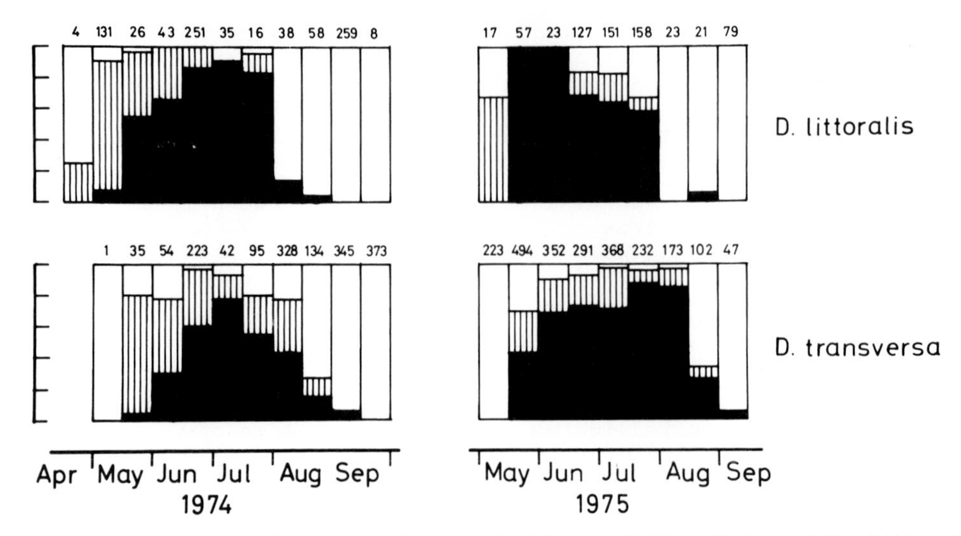

Fig. 1. The seasonal variation in the reproductive condition of *Drosophila littoralis* and *D. transversa* populations at latitude 65°N (Oulu, Finland). Black: egg-laying females, shaded: developing females, white: undeveloped (diapausing or very young) females, as proportions of total number of females studied (figures above the columns). (Lumme et al., in press).

gradually. This may be a means for avoiding catastrophal effects of late frosts (see Waldbauer and Sternburg, 1973), but this has not been studied in *Drosophila*. Temperature has a great influence on the maturation rate; in 1974 when the spring was unusually cold, some overwintered individuals were still immature after midsummer when the first generation had already started emergence from pupae. Begon (1976) has shown that adult development (maturation) is more retarded by low temperatures than larval development in *D. obscura*.

First-generation flies started eclosion soon after midsummer and continued development to reproductive maturity in both species. Because the overwintered generation is surprisingly long-lived (some *D. littoralis* individuals were found in the latter half of July), the eclosion of the first generation extends over the entire late summer. Overwintered and summer generations also interbreed to some extent (at least in many years). In August emerging adults develop no further even if the temperature remains quite high. The *D. littoralis* population stops maturation slightly earlier than *D. transversa* (compare with the spring!). This is obviously correlated with the longer developmental time from egg to adult in *D. littoralis*.

With minor modifications in timing, the seasonal development of the populations described above was found to be similar in ten out of 13 drosophilid species found in considerable numbers (Lumme et al., in press). They are *D. littoralis*, *D. montana* Stone, Griffen, and Patterson (first described in Europe as *D. ovivororum* Lakovaara and Hackman), *D. lummei* Hackman, *D. ezoana* Takada and Okada, *D. transversa*, *D. phalerata* Meigen, *D. testacea* v. Roser, *D. bifasciata* Pomini, *D. obscura*, and *D. eskoi* Lakovaara and Lankinen. In addition, observations in northern Finland show that *D. subarctica* Hackman has a similar phenology (unpublished). These species are spending their winter dormancy as reproductively immature adults, and at our latitude of study they have only a partial second generation during the summer (in fact, this is concluded only from the observation that some adults of the first generation lay eggs early enough). The onset of dormancy occurs in adults eclosing in August and thereafter, and reproduction starts again in May.

The adult is also the most common stage of overwintering in *Drosophila* species studied so far in areas outside Finland. Carson and Stalker (1948) confirmed it in *D. robusta* Sturtevant. Basden (1954) studied Scottish populations of *D. obscura*, *D. subobscura*, and *D. tristis* Fallen, and concluded that they also overwinter at the adult stage. Begon (1976) has ascertained this for *D. obscura*, but in his careful experiments *D. subobscura* showed retarded rather than arrested (adult) development during the winter. This can mean that other developmental stages, too, are present in wintertime. In fact, finding of dormant adults does not mean that other stages are absent. Possible overwintering of eggs, larvae, or pupae is very difficult to ascertain or exclude when the breeding sites remain unknown, if the adult population is not observed continuously. For *D. phalerata* and *D. transversa* hibernation in the adult has been shown in many localities (Geyspits and Simonenko, 1970; Charlesworth and Shorrocks, 1976; Muona and Lumme, in press). The adult at the stage of winter dormancy has also been found in *D. rufifrons* Loew (Bertani, 1947), *D. lowei* Heed, Crumpacker, and Ehrman (Heed et al., 1969), *D. grisea* Patterson and Wheeler, and *D. macroptera* Patterson and Wheeler (Kambysellis and Heed, 1974).

In our field collections in the years 1974 and 1975, and also in other data from Finland, three drosophilid species differ clearly from the others in their seasonal occurrence: *D. alpina* Burla, *D. subsilvestris* Hardy, and *Chymomyza costata* Zetterstedt. Their flying time begins much later than that of the other species, and the first individuals caught sometime after midsummer are all newly emerged. Because this is the time of emergence of the first generation of many other species there are two alternative explanations for this observation: the overwintered adults are not attracted by our baits, or these three species overwinter in some other stage. Some laboratory observations indicate that *D. alpina* has an obligatory pupal diapause at our latitudes (Lumme et al., in press). *D. subsilvestris* shares a similar phenology in the field, but we have never succeeded in culturing it in the laboratory, and the overwintering stage remains unknown. That *Chymomyza costata* starts flying so late is not surprising, because laboratory observations have shown that it has a larval diapause (Hackman et al., 1970; Lakovaara et al., 1972). Larval diapause is also known in *D. deflexa* Duda (Basden, 1954), and pupal diapause possibly occurs in some high-altitude populations of *D. persimilis* Dobzhansky and Epling (Dobzhansky and Epling, 1944).

Thus the overwintering in the adult stage seems to be the most common means in *Drosophila* species in temperate regions studied so far. One technical aspect, however, can bias this impression. The standard method in collecting drosophilids is the use of baits. Developmental stages other than adults are not trapped, and thus adult dormancy is the most obvious one to be found without any systematic study. Furthermore, retarded adult development during low winter temperatures can lead to a false impression of synchronized populations (the case of *D. subobscura*, Basden, 1954 *contra* Begon, 1976). Plenty of experimental work is still needed on this topic, and it is especially important to study developmental stages other than adults in their natural environments to obtain a clear picture of the life cycles.

To summarize the above description, a hypothetical scheme of the life cycle of a species overwintering as an adult is presented in Figure 2.

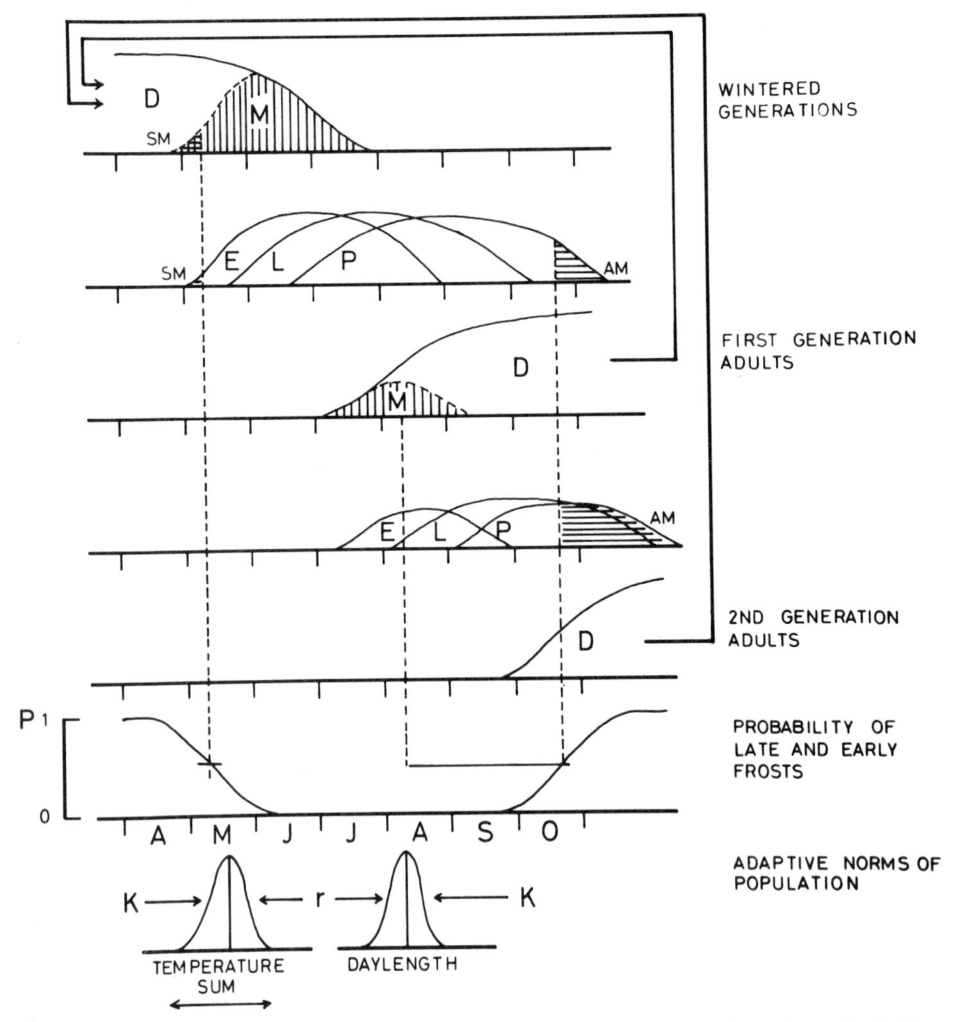

Fig. 2. Scheme of the life cycle of a northern *Drosophila* overwintering as adult. D: diapausing, M: mature, reproducing adults, E: eggs, L: larvae, P: pupae. SM: mortality due to late spring frosts, AM: mortality due to early autumn frosts. r: r-selection acting toward early beginning and late ending of reproductive period, K: K-selection, opposite to r-selection.

It must be stressed that the presentation is based mainly on field observations on adults, but it is supported by culture experiments on artificial medium. Figure 2 also summarizes the environmental variables on which primary attention must be focused (Stern and Roche, 1974), i.e., date and distribution of the temperature sum needed to initiate reproduction in the population, frequency, and distribution in time of spring frosts, and frequency and distribution in time of early frosts in the fall. Stern and Roche (1974) deal with the synchronization of forest trees with the season; a *Drosophila* population is analogous to a perennial tree. The timing of the beginning of the reproductive period seems to depend on rising temperatures only; daylength is long enough before the snow has melted in our latitude of study (Lumme et al., in press; see also Tauber and Tauber, 1976). The stopping of reproduction in August is the topic of the latter part of this paper. The frosts usually come at our latitude some two to three months later. Considering the long time needed for the development from egg to adult, which can be two months or more in autumn temperatures for instance in *D. littoralis*, such early stopping of reproduction is necessary if prereproductive (diapausing) adults are the only stages surviving the winter.

Environmental Factors Influencing the Onset of Adult Diapause

The Influence of Daylength on the Adult Reproductive Diapause

Photoperiod has long been known to be the main factor controlling diapause in insects (Danilevskii, 1965; Beck, 1968). It is surprising that the first report known to us of photoperiodic control of diapause in *Drosophila* was published as late as 1970 by Geyspits and Simonenko. They studied *Drosophila phalerata*. Thereafter photoperiodic control of diapause has been ascertained in several species in temperate regions. As in other groups of insects, it may be considered a rule at least for populations at northern latitudes. In field observations, the only sign of this is the quite early stopping of reproduction, but no final conclusions on the role of daylength can be drawn on field observations alone.

The first indication of a photoperiodic diapause, which may be seen in laboratory work oriented to other subjects, is that flies collected from the field stop breeding as first-generation adults (in the case of reproductive diapause) or do not lay eggs, if collected late in the summer. If cultures are illuminated continuously (24 hr per day instead of "normal" conditions), breeding sometimes takes place (Lakovaara et al., 1972). However, the very high temperatures used in laboratories working with, for example, *D. melanogaster* can allow culturing in all daylengths, and thus photoperiodism is not detected (this may explain the late start of studies in this field).

In the last few years, photoperiodic control of diapause has been found in a variety of species overwintering as adults, but it has not so far been observed in the few species known to overwinter at other stages. The photoperiodic species are *D. phalerata* and *D. transversa* (Geyspits and Simonenko, 1970); *D. littoralis* (containing four sibling species), *D. bifasciata*, and *D. subarctica* (Lakovaara et al., 1972); the siblings of *D. littoralis*: *D. lummei*, *D. ezoana* (the name *D. lakovaarai* was erroneously used),. and *D. montana* (a synonym *D. ovivororum* was used) (Lakovaara et al., 1973); *D. obscura* (Begon, 1976); and *D. grisea* (Kambysellis and Heed, 1974). Few species have been studied in detail, however.

Of the 11 species observed to overwinter as adults at our latitude of study (Table 1), ten species have photoperiodically controlled diapause. The eleventh, *D. eskoi*, has not been subjected to detailed experiments, but it breeds readily in continuous illumination, which is standard method in our laboratory.

In connection with our field work on phenology described above, the photoperiodic reaction curves of *D. littoralis* and *D. transversa* were determined from good population samples (more than ten females founding a strain). The critical daylength of the *D. littoralis* population was found to be 17.9 hr (at 16°C). This is the daylength in the first

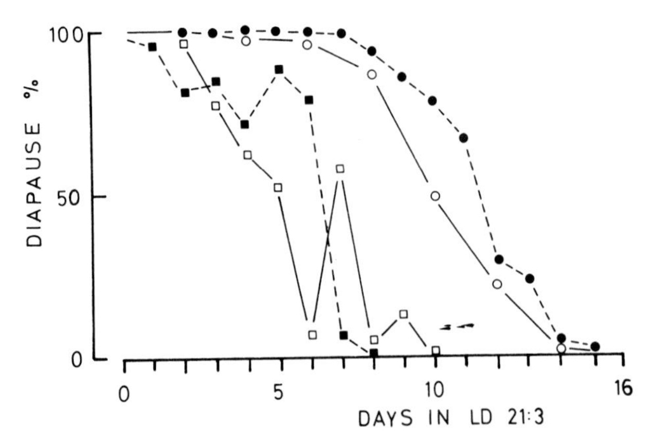

Fig. 3. Proportion of diapausing females as a function of days spent at long day (LD 21:3, 16°C) in two inbred lines of *D. littoralis*. Squares: line from Zürich (critical daylength 13.0 hr), circles: Oulu (17.6 hr). Open symbols: long-day treatment immediately after eclosion to adults; closed: long-day treatment after 15 days in LD 12:12.

days of August at the latitude 65°N, if determined from sunrise to sunset. If civil twilight is included, the corresponding time is in the last week of August, but we consider the beginning of August to be more in accordance with the field observations. The critical daylength of the *D. transversa* population sample was 16.5 hr, corresponding to 16 August (Muona and Lumme, in press). These values are satisfactorily correlated with the field observations (Fig. 1). The stopping of reproduction seems to occur slightly earlier in *D. littoralis* than in *D. transversa*, which is probably due to the longer developmental time of *D. littoralis*. The comparison of field and laboratory results in detail is, however, difficult, for the reasons given below.

The type of the photoperiodic reaction curve seems to be similar in all species studied so far: short-day diapause - long-day reproduction is the general pattern. In *D. phalerata*, however, Geyspits and Simonenko (1970) obtained another peak of diapause in long daylengths, and explained it as a sign of summer diapause (aestivation). At least in our latitudes, summer diapause is lacking in the field, but also northern *D. phalerata* and *D. testacea* flies develop slower in LD 21:3 than in LD 18:6 (Muona and Lumme, in press). At any rate, they develop to reproductive maturity within three weeks after eclosion, when subjected to daylengths longer than the critical daylength. In our opinion, Geyspits and Simonenko (1970) draw conclusions on too young flies, in which the ovaries have not had time enough to develop further from the post-eclosion immaturity (see Fig. 3).

Temperature Dependence of the Photoperiodic Reaction

Geyspits and Simonenko (1970) showed a strong effect of temperature on the photoperiodic reaction in *D. phalerata*; increasing temperature shortens the critical daylength. In *D. littoralis*, increasing temperature to 22°C (from the usual 16°C) abolishes diapause almost completely, and females mature in all daylengths studied (Lumme et al., 1974b). Charlesworth and Shorrocks (1976) mention a high temperature dependence in the photoperiodic reaction of *D. phalerata* and *D. transversa* populations in England. For these two species, it has been claimed that the influence of temperature is necessary for the local fine adjustment of northern populations because the clinal genetic variation of the photoperiodic reaction is not wide enough (see below) to explain the field results (Muona and Lumme, in press). High-temperature dependence(if within the ecological range of a species) of course diminishes the usefulness of the photoperiodic reaction, which basically is a means for obtaining temperature-independent information on the time of the year. On the other hand, it can hardly be considered disadvantageous. Further studies are needed on this problem.

Nutritional Factors Influencing Diapause?

The effect of food quality or quantity on diapause is known in some insects (Beck, 1968). In *D. obscura*, Begon (1976) observed differences in the growth rate of ovaries in females fed on "full" and "half" diets. In our laboratory we have observed that the medium used for storage of flies during field collections, pure agar-sugar, is not sufficient for the growth of ovaries in many species (the culture medium is based on malt, Lakovaara, 1969). The role of yeasts in the maturation of adults is known in some *Drosophila* species (Spieth, pers. comm.). Unfortunately, we have not studied whether the retardation due to poor diet interacts with the photoperiodic reaction. It may be possible that flies on sugar medium react to the photoperiod, but the growth of ovaries is secondarily retarded because of lack of nutrients.

Field observations on the fungivorous *D. transversa* give some hint of the possible importance of nutritional factors in the field (Muona and Lumme, in press). In late summer 1974 the fungal crop was unusually good, and the proportion of developing females in samples at that time (late July-August) was clearly greater than in the following year when the amount of fungi was much smaller (Fig. 1). This problem merits further studies.

The Developmental Stage of Determination of the Reproductive Diapause

In many insects, diapause induction (by photoperiod) precedes the stage of actual manifestation by a long time (Beck, 1968). In *D. littoralis* we found that the photoperiodic determination of diapause occurred no earlier than after the emergence of the adults (Lumme et al., 1974b). We concluded that diapause is terminated by sufficient daylength and that each eclosing individual is in diapause (or at least prediapause). This description may not be valid from the physiological point of view, but it illustrates well the situation. The termination of diapause offers some technical advantages in laboratory work when termination occurs in as short a time (and can be observed,too) as two to three weeks. Before eclosion, flies can be cultured conveniently in uniform conditions adjusted to maximize the growth rate and adult production (18°C, continuous illumination is used in our laboratory for all species). When experimental animals are taken from these stock cultures within two days after adult eclosion and subjected to light-dark regimes at 16°C, the photoperiodic reaction can be seen in three weeks.

In *D. transversa*, *D. phalerata*, *D. testacea*, *D. lummei*, *D. montana* and *D. ezoana* (in fact, in all species we have studied in detail) reproductive diapause is similar to that found in *D. littoralis*. Full incidence (up to 100%) of diapause can be seen if the adults are harvested from stock cultures kept in diapause-preventing conditions. Begon (1976) observed full diapause in *D. obscura* females transferred to short daylengths only a few days before the emergence of the adults (he tried not to do it after eclosion). Detailed studies on adult developmental physiology have not been made, and the physiological condition of females immediately after eclosion to adults has not been compared to the later superficially similar stage which continues in short daylengths.

In many species (including *D. littoralis*, Lumme et al. 1974b) flies collected from wild populations in autumn or winter develop to reproductive maturity immediately if subjected to long daylengths and/or sufficiently high temperature. This has lead to the conclusion that the overwintering stage is not a true diapause, but a reproductive arrest (Basden, 1954), also called Carson-Stalker arrest as described first in *D. robusta* (Carson and Stalker, 1948). An immediate onset of reproduction was also found in *D. macroptera* (20°C, all daylengths) and *D. grisea* (20°C, long day) by Kambysellis and Heed (1974) and in *D. obscura* (Begon, 1976), *D. transversa* and *D. phalerata* (Muona and Lumme, in press). Thus the cold period (chilling) is not needed to terminate diapause in *Drosophila* in general. According to Muller (1970, see also Thiele, 1973), the type of dormancy thus seems to be photoperiodic quiescence; it continues as long as the environmental cue (daylength) remains inducing. In the field, the dormancy has characteristics of oligopause (Begon, 1976); we prefer to call it diapause because of the limited value of these classifications (see also Tauber and Tauber, this volume).

Figure 3 presents the results of an experiment designed to determine the number of long-day cycles necessary to terminate diapause in two inbred lines of *D. littoralis*, one from Zurich (critical daylength 13.0 hr, see Figure 4) and the other from Oulu, Finland (critical daylength 17.6 hr). One group of adults was subjected to long days LD 21:3 immediately after eclosion to adults, the other was kept for 15 days in short-day LD 12:12, and thereafter exposed to LD 21:3 for

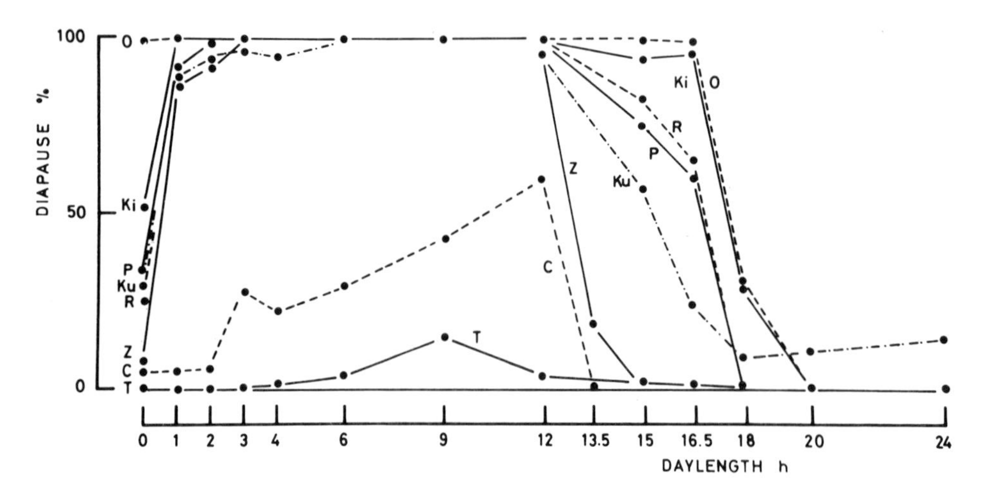

Fig. 4. Photoperiodic reaction curves of eight inbred lines of *D. littoralis*. O: originates from Oulu, Finland (65°N), Ki: Kilpisjärvi, Finland (69°N), R: Rovaniemi, Finland (66°N), P: Paltamo, Finland (64°N), Ku: Kuopio, Finland (63°N), Z: Zurich, Switzerland (47°N), C: Caucasus, USSR (42°N), T: Ticino, Switzerland (46°N) (Lumme and Oikarinen, 1977).

a given number of days. After the long-day exposure, the flies were kept at LD 12:12 until they were three weeks old, or until at least one week after the exposure (the growth of ovaries takes some time before it is visible). No great differences can be seen between the age groups (the short day does not deepen the diapause), but the difference between the lines is great. On the average, the breaking of diapause takes some five days more in northern flies than in southern ones. The relationship of this observation to the different critical daylengths of the lines remains to be studied. The variation between individuals is quite high, which makes the curves smooth. One conclusion is obvious: the onset of reproduction is irreversible (switch). The time lag between emergence and diapause termination must of course be considered in the interpretation of field observations and laboratory curves of photoperiodic response.

Summarizing, in all *Drosophila* species studied in detail, the photoperiodic control of winter dormancy is expressed as the termination of the posteclosion immaturity within three weeks or less in suitable (summer) conditions. This is advantageous since the late "prediction" of the coming winter has a high probability of being correct. If the daylength should allow for reproduction, there is, however, a delay before starting egg-laying. This delay seems to be shorter in southern populations, in which reproduction is the more possible alternative for most flies (several generations per year).

Geographical Variation in Photoperiodic Diapause

In the first paper published on the photoperiodic diapause of *D. phalerata*, the genetic differences between two strains originating from different areas were studied (Geyspits and Simonenko, 1970). The general rule among insects is that the critical daylength increases some 1 to 1.5 hr for each 5° of latitude (Danilevsky et al., 1970). and corresponding variation is found in almost all insects studied (e.g., Beck, 1968; Saunders, 1976). Of course latitude alone is not responsible for this variation, for climate also exerts an influence. Very accurate latitude-altitude dependence, for example, has been found in some insects (Bradshaw, 1976).

We have studied the critical daylengths of diapause termination in some Finnish populations of *D. transversa* (Muona and Lumme, in press) from 60° to 69° N lat. Their critical daylengths varied between 15.9 and 16.5 hr light per day at 16° C. In northern England, a value of 12.7 hr (15° C) has been measured for this species (Charlesworth and Shorrocks, 1976) Strains of *D. phalerata* from latitudes 60°, 61°, and 65° N were also studied; their critical daylengths were 15.7 hr, 16.0 hr, and 17.0 hr (16° C), respectively (Muona and Lumme, in press). Charlesworth and Shorrocks (1976) give a value of 14.8 hr (15° C) for a population from northern England (54° N lat.). Geographical variation thus is present in these two fungivorous species, even if it is surprisingly small in the Finnish populations of *D. transversa* (see Fig. 5).

In *D. littoralis*, the photoperiodic response of some inbred lines from different geographical origins has been studied (Lumme and Oikarinen, 1977). Due to inbreeding, they do not represent the populations of origin well, but are presented in Figure 4 as examples of possible modes of response in this species. A wide variation is seen, includ-

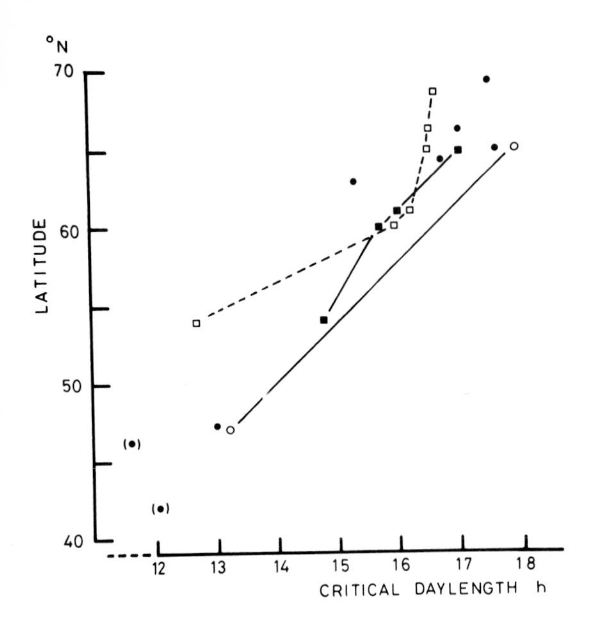

Fig. 5. Geographical variation in the photoperiodic reaction in *Drosophila* species: critical daylengths as function of the latitude. Open circles: *D. littoralis*, population samples; closed circles: *D. littoralis*, inbred lines (Lumme and Oikarinen, 1977); open squares *D. transversa*, and closed squares: *D. phalerata* (Muona and Lumme, in press; Charlesworth and Shorrocks, 1976). Points within brackets: incomplete diapause.

ing almost no photoperiodic reaction in a line coming from Ticino, Switzerland (45°N lat.), in which the maximum percentage of diapausing females was found to be 14.4% at LD 9:15 (16°C). The longest critical daylength was 17.6 hr in a line originating from Oulu, Finland (65°N lat.), which is quite close to the critical daylength determined from a population sample from the same locality, 17.9 hr (see above).

In Figure 5 we summarize the information available on the latitude-dependence of critical daylengths in *Drosophila*. The observations are roughly in accordance with the general rule of a 1- to 1.5-hr increase of critical daylength per 5° of latitude. The cline is, however, not as steep as supposed in the Finnish populations of *D. transversa*. The reasons for this are not apparent (some speculations are presented in Muona and Lumme, in press), but the phenomenon is not unknown in other insects (e.g., *Pieris brassicae*, Danilevsky et al., 1970). Studies on the reasons for the relaxation of the cline may be valuable for understanding the development of clines in general.

For summarizing the clinal variation, and for connecting it with the ecological basis, a schematic model of a cline in the photoperiodic response is presented in Figure 6. The theoretical model of Cohen (1970) for optimal timing of diapause is especially applicable to the *Drosophila* species with a photoperiodically "terminated" adult reproductive diapause. In a given locality, the photoperiodic reaction of a population is adjusted by natural selection to maximize the long-term growth of the population. The optimum reaction depends on the ratio of two probabilities: the probability of survival over the winter in diapause and the probability of successful reproduction. This is scored by natural selection independently in each cohort of eclosing adults. The mechanism in *Drosophila* allows the full use of the growing season, because the eclosion of adults continues from midsummer to late autumn and the number of generations is a continuous variable (Fig. 2).

Because in the north the autumn frosts come earlier on the average and daylength on a given date is longer, the daylength "predicting"

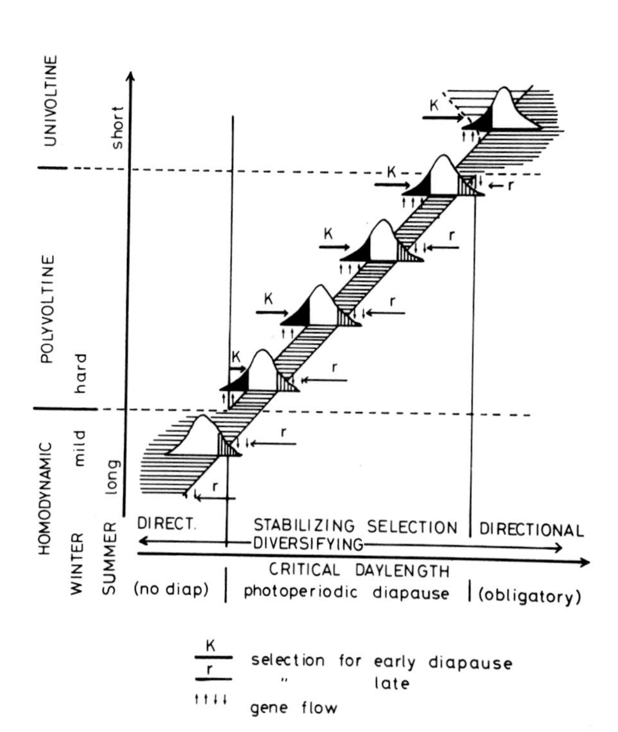

Fig. 6. A schematic model of the cline of photoperiodic diapause.

the average time of frosts greatly increases toward the north. Populations must follow this change in their genetic constitution for maximal exploitation of the growing season. The selective forces in this situation can be simplified as follows. Factors related to the short-term growth of the population, regardless of environment (r-selection, Pianka, 1974), select against early entry into diapause, i.e., they select toward shorter critical daylength for diapause induction. The extreme r-strategist has no diapause at all (for instance, the domestic species in the north). The environmental factor, autumn frost, selects against those genotypes which try to reproduce too late (K-selection, also hard selection *sensu* Wallace, 1975). Thus, local populations are subjected to stabilizing selection; the fitness of extreme phenotypes is low (Cohen, 1970; Levins, 1968). On the other hand, due to geographical variation in climate and daylength, this stabilizing selection implicitly leads to disruptive or diversifying selection along a continuous cline (Stern and Tigerstedt, 1974:172). Gene flow between locally adapted populations is an important factor; its main influence obviously is making adaptation worse. On the other hand, it increases the intrapopulation variation, which may be useful (and therefore balanced!) when successive years differ in autumn weather.

This selective framework can relax at both ends of the cline. In the very south winters can be so mild that specific overwintering mechanisms are not necessary (e.g., inbred line of *D. littoralis* originating from Ticino, southern Switzerland, in Fig. 4). The photoperiodic reaction is then no longer useful; it can either disappear or not develop at all in such populations (the genus originates from the tropics). However, these southern populations exchange genetic material with other populations, and the species must resist this load. In the very north the photoperiodic reaction can also be neutral if the summer is always so short that only the first generation (or even a part of it) can develop. Selection against longer critical daylengths

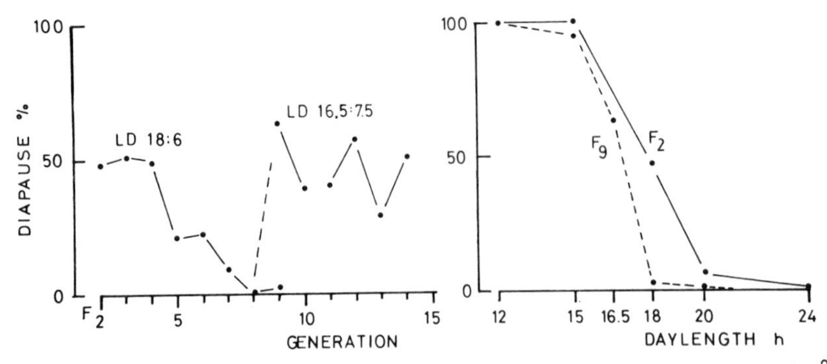

Fig. 7. Selection against diapause in *D. littoralis* population from Oulu (65° N, Finland). Population was founded by ten inseminated females; selection was started in the second laboratory generation. Photoperiodic reaction curves of the population before and after the selection are presented on the right (Oikarinen and Lumme, in preparation).

(r-selection) relaxes. Furthermore, low temperature can be sufficient for the prevention of reproduction, or diapause can develop to be obligatory (Danilevsky et al., 1970). Directional selection in the ends of a cline postulated by Stern and Tigerstedt (1974) occurs during the colonization of new areas of distribution, when the optimum adaptation is not yet achieved (perhaps the cline of *D. transversa* in Fig. 5 looks like this?). Similar arguments are made with respect to the maintenance of high additive genetic variance for diapause in *Oncopeltus fasciatus* (Dingle et al., 1977).

Genetic Variation within Local Populations

Implicit in the above model of the cline is that local populations contain genetic variation, at least to some degree, due to random variation in the environment, i.e., time of autumn frosts (balancing selection, Levins, 1968) and gene flow between local populations. These factors of course vary greatly from species to species. The within-population variation is most easily demonstrated by selection experiments. Figure 7 presents the selection response of a *D. littoralis* population sample taken from Oulu, Finland (65° N lat.), which was founded from ten inseminated wild females (Oikarinen and Lumme, in preparation). The population was selected against diapause by keeping the adults of each generation three weeks in a given daylength (16° C). The eggs laid by them during the third week were used to start the following generation.

The original critical daylength of the population was 17.9 hr light per day. Selection was started at LD 18:6 where the percentage of nonreproducing females was initially 47%. After six generations of selection, all females of the population were reproducing at LD 18:6. After this, selection was continued at LD 16.5:7.5 where the diapause percentage was now 62%. Selection was no longer successful; after five generations no statistically significant decrease in diapause percentage was observed. Unfortunately the selected line died after a total of 12 generations of selection. Very similar results were obtained with other strains (Oikarinen and Lumme, in preparation). It is worth mentioning here that nothing happened in control lines cultured similarly in LL, 18°C, or in those selected for fast reproduction in LL, indicating that the fecundity in diapause-preventing conditions is not correlated or linked with the photoperiodic reaction in this species (compare with results of Dingle et al., 1977; and Istock, this volume). The slow response to selection in first generations may indicate that the advantageous genetic material was quite rare at the beginning of the experiment (Fig. 7).

After the successful selection at LD 18:6, the response curve of the strain was determined again (Fig. 7). The critical daylength was found to be about 1 hr shorter, now 16.9 hr and the response curve was slightly steeper. The steepness of the photoperiodic response curve of a population indicates the phenotypic variation in the critical daylengths of individuals; clearly selection had decreased this.

The strong response to selection in the first generations is similar to that found in many other insects (e.g., Danilevsky et al., 1970; Honek, 1972; Herzog and Phillips, 1976; Dingle et al., 1977, and references therein; Hoy, this volume). It agrees with the genetic system in *D. littoralis* described below: selection against dominant "alleles" in a single "locus" is very efficient. What the selection response does show is that there is high additive genetic variance within the population, and that it can easily be changed in the laboratory, a phenomenon which must be considered in studies of the photoperiodic behavior of populations (Lumme and Oikarinen, in press; Dingle et al., 1977). If selection can alter the critical daylength of an experimental population, it can also be altered by random genetic drift during laboratory culturing. The seasonal changes in the reactions of the *D. phalerata* laboratory population found by Geyspits and Simonenko (1970) can quite easily be interpreted by random drift or uncontrolled selection, rather than by a circannual clock.

To further evaluate the genetic variation within and between populations of *D. littoralis* the following experiment was done (Lumme and Oikarinen, 1977). From each of eight strains of different geographical origin (42°N to 69°N lat.) five lines were founded and full-sib inbred for five generations. Variation was studied in the sixth generation by taking five pairs from each line (= 8 X 5 X 5 planned) and studying the diapause percentages in their progenies at LD 16.5:7.5 (16°C). About 91% of the total variation was due to differences between the original strains. Of the remaining variation, more than half (5.5% of the total) was due to differences between lines, within strains. The differentiation between lines was highly significant, although the inbreeding was conducted only through five generations. This again supports the possibility of a few-factor genetic basis for the inheritance of the photoperiodic reaction in this species.

Genetics of the Photoperiodic Diapause in *Drosophila littoralis*

In a preliminary study (Lumme et al., 1975) three strains of *D. littoralis* were crossed, and the results indicated that the control of photoperiodic diapause was inherited as a single mendelian unit. This observation is just the opposite of the results of most studies on the genetics of photoperiodism (e.g., Danilevskii, 1965), even those done on this same species (Lakovaara et al., 1972), in which control of the photoperiodic response is found or thought to be polygenic. Therefore it was necessary to analyze our results in more detail (Lumme and Oikarinen, 1977).

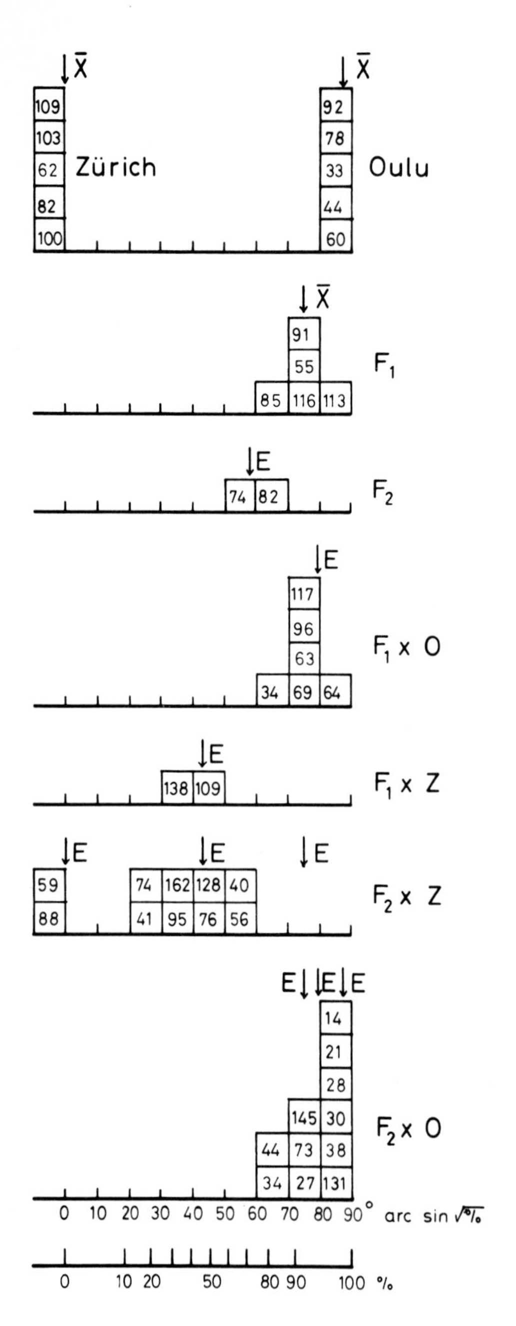

Fig. 8. Distribution of progenies of single pairs in inbred line Zürich (Z) and Oulu (O) of *D. littoralis*, and in their crosses, according to the diapause percentage in LD 16,5:7.5, 16°C. The number of females studied from each progeny is given within the squares. F_1's contain reciprocals; F_2 is from a cross O X Z; F_1 X O contains reciprocals O X (O X Z) and (O X Z) X); F_1 X Z is made in a form (Z X O) X Z; in the third generation F_2 females from a cross O X Z are crossed to O or Z males. E: expected diapause percentages on the basis of single segregational unit model of inheritance, calculated from the P and F_1 averages (x) (Lumme and Oikarinen, 1977). Arrows: exact points of averages and expectations.

To illustrate the methods used, an example of crosses between two inbred lines is given in Figure 8. Altogether the offspring of 46 pairs in two inbred lines, Zürich and Oulu, and their crosses were evaluated; the total number of females dissected for this was 3493. At the individual level, the photoperiodic response is an all-or-none reaction. To assess variation, populations must be studied, and to determine the genotype of a fly, its progeny (hopefully with a known partner) must be analyzed. This makes the genetic analysis of photoperiodism a tedious job (in some insects, the situation is different: single individuals can be "measured" quantitatively, for example, according to the time of first eggs laid). Therefore it is also necessary to restrict the environmental variables to a minimum; we made all crosses in one condition only, LD 16.5:7 at 16°C (compare Fig. 4).

There was no diapause in the Zürich line under the experimental conditions; in Oulu diapause was almost complete (99.8% on the average). The five F_1 families studied (reciprocals included) were all closer to Oulu than Zürich, the weighted mean of diapausing females being 93.3%. In the following generation (F_2 and backcrosses to F_1) segregation cannot be seen directly; the diapause percentage seen is an average of the diapause percentages of different genotypes weighted by their proportions. Thus the generation mean is subject to two independent sources of error variation: variation in the diapause percentage of a given genotype, and variation in the genotype frequencies. Thus the samples (progenies) must be quite large to be of any value. In Figure 8, the distribution of offspring from single pairs is given, and the number of females obtained from each pair is given within the squares. The distribution is presented for *arc sin* transformed percentages

Because the hypothesis during this study was the control of inheritance by a single mendelian unit, the expected values in Figure 8 are calculated according to this model on the basis of diapause percentages observed in the parental lines and in the F_1 (averages). Thus, for example, the expected value for the F_2 generation is (0% + 2 X 93.3% + 99.8%)/4 = 71.6%, and for a cross 0 X (0 X Z) it is (99.8% + 93.3%)/2. The observed diapause percentages in the second generation are well in accordance with the model.

The segregation of "alleles" in the diapause "locus" can be seen directly in the backcrosses to F_2 generation females. Of these progenies, one part is expected to be pure homozygotes, two parts homozygotes and heterozygotes in a 1:1 ratio, and one part of progenies is expected to contain only heterozygotes. The results in Figure 8 are roughly in accordance with these calculations, but the middle group in backcrosses to the Zürich line is quite variable; it may indicate that some recombination within the unit controlling diapause has occurred. On the other hand, the pure heterozygote group is lacking, but this could be due to the small number of pairs studied.

Similar indication of the linkage of diapause factors within one segregating unit was obtained in larger samples, too. All eight inbred lines given in Figure 4 were crossed with each other (i.e., a diallel cross), and the data from F_1, F_2, and backcrosses to F_1 are presented in Figure 9. In Figure 9A the F_1 diapause percentages are displayed as a function of the average of the parents. It can be seen that the progenies are not intermediate, but a significant dominance deviation is present. In a more detailed analysis it was obvious that longer critical daylength is in most cases dominant over the shorter one (points above the diagonal). In Figure 9A also the points below the diagonal could represent such cases, because the reaction was studied in a long day in which the southern lines were "out of scale" (at LD 16.5:7.5 they are all reproducing, see Fig. 4). The important point considering the analysis of the following generation is that the F_1 phenotype can be separated from the average of the parents, a necessary prerequisite for analyzing segregation by present methods.

The distribution of the F_2 points in Figure 9A is quite similar to that of the F_1's. They were, however, not identical. In Figure 9B,

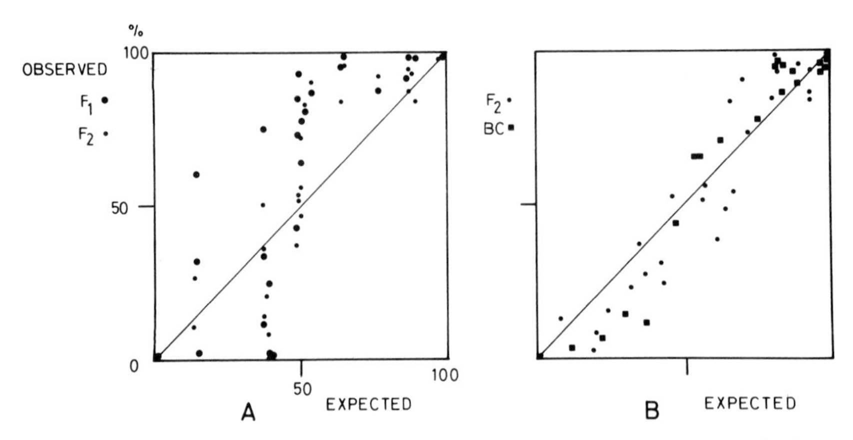

Fig. 9. A. F_1 and F_2 diapause percentages of eight line diallel cross (weighted means over replicates and reciprocals) as a function of the average of diapause percentages in the parents (expected). B. F_2 and backcross diapause percentages from the diallel as a function of the expected diapause percentages based on single segregational unit model of inheritance (Lumme and Oikarinen, 1977).

the F_2 diapause percentages are drawn as a function of the expected values obtained by a simple mendelian model (see above). The observed points fit well with the expected ones; the correlation coefficient was 0.951 (N = 27), and the linear regression was $y = 1.08x - 4.80$. Similarly, the backcross results were compared with the model; here $r = 0.973$ ($n = 29$), and regression equation $y = 1.08x - 5.56$ was obtained. The backcrosses were made reciprocally to the female parent, which indicated that the controlling unit was autosomal, because the reciprocals were identical with each other. From all of the second generation, the comparison of expected diapause percentages with those observed gives $r = 0.964$ ($n = 56$), and the slope of the regression was 1.080. The results from crossing experiments thus confirmed the hypothesis of a single mendelian, autosomal unit as the genetic basis of the photoperiodic reaction. At least within the first few generations, the whole control unit is segregating like alleles in a single locus, and any possible recombination within it has such a small effect that it cannot be seen. In backcrosses to F_2 some indication of possible recombination was seen, but the ultimate structure of the controlling unit remains to be studied.

At the level of population genetics and phenology, the genetic system controlling the geographical variation in the photoperiodic reactions of *D. littoralis* has some interesting implications. It was shown that the longer critical daylength was in general (incompletely) dominant over the shorter one. The long critical daylength is a "northern" character and safe with regard to the environment (it is also evolved later); its dominance can be important in maintaining genetic variation within populations. Implicitly, the gene flow northwards is easy. If a southern immigrant interbreeds within the northern population, the progeny are partially protected from the autumn frosts by this dominance. Thus the southern gene pool can easily disperse to the north. On the other hand, the gene flow in the opposite direction is restricted; in the southern population the F_1 of a northern immigrant stops reproduction too early and suffers a reproductive deficit. This speculation remains to be verified quantitatively.

The reasons for the linkage observed are also unknown (to us). One attractive possibility is that there is only one locus responsible for photoperiodic time measurement (see the chronon hypothesis for biological clocks, Ehret and Trucco, 1967). The more obvious alternative is that there are many genes arranged close to each other to form a supergene (no inversions in this species). The supergene hypothesis also fits better with observations on other insects with polygenic control of the photoperiodic reaction. The selective framework presented in Figure 6 perhaps contains qualitatively the essential factors for understanding supergene formation. Disruptive selection with a system of random mating can lead to linkage between many loci responsible for the character selected (see Ford, 1975, and references therein). However, the situation along a continuous cline is different; the intermediates between different phenotypes are not selected against if they are situated in the intermediate area. In any case, stabilizing selection in local populations is against extreme phenotypes, and this can lead to linkage for avoiding segregational load. The effect of migrational load is similar; it broadens the phenotypic variation of a population and is probably easily eliminated (with minimum disturbance to the coadapted gene pool) in the presence of linkage. Dingle et al. (1977) reduce the situation into a few words: the degree of local adaptation implicitly is a measure of dispersal, selection pressure, etc. The theory of clines in quantitative characters is, however, not in final form.

Further studies are needed on different species concerning the genetic system of photoperiodic clines and "genetic rheostat" vs. "switch" strategies (Dingle et al., 1977). The opportunity for using *Drosophila* species is present, and the generality of the single segregating unit model of *D. littoralis* can be tested with other species. It may be possible to find free recombination in some species; according to the literature, it is usual in insects. The relaxation at the northern end of the cline in *D. transversa* (Fig. 5) is also genetically interesting in this connection. Unfortunately, the ecologic basis for studying population genetic problems like the clines in *Drosophila* is quite weak so far; the essential knowledge on migration, population structure, etc., is almost completely lacking at least in the species showing photoperiodic reactions, and knowledge of the phenology, too, can be considered superficial at the best. In addition, none of the really well-known species used by geneticists have so far been observed to have any photoperiodic reactions. One possibility, however, is worth mentioning here. *D. lummei*, which has photoperiodic reproductive diapause similar to *D. littoralis*, can be easily crossed with *D. virilis* Sturtevant, which is one of the genetically well-known species of the genus but has no sign of a photoperiodic reaction (unpublished observations). This species pair could allow further analysis of the genetics (and also physiology) of diapause. Although the difficulties arising from the use of crosses between species are obvious, interspecific crosses can provide interesting data on the underlying mechanism (Tauber and Tauber, this volume).

Evolutionary Aspects

At the present stage of knowledge, not very much can be said about the evolution of "synchronization" to northern environments in *Drosophila*. The number of species studied in detail is very low, or perhaps more accurately no species is known in enough detail. Some observations, however, suggest future studies.

In Table 1 we list the *Drosophila* species recorded so far in Scandinavia and Finland (Lakovaara et al., 1974), with some remarks concerning their relation to the overwintering problem, if anything is known.

Three species groups, the *D. virilis-*, *D. obscura-* and *D. quinaria*-groups are well represented in the northern fauna. Eleven species out of 23 found northwards of the latitude 65°N in Europe belong to one of the three (five are unestablished immigrants, not surviving outdoors over the winter). In our collection data from the years 1974 and 1975 from Oulu (65°N lat.) these 11 species were all recorded. The proportion of the three species groups mentioned out of the total number of individuals was 23090/23802 (97%), which indicates that these groups can be considered as "successful colonizers" of our quite northern study area (Lumme et al., in press).

The position of these three species groups in the phylogenetic tree of Drosophilidae is not the same; they represent very different lineages and belong to different subgenera (Throckmorton, 1975). The *D. obscura* -group belongs to the subgenus *Sophophora*. The *D. virilis*- and *D. quinaria*- groups are included in the subgenus *Drosophila*, but they are not close relatives within this extremely diverse subgenus. However, in all these species groups an essentially identical photoperiodic adult dormancy (photoperiodic quiescence) has been found among the northern species, as at least one means of synchronization. Within the groups, and also within a species, there is variation in the seasonal strategy.

The *D. obscura*-group seems to be especially variable with respect to the overwintering mechanism. Of the species studied, *D. pseudoobscura* Frolova is widely used in studies on circadian rhythms (which are possibly connected with photoperiodism), but no photoperiodic reaction of any kind has been found in it (Pittendrigh and Minis, 1971). The same is true for *D. subobscura* (Begon, 1976), which is the favorite of European population geneticists. On the other hand, *D. obscura* (Begon, 1976) and *D. bifasciata* (Lakovaara et al., 1972) have a photoperiodic adult diapause, and this may also be the means of overwintering in *D. eskoi* (Lumme et al., in press), *D. lowei* (Heed et al., 1969) and *D. tristis* (Basden, 1954). Furthermore, *D. alpina* (Lumme et al., in press) and possibly *D. persimilis* (Dobzhansky and Epling, 1944) have a pupal diapause. *D. subsilvestris* also overwinters in some other stage than the adult (Lumme et al., in press). Unfortunately, only a few of these species have been studied in detail. This species group seems to be an especially interesting object for further studies on the evolution of seasonal synchronization in particular, because the phylogenetic relationships between the species are quite well known (Lakovaara et al., 1976), and it has long been the favorite group of population geneticists.

In contrast to the *D. obscura*-group, only adult diapause has been found in species of the *D. virilis*- and *D. quinaria*-groups (or anywhere, with the exception of *D. deflexa*, Basden, 1954). This photoperiodic diapause expresses itself as a continuous variable along a climatic gradient; the critical daylength of a population depends on the climate and latitude. Diapause can even be absent in southern populations of those species such as *D. littoralis* which are distributed to the northern extreme of the range of *Drosophila*.

Summary

Phenology of drosophilids in northern latitudes is predominantly determined by the winter. In most species (possibly in all), the necessary cold-resistance is restricted to only one developmental stage, which in most species is the adult before reproduction; few species are dormant as pupae or larvae.

The coming winter must be forecast long before, because all living individuals expected to survive must have time enough to achieve the diapause stage. In all *Drosophila* species having an adult diapause and studied in detail, the photoperiod has been found to be the main environmental cue for onset of dormancy. All drosophilids in the very north are long-day insects. The type of dormancy seems to be "photoperiodic quiescence"; it is not induced earlier than after the emergence of adults, and it can be terminated in the laboratory at any time by long day. In the wild, however, reproduction starts when the temperature has been high enough in spring. This type of dormancy has been found in many independent lineages of the phylogenetic tree of drosophilids, and it has obviously evolved several times during their evolution. The similarity of different species in the (superficially known) physiological characters of diapause may indicate a conservative Dipteran neuroendocrine system.

Geographical and local variation is found in species studied in this respect; the increase in critical daylength northwards seems to follow the same trend as in other insects. The genetics of this variation has been studied in one species only; in *Drosophila littoralis* the genetic system controlling the clinal variation is a supergene, which varies greatly across latitudes. This differs clearly from the observations made in most other insects.

The opportunity for using *Drosophila* species in studies concerning the evolution of "escape in time" is quite recent. Although basic knowledge is still inadequate, some promising possibilities for comparative studies have already been found.

Acknowledgments This review is based on the long-lasting cooperation in the Department of Genetics, University of Oulu, and I express my sincere thanks to the head of Department, Professor Seppo Lakovaara, and to my colleagues working with me, Aila Oikarinen, Outi Muona, Liisa Pohjola, Ritva Pöllänen, P. Lankinen, M. Orell, R. Alatalo, A. Saura, and J. Lokki. Financial support from The Finnish Cultural Foundation, Alfred Kordelin Foundation, The University of Oulu, and The National Research Council for Sciences of Finland is acknowledged.

References

Basden, E.B.: The autumn flush of *Drosophila* (Diptera). Nature (Lond.) *172*, 1155-1156 (1953).

Basden, E.B.: Diapause in *Drosophila* (Diptera: Drosophilidae). Proc. R. Entomol. Soc. Lond. *29*, 114-118 (1954).

Basden, E.B.: Drosophilidae (Diptera) within Arctic Circle. I. General survey. Trans. Roy. Ent. Soc. *108*, 1-20 (1956).

Basden, E.B., Harnden, D.G.: Drosophilidae (Diptera) within Arctic Circle. II. The Edinburgh University expedition to subarctic Norway, 1953. Trans. Roy. Ent. Soc. *108*, 147-162 (1956).

Beck, S.D.: Insect Photoperiodism. New York and London: Academic Press 1968.

Begon, M.: Temporal variations in the reproductive condition of *Drosophila obscura* Fallén and *D. subobscura* Collin. Oecologia (Berl.) *23*, 31-47 (1976).

Bertani, G.: Artificial "breaking" of the diapause in *Drosophila nitens*. Nature (Lond.) *159*, 309 (1947).

Bradshaw, W.E.: Geography of photoperiodic response in diapausing mosquito. Nature (Lond.) *262*, 384-386 (1976).

Burla, H.: Systematik, Verbreitung und Ökologie der *Drosophila*-Arten der Schweiz. Rev. Suisse Zool. *58*, 53-175 (1951).

Burla, H.: Jahreszeitliche Häufligkeitsänderungen bei einigen schweizerischen *Drosophila*-Arten. Rev. Suisse Zool. *68*, 173-182 (1961).

Bächli, G.: Faunistische and ökologische Untersuchungen an Drosophiliden-Arten (Diptera) der Schweiz. I. Fangort Zürich. Bull. Soc. Entomol. Suisse *45*, 49-53 (1972a).

Bächli, G.: Faunistische und ökologische Untersuchungen an Drosophiliden-Arten (Diptera) der Schweiz. III. Fangort Aigle VD. Bull. Soc. Entomol. Suisse *45*, 255-259 (1972b).

Carson, H.L., Stalker, H.D.: Reproductive diapause in *Drosophila robusta*. Proc. Natl. Acad. Sci. U.S.A. *34*, 124-129 (1948).

Charlesworth, P., Shorrocks, B.: Overwintering strategies of fungal feeding *Drosophila*. Abstr. 5th European Drosophila Res. Conference, Louvain-la-Neuve, Belgium (1976).

Cohen, D.: A theoretical model for the optimal timing of diapause. Amer. Natr. *104*, 389-400 (1970).

Cooper, D.M., Dobzhansky, T.: Studies on the ecology of *Drosophila* in the Yosemite region of California. I. The occurrence of species of *Drosophila* in different life zones and at different seasons. Ecology *37*, 526-533 (1956).

Danilevskii, A.S.: Photoperiodism and Seasonal Development of Insects. Edinburgh: Oliver & Boyd 1965.

Danilevsky, A.S., Goryshin, N.I., Tyshchenko, V.P.: Biological rhythms in terrestrial arthropods. Ann. Rev. Entomol. (Wash.) *15*, 201-244 (1970).

Dingle, H., Brown, C.K., Hegman, J.P.: The nature of genetic variance influencing photoperiodic diapause in a migrant insect, *Oncopeltus fasciatus*. Amer. Natur. (1977).

Dobzhansky, T., Epling, C.: Contributions to the genetics, taxonomy, and ecology of *Drosophila pseudoobscura* and its relatives. Carnegie Inst. Wash. Publ. *554*, 1-183 (1944).

Dobzhansky, T., Pavan, C.: Local and seasonal variations in relative frequencies of species of *Drosophila* in Brazil. J. Animal Ecol. *19*, 1-14 (1950).

Ehret, C.F., Trucco, E.: Molecular models for the circadian clock. I. The chronon concept. J. Theor. Biol. *15*, 240-262 (1967).

Ford, E.B.: Ecological Genetics (fourth edition). London: Chapman and Hall 1975.

Geyspits, K.F., Simonenko, N.P.: An experimental analysis of seasonal changes in the photoperiodic reaction of *Drosophila phalerata* Meig. (Diptera, Drosophilidae). Entomol. Rev. Wash. *49*, 46-54 (1970).

Hackman, W., Lakovaara, S., Saura, A., Sorsa, M., Vepsäläinen, K.: On the biology and karyology of *Chymomyza costata* Zetterstedt, with reference to the taxonomy and distribution of various species of *Chymomyza* (Diptera, Drosophilidae). Ann. Entomol. Fenn. *36*, 1-9 (1970).

Heed, W.B., Crumpacker, D.W., Ehrman, L.: *Drosophila lowei*, a new American member of the *obscura* species group. Ann. Entomol. Soc. Amer. *62*, 388-393 (1969).

Herting, B.: Untersuchungen über die Ökologie der wildlebenden *Drosophila*-Arten Westphalens. Zeitschr. Morph. Ökol. Tiere *44*, 1-42 (1955).

Herzog, G.A., Phillips, J.R.: Selection for a diapause strain of the bollworm, *Heliothis zea*. J. Heredity *67*, 173-175 (1976).

Honěk, A.: Selection for non-diapause in *Aelia acuminata* and *A. rostrata* (Heteroptera, Pentatomidae) under various selective pressures. Acta entomol. bohemoslov. *69*, 73-77 (1972).

Hunter, A.S.: High altitude *Drosophila* of Colombia (Diptera, Drosophilidae). Ann. Entomol. Soc. Amer. *59*, 413-423 (1966).

Kambysellis, M.P., Heed, W.B.: Juvenile hormone induces ovarian development in diapausing cave-dwelling *Drosophila* species. J. Insect Physiol. *20*, 1779-1786 (1974).

Kimura, M.T.: Microdistribution and seasonal fluctuation of drosophilid flies dwelling among the undergrowth of plants. J. Fac. Sci. Hokkaido Univ. VI Zool. *20*, 192-202 (1976).

Lachaise, D.: Les Drosophilidae des savanes preforestières de la region tropical de Lamto (Côte d'Ivoire). I. Isolement écologique des éspecès affines et sympatriques; rythmes d'activité saisonnière et circadienne; role des feux de brousse. Ann. Univ. Abidjan (Ecologie) 7, 7-152 (1974).

Lakovaara, S.: Malt as a culture medium for *Drosophila* species. Drosophila Inform. Serv. *44*, 128 (1969).

Lakovaara, S., Hackman, W., Vepsälainen, K.: A malt bait in trapping drosophilids. Drosophila Inform. Serv. *44*, 123 (1969).

Lakovaara, S., Saura, A., Koref-Santibañez, S., Ehrman, L.: Aspects of diapause and its genetics in northern drosophilids. Hereditas *70*, 89-96 (1972).

Lakovaara, S., Lumme, J., Oikarinen, A.: Genetics and evolution of diapause in European species of the *Drosophila virilis* group (Abstract). Genetics *74*, s143 (1973).

Lakovaara, S., Lankinen, P., Lokki, J., Lumme, J., Saura, A., Oikarinen, A.: The Scandinavian species of *Drosophila*. Drosophila Inform. Serv. *51*, 122 (1974).

Lakovaara, S., Saura, A., Lankinen, P., Pohjola, L., Lokki, J.: The use of isoenzymes in tracing evolution and in classifying Drosophilidae. Zoologica Scripta *5*, 173-179 (1976).

Lees, A.D.: The Physiology of Diapause in Arthropods. London: Cambridge Univ. Press 1955.

Levins, R.: Evolution in Changing Environments. Monographs in Population Biology. Princeton: Princeton Univ. Press 1968.

Lumme, J., Oikarinen, A.: The genetic basis of the geographically variable photoperiodic diapause in *Drosophila littoralis*. Hereditas *86*, 129-142 (1977).

Lumme, J., Lakovaara, S., Lankinen, P., Saura, A.: *Drosophila* fauna in northernmost Scandinavia. Abstr. 4th European Drosophila Res. Conference, Umea, Sweden (1974a).

Lumme, J., Oikarinen, A., Lakovaara, S., Alatalo, R.: The environmental regulation of adult diapause in *Drosophila littoralis*. J. Insect Physiol. *20*, 2023-2033 (1974b).

Lumme, J., Lakovaara, S., Oikarinen, A., Lokki, J.: Genetics of photoperiodic diapause in *Drosophila littoralis*. Hereditas *79*, 143-148 (1975).

Lumme, J., Muona, O., Orell, M.: Phenology of northern drosophilids. Ann. Ent. Fenn. (in press).

Mather, W.B.: The genus *Drosophila* (Diptera) in Eastern Queensland. II. Seasonal changes in a natural population 1952-1953. Aust. J. Zool. *4*, 76-89 (1956).

Muona, O., Lumme, J.: Geographical variation of the reproductive cycle and photoperiodic diapause in *Drosophila transversa* Fallen and *D. phalerata* Meigen. Oikos (in press).

Muller, H.J.: Formen der Dormanz bei Insekten. Nova Acta Leopoldina *191*, 1-27 (1970).

Patterson, J.T.: Studies on the genetics of *Drosophila*. III. The Drosophilidae of the Southwest. Univ. Texas Publ. *4313*, 7-216 (1943).

Pianka, E.R.: Evolutionary Ecology. New York: Harper & Row 1974.

Pipkin, S.B.: Seasonal fluctuations in *Drosophila* populations at different altitudes in the Lebanon mountains. Z. Indukt. Abstamm. Vererbungs. *84*, 270-305 (1952).

Pittendrigh, C.S., Minis, D.H.: The photoperiodic time measurement in *Pectinophora gossypiella* and its relation to the circadian system in that species. In: Biochronometry. Menaker, M. (ed.). Washington, D.C.: National Academy of Science 1971.

Rocha Pité, T.M.: An introduction to the study of Portuguese Drosophilidae. Arq. Mus. Bocage *3* 367-384 (1972).

Saunders, D.S.: Insect Clocks. Oxford: Pergamon Press 1976.

Shorrocks, B.: *Drosophila*. London: Ginn & Co. Ltd. 1972.

Shorrocks, B.: The distribution and abundance of woodland species of British *Drosophila* (Diptera: Drosophilidae). J. Animal Ecol. *44*, 851-864 (1975).

Stern, K., Roche, L.: Genetics of Forest Ecosystems. Berlin: Springer Verlag 1974.

Stern, K., Tigerstedt, P.M.A.: Ökologische Genetik. Stuttgart: Gustav Fischer Verlag 1974.

Tauber, M.J., Tauber, C.A.: Insect phenology: Criteria for analyzing dormancy and for forecasting postdiapause development and reproduction in the field. Search

(Agriculture), Cornell Univ. Exp. Sta., Ithaca, N.Y. *3*, 1-16 (1973).

Tauber, M.J., Tauber, C.A.: Physiological responses underlying the timing of vernal activities in insects. Int. J. Biometeor. *20*, 218-222 (1976).

Thiele, H.U.: Remarks about Mansingh's and Müller's classifications of dormancies in insects. Can. Ent. *105*, 925-928 (1973).

Throckmorton, L.H.: The phylogeny, ecology, and geography of *Drosophila*. In: Handbook of Genetics. Vol. 3. Invertebrates of Genetic Interest. King, R.S. (ed.). New York: Plenum Press 1975, pp. 421-469.

Toda, M.J.: Seasonal activity and microdistribution of drosophilid flies in Misumai in Sapporo. J. Fac. Sci. Hokkaido Univ. VI Zool. *18*, 532-550 (1973).

Wakahama, K.: Studies on the seasonal variations of populations in *Drosophila*. II. The effect of altitude on seasonal activity of *Drosophila*, with a note on the monthly numerical variation of species. J. Fac. Sci. Hokkaido Univ. VI Zool. *15*, 65-73 (1962).

Waldbauer, G.P., Sternburg, J.G.: Polymorphic termination of diapause by *Cecropia:* genetic and geographical aspects. Biol. Bull. *145*, 627-631 (1973) .

Wallace, B.: Hard and soft selection revisited. Evolution *29*, 465-473 (1975).

Wheeler, M.R.: A nomenclatural study of the genus *Drosophila*. Univ. Texas Publ. *5914*, 181-205 (1959).

Wheeler, M.R., Hamilton, N.: Catalog of *Drosophila* species names, 1959-1971. Univ. Texas Publ. *7213*, 257-268 (1972).

Wheeler, M.R., Throckmorton, L.H.: Notes on Alaskan Drosophilidae (Diptera), with the description of a new species. Bull. Brooklyn Entomol. Soc. *LV*, 134-143 (1960).

Fitness Variation in a Natural Population

CONRAD A. ISTOCK

When Fisher (1958, but originally 1930) derived the fundamental theorem, he created an impressive but problematical view of natural selection. Some major conceptual difficulties intrinsic to the theorem were not noticed by Fisher and have gone largely unnoticed by subsequent students of evolution. The theorem is typically recited without comment or qualification (Feldman and Lewontin, 1975; Ayala, 1976). Empirical findings reviewed here strongly argue that these conceptual problems may be severe.

The theorem states that the rate of change of average fitness in a population at any time equals the additive genetic variance in fitness at that time. If, following Fisher, we take fitness to be the expressed average Malthusian parameter, called $\bar{r}$, or the intrinsic rate of increase in much of the ecological literature, we can write:

$$\frac{d\bar{r}}{dt} = V_A = h_N^2\ V_P$$

after Istock et al. (1976a), where $\bar{r}$ is the average fitness, V_A is the additive genetic variance of r, V_P is the phenotypic variance of r, and h_N^2 the heritability of r in the narrow sense (Falconer, 1960: Mather and Jinks, 1974). The theorem posits continuous directional selection for higher expressed $\bar{r}$ while h_N^2 and V_A go to zero. It does indeed suggest a fundamental mechanism of microevolution. Kimura (1958, or Crow and Kimura, 1970) has made interesting extensions of the theorem to cover nonrandom mating, environmental change, linkage, and epistasis.

Confusion over the precise meaning of the Malthusian parameter, the symbol m used by Fisher (1958), may arise because Fisher used m in two ways. At some points m is a measure of whole population increase, which is the same as average individual reproductive rate (pages 26-30, or page 39 of Fisher, 1958). At other points m is the reproductive, or replication, rate of an allele averaged over all possible genotypic backgrounds and environments (e.g., at the beginning of the derivation of the fundamental theorem on page 37, and as a consequence of his definition of genetic variance, or see page 47, Fisher, 1958). Both uses of m come together in the derivation of the theorem itself. The double summation of additive effects of fitness (α) weighted by the associated changes in allelic frequency (dp) over all alleles at all loci yields the change in average individual fitness, thus making direct contact with the usual definition of $\bar{r}$ in contemporary ecology. This interpretation agrees with the careful distinctions made by Crow and Kimura (1970) in their treatment of the theorem.

Conceptual problems with Fisher's theorem arise from its directional selection postulate. This postulate implies that: (i) the mechanism of the theorem can only proceed in populations not in ecological, that is demographic, equilibrium, where $\bar{r} \neq 0$ and (ii) in microevolutionary time, V_A becomes very small and evolution all but ceases except for the slow rate dependent on the recharging of V_A by recurrent mutation. Larger increases in V_A could arise with drastic environmental changes evoking new and strong genotype-environment interactions. The latter no doubt occurs, but lies outside the theorem's original setting.

Many ecologists would assert that natural populations are seldom in perfect equilibrium, so the condition that $\bar{r} \neq 0$ may not seem too bothersome. During repeated or incessant departure from equilibrium, the mechanism of the theorem accomplishes selective change and adaptation. But what of the adaptive properties being selected in each such episode? Will there be a consistent pattern of change in the features of life history which lie behind fitness and its variation? Departures from equilibrium vary both up and down creating different population states, as will departures varying in degree and background circumstance. Conflicting selective patterns should follow and alternate with each other through time favoring different genotypes with differing life histories. In microevolutionary time the result may be a conservation of the genetic variance of fitness (Haldane and Jayakar, 1963). From this conservation property, one of the cherished objects of evolutionary ecologists, an optimal life-history pattern (Leon, 1976), may emerge, but it will likely be an average, an abstraction, surrounded by ample, heritable, life-history variation (Mertz, 1970). Surely conflicting and alternating selective patterns will not produce the consistent rise of $\bar{r}$ predicted by Fisher's theorem, though artificial selection should do so. Hence the theorem contains a solid contradiction: for ecological reasons it describes evolution by "fits and starts," but this will not likely or generally yield the simple outcome predicted.

At a much broader and more intuitive level Fisher's theorem clashes with many modern notions about evolution; and *the issue is whether* V_A *for fitness tends near zero in microevolutionary time*. We readily entertain notions of continuous, perhaps cyclic, genetic tracking of biotic and abiotic change by a population (Krebs, 1970; Felsenstein, 1976). We think, or imagine, that adaptive change can proceed in populations under ecological, but not genetic equilibrium (Haldane, 1964). Populations are supposed to have reserves of genetic variation to meet progressively changing environments and to bridge the expanse from micro- to macroevolution (Levins, 1968). We believe that species divergence with progressively strengthened genetic isolation follows from geographical isolation (Mayr, 1963; Dobzhansky, 1970), and perhaps sometimes from prolonged disruptive selection (Thoday and Boam, 1959). All these ideas assume persistent variation in fitness tied to specific adaptive traits. The microevolutionary machine seemingly runs unstopped over enormous time spans and generally has a reserve of raw material for continuing adaptation in most natural populations most of the time. Conserving, shifting, exchanging, or reallocating processes would then predominate over outright expenditures in the economy of fitness variation under selection. This is the alternative to Fisher's theorem and a probable basis for long-term evolutionary change. This requires a mechanism for conservation of fitness variation in general, over and above the new fitness variation exposed by recombination as local populations work their ways over Wright's (1932) adaptive landscape. I will propose such a mechanism at the end of this paper.

Kimura (1958, and also Crow and Kimura, 1970) recognized the problem created if V_A tends to vanish. He solved for the rate of change in V_A as:

$$\frac{dV_A}{dt} = 2\sum_i p_i \, (r_i - \bar{r})^3$$

for multiple alleles and multiple loci, and neglecting dominance deviations by assuming they cancel when many loci influence fitness. The summation is over all genotypes i. The expression means that V_A changes as twice the third moment of the distribution of r. V_A will decrease if negative deviations dominate. Kimura concluded that V_A will decrease slowly if the distribution of r is close to normal, with the third moment (skewness) consequently close to zero. This does not solve the problem intrinsic to Fisher's theorem. If the distribution of r is or becomes normal with genetic variance greater than zero, V_A is conserved under Kimura's solution and $\bar{r}$ cannot increase except by addition of new variation which simultaneously alters the symmetry of the distribution. The possibility that V_A will be stabilized by "normalizing" selection and the subsequent requirement for additional genetic variation for r if $\bar{r}$ is to again increase are both inconsistant with Fisher's original model. Fisher's model may only be applicable to instances of steady directional selection in nearly constant environments, a process akin to artificial selection. Kimura's model is really an independent derivation and may be more interesting than Fisher's theorem as a simple but highly general representation of natural selection in the microevolutionary sense.

Fisher (1958) also thought that selection for increasingly higher mean fitness would always favor tighter linkage and reduced recombination among the loci influencing fitness characters. This could lock-in the most fit allelic combinations giving consistently and persistently superior genotypes. This contention is a corollary to the theorem. If chromosomal recombination continuously disrupts the most fit linkage groups, it will generate variance in fitness. Hence, full realization of the theorem requires full suppression of recombination. Turner (1967) argued the opposite: when many loci are involved, selection would favor intermediate recombination levels and the genotype would not "congeal," to use Turner's phrase. Lewontin (1971) agreed with Fisher, and rejected the idea of maximization of fitness by intermediate recombination frequencies. He suggested that the perpetuation of recombination among fitness loci might "be sought in some more general long-term advantage of recombination for adaptation to a varying environment." The latter suggestion, while clearly of interest, skirts the problem of connecting fitness variation in its micro- macroevolutionary roles and also skirts the implicit questions of individual and group selection. Data presented here will support Turner's contention, at least in a hypothetical sense, and will conform as well to Maynard Smith's (1974) claim that an evolutionarily stable strategy (ESS) can emerge from the mixing (polymorphism) of two or more pure strategies within the same population.

Lewontin (1974) subsequently dropped the requirement for a long-term advantage from recombination and stated that "one possible answer is that genetic flexibility for adaptation to a fluctuating environment demands recombination to generate adaptive combinations different from those previously selected." This seems closer to my results, but it would be better to say that one possible answer is that genetic recombination, along with segregation and reassortment of homologous chromosomes, allows presently successful individuals, different in fitness characters among themselves, to generate clusters of offspring varying around the central tendency of the whole population of successful individuals - the region of central tendency being the best predictor of future success in a potentially shifting, but continuous and autocorrelated sequence of environment.

Review of Empirical Results

For the remainder of this paper I will rely on results obtained during my studies of the ecology and genetics of the autogenous (nonbloodfeeding), pitcher-plant mosquito *Wyeomyia smithii* (Istock et al., 1975, 1976a, b). These results along with those of other authors (Smith, 1902;

Smith and Brust, 1971; Evans and Brust, 1972; and Bradshaw and Lounibos, 1972) give a clear picture of the phenotypic and genotypic expression of two major fitness components: the rate of larval development (egg to pupa development time) and the fraction of f_1 diapausers under warm-season conditions. The relation of these two developmental characters to expressed fitness requires a little more elaboration.

Larval development time controls the short-run expression of population or genotypic increase within a single generation. It influences the time to first reproduction and the mean generation time. Hence, the "compounding of interest" aspect of population change is increased or decreased as larval development time is shorter or longer, respectively. The relationship is shown in the Lotka equation for species with overlapping generations:

$$1 = \sum_{\alpha}^{\omega} l_x m_x e^{-rx} = \sum_{\alpha}^{\omega} V_x e^{-rx}$$

where x is the index of age, α is the age of first production, ω is the age of last reproduction, $V_x = l_x m_x$, and r is fitness or the instantaneous reproductive rate. Hence, fitness is a function of several variables of which development time is one. Over one generation the net reproductive rate R_0 is

$$R_0 = \sum_{\alpha}^{\omega} V_x$$

and the V elements will be positioned not only with respect to age but also over the reproductive season. Development time influences both the seasonality of reproduction and the number of generations in a given year. For the northern *W. smithii* population I have studied, the shortest development times will permit three successive generations per year followed by a long winter diapause. Longer development times permit only one or two generations per year. Thus, development time is a fitness component both in the strict sense of Fisher, and additionally through its timing and fitting of the generations within a single summer. *It controls the statistical distributions describing the temporal spacing and number of successive generations per year for all the progeny in the lineage of each individual.*

The length, seasonality, and number of generations per year are all variable in *W. smithii* populations as a consequence of the fundamental variation in larval development time. Variation in the seasonal spacing and number of generations per year introduces an ecological realism which will be important to my conclusions about the realism of Fisher's theorem. Figure 1 displays the phenotypic variation in development time for nondiapausing larvae in laboratory cohorts grown at 15 hr light and a fluctuating temperature regime of 17°-35°C daily (see Istock et al., 1976a, for details). These environmental conditions mimic spring and late summer conditions of the larval environment at Kennedy Bog, Rochester, New York, where most of my field studies have been conducted. Figure 1 also shows the development time difference between the sexes.

At a 15 hr photoperiod about 43% of the larvae of *W. smithii* from Kennedy Bog will enter a stable third instar diapause, which can only be terminated by a longer photoperiod. There are four larval instars in all. This warm season diapause thus serves as another control over the number of generations per year. With increasing time past the summer solstice (16 hr photoperiod) increasing numbers of larvae enter diapause thus shortening the number of generations per year for their lineages. By late August and early September large numbers of larvae

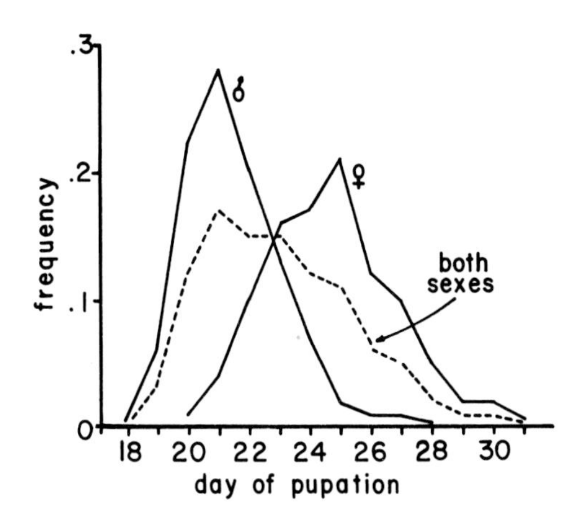

Fig. 1. Distribution of larval development time for the 1973 stock from Kennedy Bog which served as the parental generation for the selection lines of Figure 4. These larvae grew at 15 hr light, 17°-35°C daily temperature cycles and 0.5 ml of std. food suspension per day. Typically the males develop faster than females, but overlap broadly with them.

in nature have entered diapause, thus opting for fewer generations per year. The fraction of individuals diapausing is a variable trait within cohorts, and among the f_1 offspring of single females. It allows the offspring of a given lineage to delay reproduction until the following year after varying numbers of completed generations in the current year.

With the foregoing as background I next want to review six of the main results of my work.

(1) *Expressed r is a realistic and interpretable measure of fitness.* Table 1 summarizes a laboratory analysis of the response of r to varying food and density levels along with the associated fractions of diapausing larvae (from Istock et al, 1975). Within the table the allotment of food per individual was initially constant over each diagonal. These cohorts were grown at 15 hr light and a 17°-35°C diurnal temperature cycle. A few conclusions emerge. (i) The fraction diapausing is a function of food supply as well as photoperiod; at very low food all individuals diapause. (ii) There is a good relation between r and food supply, i.e., r behaves as a reasonable measure of expressed fitness. (iii) Quite surprisingly, r is positive as soon as both females and males commit themselves to completion of the life cycle. This suggests that the interrelations of responses to food, photoperiod, and diapause are tuned to make r a good indicator of potential population replacement. If $\bar{r} = 0$, a population is in numerical equilibrium. Negative r would mean less than simple replacement. Apparently one cannot force individual larvae to consistently realize fitness below replacement; instead they enter diapause.

(2) *Much of the variation in development time and tendency to diapause under warm season conditions is genetic.* Figure 2 shows the parent-offspring resemblance in development time. The observed h^2_N is 0.33 showing that 33% of the observed phenotypic variation in development time (Fig. 1) is due to underlying additive genetic variation (Istock et al., 1976a). After 15 generations of artificial directional selection for shorter development time h^2_N is zero as shown in Figure 3. This demonstrates the expected outcome from Fisher's theorem under directional, artificial selection. I do not know to what extent recombination might recover genetic variation in development time if selection were relaxed for some generations. It is an interesting question. What is most interesting here is the contrast between the laboratory selection and nature. Additive genetic variance was reduced *à la* Fisher in a few

Table 1. Responses of laboratory cohorts of *W. smithii* to different levels of food and larval density; temperature regime a 17°-35° C daily cycle, photoperiod 15 hr light. From Istock et al. (1975)

Larval density (#/60 ml water)	Food level (ml. std. food suspension per day)				
	0.05	0.1	0.2	0.4	0.8
a. Percent diapausers					
10	100	68.8	46.2	15.4	23.5
20	-	88.0	76.5	52.8	31.3
40	100	100	85.5	71.1	46.6
b. Mean total no. eggs produced per female					
10	n.p.	12.7	56.5	72.7	76.5
20	-	n.f.	18.8	44.8	41.6
40	n.p.	n.p.	37.8	7.7	23.6
c. Mean generation time in days					
10	n.p.	52.7	53.7	42.1	40.8
20	-	n.f.	57.4	49.5	46.4
40	n.p.	n.p.	57.8	51.3	48.1
d. Net reproductive rate					
10	n.p.	1.55	9.16	8.80	12.37
20	-	n.f.	1.52	4.53	6.66
40	n.p.	n.p.	2.00	0.25b	3.05
e. Intrinsic rate of increase (r)c					
10	n.p.	*.008*a	.041	.052	.062
20	-	n.f.	*.007*a	.031	.041
40	n.p.	n.p.	*.012*a	-.027b	.023

n.p. No pupae produced, all survivors entered diapause.
-- This treatment not done.
n.f. No female pupae appeared.
a These values form a boundary between no completion of life cycle and metamorphosis by some individuals. At this boundary all three values of r are slightly positive.
b This surprisingly low value is probably an aberrant result from the low yield of females and low egg production for this treatment.
c r varies significantly with crowding on the food resource (correl. coef. = 0.647, $p < .05$, 8 d.f.).

generations while the continuance of such variation in nature has been widely observed in *W. smithii*. It is clear then that retention of variation in development time has to mean retention of variation in fitness.

By selecting differentially for fast direct development, slow direct development, and diapause, the genetic nature of these was demonstrated. Figure 4 shows the responses of these characteristics to artificial selection. It also shows that selection for fast direct development elicits a correlated increase in the nondiapausing fraction, while selection for diapause elicits slower and slower development. Selection for slow direct development elicits a weaker, but noticeable response in tendency to avoid diapause. The genetic nature of the correlated responses is well demonstrated in Table 2 for mass reciprocal crosses of the resultant populations in the above selection experiment. The correlation of fast development with nondiapause and dia-

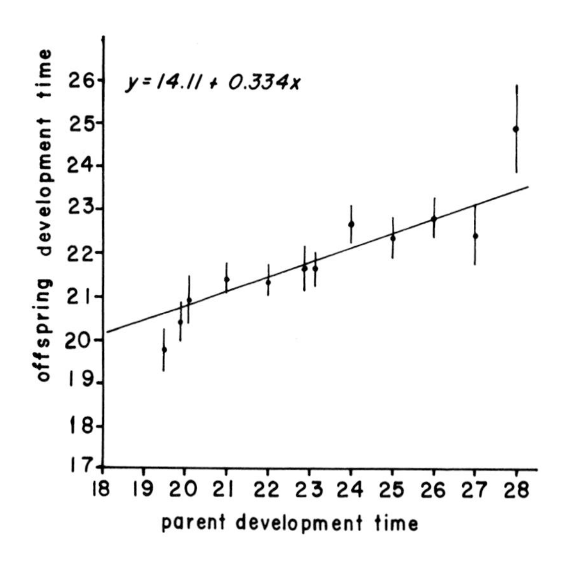

Fig. 2. Parent-offspring relation for variation in development time for the unselected 1973 stock from Kennedy Bog. The parent-offspring regression equation is printed above and graphed as the solid line. The vertical bars give exact 99% confidence limits. The estimate of h_N^2 is 0.33, standard error = 0.028.

Table 2. Larval development time (days from egg to pupa) and % diapause in mass reciprocal crosses between fast (after 15 generations of selection) and diapause (after 7 generations of selection) lines of Figure 4. All crosses were done simultaneously. The error terms are 95% confidence intervals.

Cross	Development time	% Diapause
fast X fast	19.6 ± 1.45	0.7 ± 0.2
fast ♀ ♀ X diapause ♂ ♂	23.0 ± 2.82	45.8 ± 1.5
diapause ♀ ♀ X fast ♂ ♂	24.7 ± 2.91	61.0 ± 2.3
diapause X diapause	27.2 ± 3.64	93.8 ± 0.8

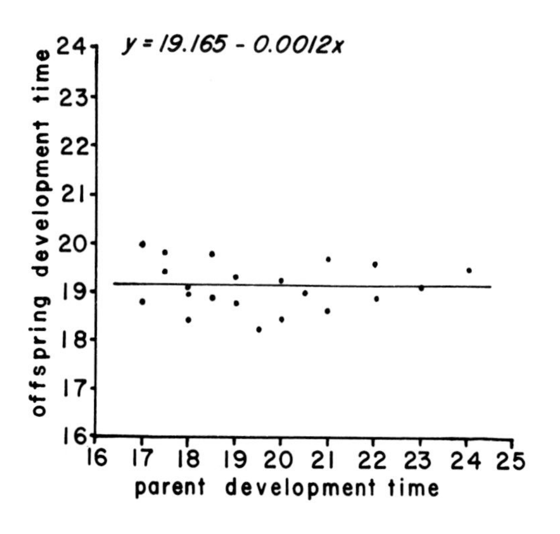

Fig. 3. Parent-offspring relation for variation in development time after 15 generations of mass selection for fast development time as shown in top graph of Figure 4. Fast development was operationally defined as a 17-20-day phenotype. The regression equation is printed above and graphed as the solid line. The estimate of h_N^2 = -0.001, standard error = 0.013, hense h_N^2 is not significantly different from zero.

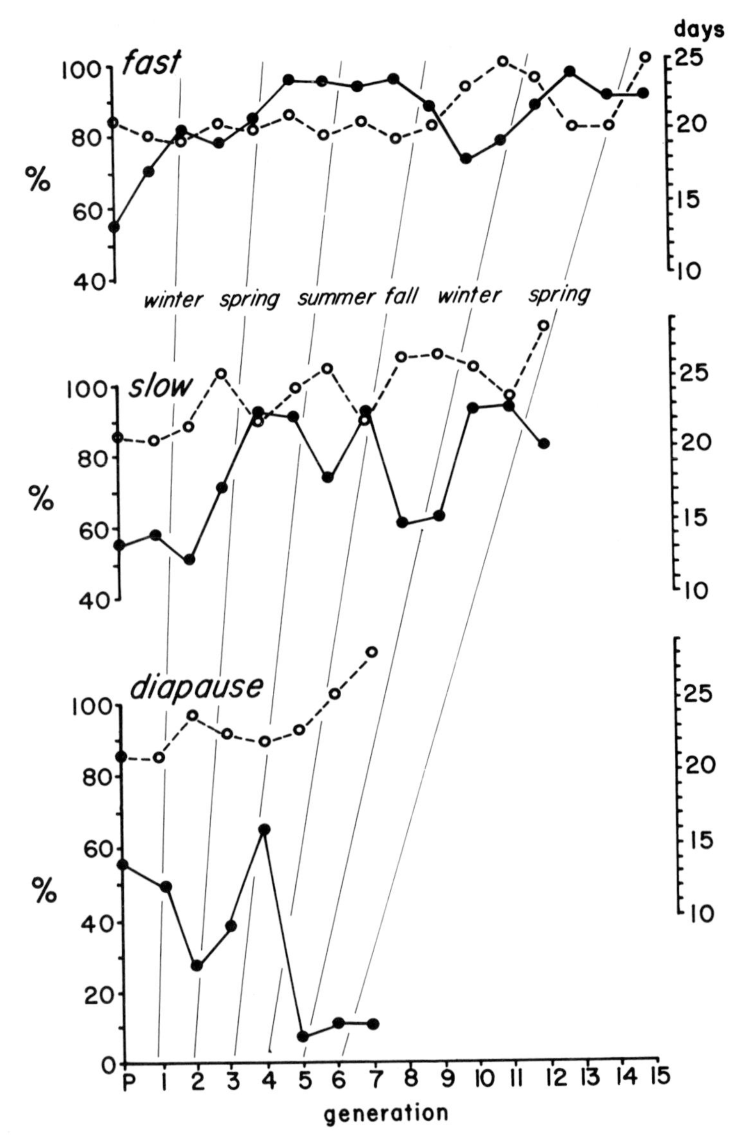

Fig. 4. Responses of *W. smithii* to mass selection in fast (15 generations), slow (13 generations), and diapause (7 generations) lines. All lines were split off from the same large parent population (P) of 600 individuals, and all extend over the same real time period. The differences in number of generations are due to differences in generation times among the lines. The fast line was perpetuated only by individuals expressing 17-20-day lapse from egg to pupa. The slow line was perpetuated by phenotypes displaying 24-35 days development. The diapause line was selected directly for diapause. All lines were maintained under the same condition; 15 hr light and 17°-35°C cycle, with 0.5 ml std. food suspension per day, except when a longer (16 hr) photoperiod was used to break diapause in order to continue selection in the diapause line. The dashed line for each type of selection plots the course of mean development time (right-hand scale). The solid lines follow the path of % nondiapause (direct developers, left-hand scale). A small fraction, 1-3% die in each generation of each line. The thin lines radiating up from the bottom give the true season outdoors. The seasons are indicated under the fast figure. For further discussion of this experiment see Istock et al. (1976a).

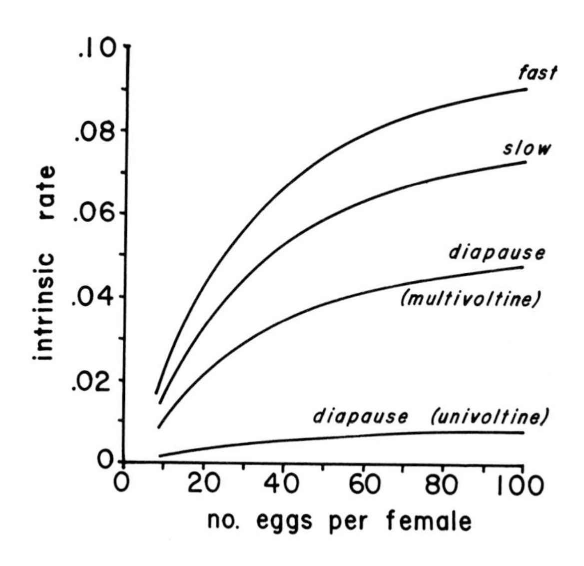

Fig. 5. Rise of potential intrinsic rate, or fitness, with egg production for each of the selection line types isolated in the experiment of Figure 4. The upper three curves represent multivoltine life cycles. This is hypothetical in the case of the multivoltine diapause curve because we artificially broke the diapause within a very short time using a long photoperiod. The lowest curve illustrates the consequence of a univoltine (extreme diapause, annual) life cycle. The differences between the curves along the ordinate describe potential fitness phenotype variation for a given level of egg production.

pause with slow development is very high (r = 0.998, $p < .01$, 2 d.f.). There is also a small maternal effect influencing the fraction diapausing.

To further test the correlation of development time and diapause a second selection experiment was performed at 16 hr light and a 17°-35°C cycle. At a daylength of 16 hr all individuals complete development, i.e., no diapause is expressed. Replicated fast and slow selection lines were maintained for five generations and then tested for a correlation of development time and percent diapause at 15 hr light. Again, fast development and low diapause were significantly correlated as were slow development and diapause. This correlation presumably arises from strong pleiotropic effects.

It is clear that considerable genetic variation underlies these two developmental fitness components and that a continuum from fast-direct development to slow-diapause development exists. This defines a distribution of fitness phenotypes whose extremes might be though of as pure strategies in Maynard Smith's (1974) sense.

(3) *Potential r differs for the extreme modalities (pure strategies) obtained by artificial selection.* From laboratory estimates of mean generation time, survival, and egg production potential values of r as a function of average lifetime egg production were calculated for the fast, slow, and diapause modalities of Figure 4. These calculations are graphed in Figure 5, where the ordinate shows the maximum expected range in fitness for the fitness phenotype variation exposed by laboratory selection. A simple result emerges. *The univoltine (one generation per year) diapause extreme will always be at a selective disadvantage if any other genotypes successfully express a multivoltine life cycle.* Hence the genetic variation will be bounded away from a pure diapause strategy by natural selection. Figure 5 also shows that potential r is not far above equilibrium replacement over the whole range of fertility for the pure diapause type, hence it is unlikely that the pure univoltine pattern could alone be successful in nature. Values for egg production of wild *W. smithii* range from very low to highs of 50-80 eggs per female lifetime (Fig. 12). Values as high as 130 eggs per female have been obtained in the laboratory, though values around 80 are more typical. It should be recalled that such egg production depends on stored larval proteins and lipids, because *W. smithii* is not a blood-feeding mosquito.

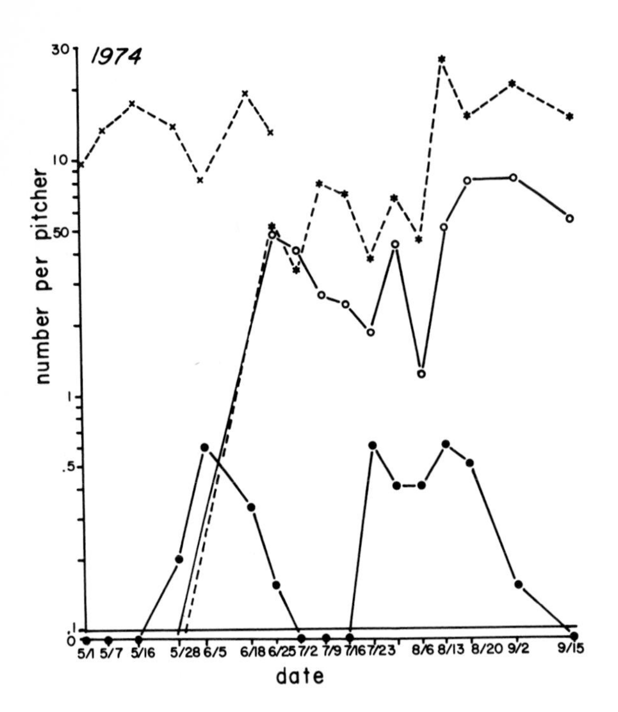

Fig. 6. Population dynamics of *W. smithii* at Kennedy Bog in 1974. The upper dashed curve shows larval density with time, X is for old pitchers and * is for new pitchers of the year. The solid line with open circles given egg densities and the solid lines with closed circles show the two pupal pulses characteristic of this population.

(4) *Larval food resource levels are seasonally patterned in nature.* A series of experiments at Kennedy Bog, principally in 1973 and 1974, have demonstrated a sequence of density-dependence -- density-independence -- density-dependence through spring-summer-fall, respectively. This sequence was determined using sequential convergence experiments with the natural population (Istock et al., 1976b). Figure 6 displays the dynamics of the natural population in 1974. The uppermost, dashed line traces mean larval density through the warm season; other details of Figure 6 are explained in the caption. A convergence experiment asks whether the observed larval density, expressed as biomass, is the maximum supportable at that time. The experiment involves introducing newborn larvae from large laboratory stocks into the field under a stratified and replicated design of treatments and controls (for further details see Istock et al., 1975; Istock et al., 1976b). Numerical convergence over the course of three successive experiments in 1974 occurred as seen in Figure 7 (for full explanation, see caption). The biomass trajectories for the three experiments appear in Figure 8. Experiment 2 in Figure 8 demonstrates the general pattern: there is a midsummer period, approximately two months long, in which biomass can be many times higher than that actually synthesized by the population. This is a period of excess resources or density-independence. It is preceded in the spring and followed in the fall by density-dependent periods as demonstrated in the composite Figure 9 (for further explanation, see figure captions and Istock et al., 1976b). It is clear the seasonal cycle is ecologically patterned and that the period of greatest larval food availability occurs in midsummer. Data obtained earlier (Istock et al., 1975) as well as new data of the next section demonstrate a positive relation between food and egg production. Hence the seasonal pattern of density effects can strongly influence fitness expression. Midsummer food abundance could also raise survival rates, thereby further elevating the expressed fitness of individuals growing during this period. It seemed (Istock, et al., 1976a,b) that the genotypes tending toward slower development and dia-

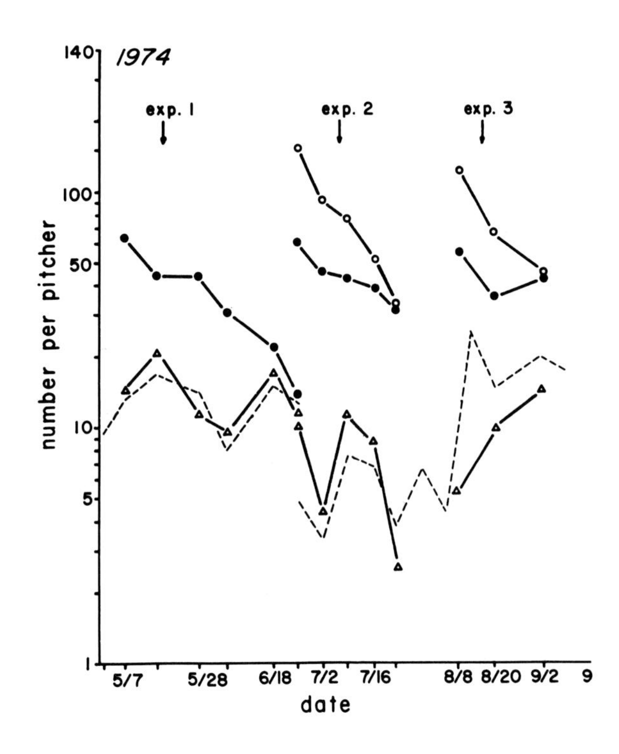

Fig. 7. Three convergence experiments at Kennedy Bog in 1974. The upper lines with open circles (O——O) show treatments with 100 newborn larvae added per pitcher, the solid lines with closed circles (●——●) show treatments with 50 newborn larvae added per pitcher, the solid lines with open triangles (△——△) are controls with no larvae added and the dashed line (----) traces the natural larval densities shown in Figure 6. For more explanation see Istock et al. (1976b).

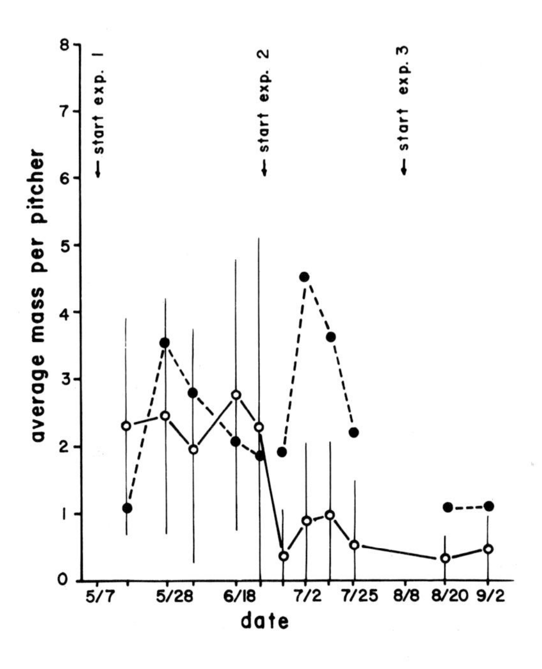

Fig. 8. Successive trajectories (●----●) of biomass in the experimental treatments of Figure 7 in contrast to the controls (O——O). The vertical lines show 95% confidence intervals for the controls. Early in the season experimentals and controls are the same (density-dependence), in midsummer they differ (density-independence) and in the fall there is density-dependence again. For data from further experiments of this type, see Istock et al. (1975 and 1976b) and Figure 9.

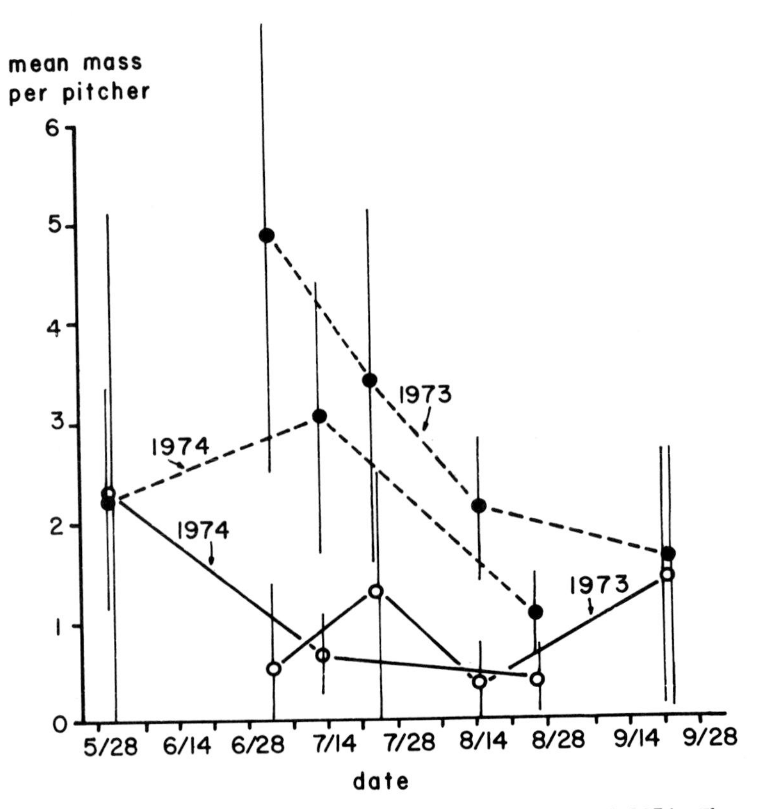

Fig. 9. A composite of the convergence experiments of 1973 and 1974. The upper dashed lines (●----●) show estimates of the carrying capacity through the spring, summer, and fall, and the lower solid line (O——O) gives the actual time course of aggregate larval mass in the controls. The thin vertical lines are 95% confidence intervals about their respective means.

pause might compensate for fitness loss from fewer generations per year by adjusting the seasonality of their growth and reproduction to avoid the density-dependent periods of spring and fall. This is part of the mechanism which preserves fitness variation, but not the essential aspect.

(5) *The fitness phenotype distribution exposed in the laboratory is also expressed in nature.* During the summer of 1976, the Kennedy Bog population was sampled for the joint expression of development time and percent f_1 diapause. The samples were groups of individuals in the same developmental stage (instars 1, 2, 3, 4, and pupa) at the same time. In all, 26 such groups were analyzed throughout the summer. The data are plotted as open circles in Figure 10. The correlation of the two fitness characters is statistically significant (correl. coeff. = 0.68, $p < .001$, 24 d.f.). The two characters are not randomly associated in nature, suggesting that the genetic correlation between these characters found in the laboratory is also manifested in nature. The regression of f_1 development time on percent f_1 diapause is the line of lower slope in Figure 10.

When only the first nine groups of contemporaneous pupae (open circles with vertical bars in Figure 10 and treated extensively in Figure 12)

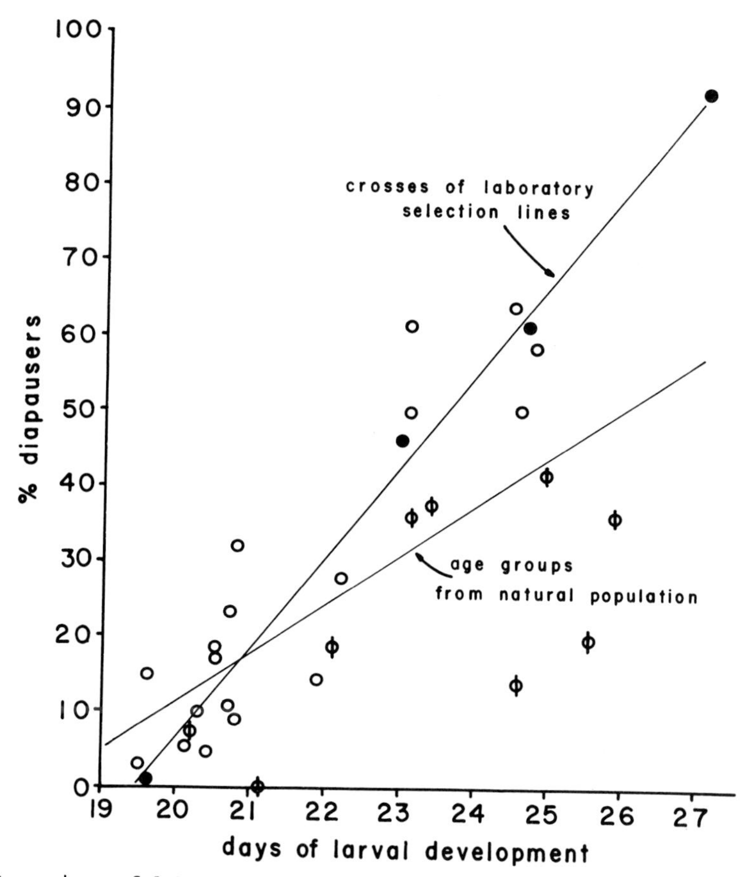

Fig. 10. Comparison of laboratory and field expression of f_1 development time and f_1% diapause. Each open circle with or without a vertical bar is the mean for an even-staged (first, second, third, fourth, or pupal instar) group isolated from nature and tested in the laboratory at 15 hr light and 17°-35°C cycle daily with 0.5 ml std. food suspension per day. The lower line is the significant regression of development time on % diapause for the open circles. The upper regression line for the closed circles is from the laboratory crosses of Table 2. It appears that the artificial selection which led to the upper line has merely tightened the genetic tendencies of the natural population and expanded the variation in the direction of greater % diapause and slower development time. As the open circles fall short of the laboratory diapause extreme they illustrate the sense in which the variation in nature is bounded away from this extreme. The open circles with a vertical bar are for the first nine pupal samples treated in Figure 12.

were tested, the correlation is similar to that for all 26 groups, but slightly short of statistical significance (correl. coeff. = 0.62, $p > .05$, 7 d.f.). Since the pupal duration is nearly invariant between individuals at a given temperature, each of the pupal samples represents the new reproductives flying a week to ten days later. Hence, such individuals are maximally likely to mate together. The variation displayed by the nine pupal samples (Figures 10 and 12) demonstrate the intermediate solution struck between intense assortative mating for development time and percent diapause (as illustrated by the laboratory crosses of selection lines) and complete panmixia with respect to these characters. Both pure strategies are mixing to some extent throughout the mating season. The mixing may be considerably greater than indicated by direct analysis of these nine samples. Each was swarmed, mated, and tested separately, but because adults can survive two to three weeks in nature, individuals emerging over periods of weeks may intermate. Unfortunately, it is not yet known if females of *W. smithii* mate once or many times.

The steeper regression line in Figure 10, along with the four solid circles, plot the laboratory crosses of Table 2. This falls at the upper side of the distribution of values from nature. It is clear that the laboratory selection lines represent a purification of the expression of the two fitness characters in nature. As predicted from the laboratory measurements the variation in nature is bounded away from the pure diapause extreme.

On Figure 10 we also note that the fast-direct development extreme is registered at the lower left boundary of the distribution of values from nature. This supports, in one sense, Lewontin's (1965) contention that selection will shorten the development time of a continuous colonizer, which *W. smithii* certainly must be in order to continuously inhabit each new generation of pitcher-plant leaves. In another sense, however, the data contradict Lewontin's contention that the mean development time would shorten and the genetic variance for development time disappear (a corollary to Fisher's theorem). The variation in development time has not vanished in the natural population and is subject to stabilizing influences (see below).

Studies of quantitative genetics and heritability are fraught with difficulty if one wishes to extrapolate laboratory results to natural populations. The effect of environmental differences could make the most careful laboratory measurements useless (Feldman and Lewontin, 1974). However, the data of Figure 10 show sufficient agreement between field and laboratory analyses that it seems safe to conclude that a genetically based, fitness-phenotype spectrum, ranging from fast-direct development to slow-diapause development, is appropriate for interpreting subsequent field observations.

(6) *There is a seasonal pattern to the expression of fitness phenotypes in nature.* Figure 11 displays the dynamics of the Kennedy Bog population in 1976. A full explanation is given in the caption, but of immediate interest are the two pulses of pupae shown with solid lines and circles at the bottom of the graph. Two such pupal pulses per season have been a consistent aspect of this population for each of the last seven years. They serve as convenient markers for the collective seasonal life-history progression of *W. smithii*. The first pulse represents pupae emerging from older overwintering leaves of the pitcher-plant; the second pulse is from new leaves of the year.

Figure 12 is set on the same abscissa as Figure 11 and displays data for pupal sex ratio, egg production, and percent f_1 diapausers over the 1976 season. There are several clear patterns.

Recall that males develop faster than females. Hence a shift in the pupal sex ratio toward males in nature indicates a coming wave of adult emergence. The upper graph of Figure 11 then roughly indicates the timing of three generations. These represent the expression of about the fastest development possible for this species and, by the correlation with low diapause, we expect these same pupal samples to transmit low diapause frequencies. Glancing down to the lowest graph we see that this is true.

Concentrating now on the lowest graph of Figure 12, and using % f_1 diapause as a quantitative genetic marker we see the following. (i) There

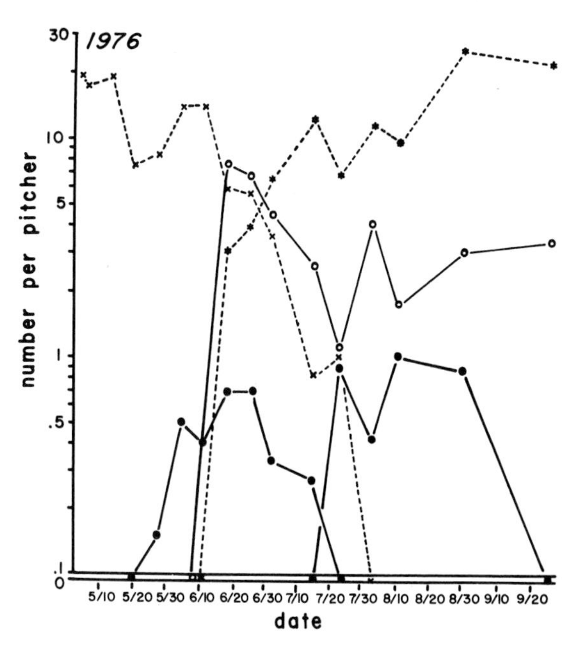

Fig. 11. Population dynamics of *W. smithii* at Kennedy Bog in 1976. All the symbols are the same as those for Figure 6. The two conspicuous pupal pulses (●——●) serve as temporal indicators for Figure 12.

is a seasonal sequencing of low diapause followed by higher in each of the pupal pulses (pulses shown from Figure 11 as dashed lines). (ii) The peaks for % diapause in both pupal pulses represent only about 40% transmission of this trait suggesting strong mixing of strategies. (iii) The two peaks of diapause transmission are separated by about one average (not short, see Table 1) generation suggesting that this is the modal expression in time of the genotypes intermediate along the fitness spectrum. (iv) There is little tendency for % diapause to shift upward with the season. This may reflect the strong pressure from fast-low diapause genotypes as they are amplified in numbers by repeated generations. (v) After two average generations (shown on lowest graph) any nondiapausers would find themselves attempting to fly and mate in the mid to late fall. In fact there are individuals that resist diapause to this degree and we note from the middle graph of Figure 12 that such pupae (9/24/76) had an extremely low egg production. These individuals also had a very low mean life span when tested under laboratory conditions. I have occasionally seen adults flying and ovipositing as late as mid-October, but have not seen these eggs successfully hatch in nature. By early to mid-September most larvae in nature are in the third instar diapause in preparation for overwintering.

The middle graph of Figure 12 records the lifetime egg production of adults from native-grown pupae of the respective sample dates. Egg production is on a logarithmic scale and the fluctuations, even in the early and middle part of the season are quite large. Egg production of 20-30 eggs per female are typical of the spring density-dependent period. Higher values reaching 40-50 eggs per female can occur in the midsummer density-independent period, although there is a dip in midsummer as well. Egg production falls quite low in the fall density-dependent period. It is clear that the higher fertilities of midsummer give some benefit in fitness to individuals completing larval growth and reproducing at this time. Larvae which grow in this period, but diapause, ought to benefit from higher winter survival and higher fertility the next spring in comparison with larvae completing develop-

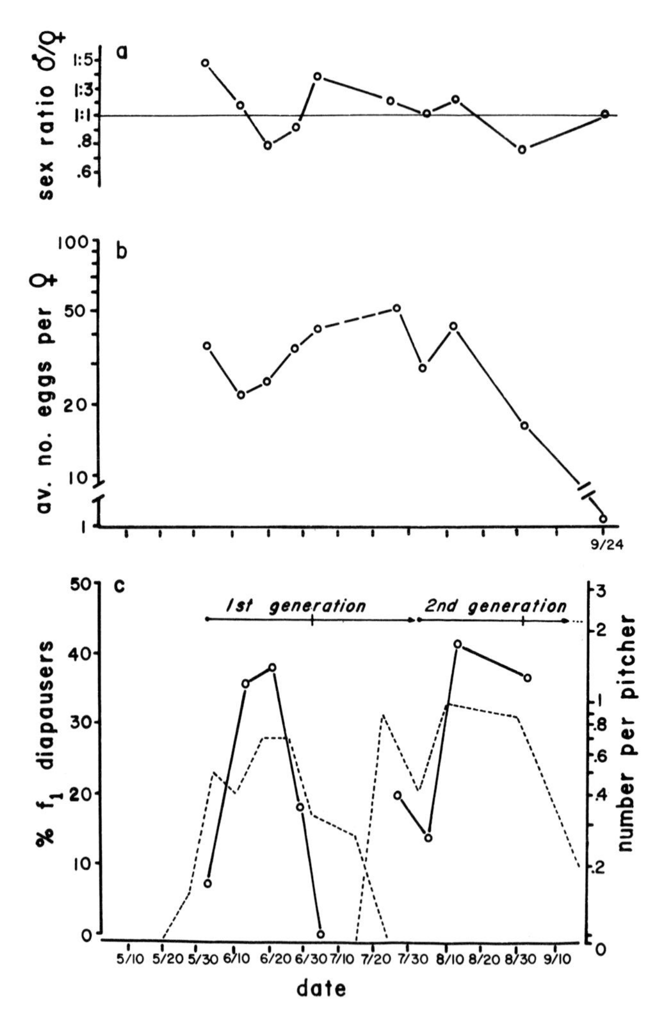

Fig. 12. Three aspects of the *W. smithii* population at Kennedy Bog in 1976. All three graphs depict properties of 10 pupal samples (open circles) taken as pupae directly from nature. Hence these samples reflect the properties of individuals experiencing larval growth conditions in nature at various times during the spring, summer, and fall. An interpretation of these graphs is given in the text.

ment and diapausing during the fall. The most important observation, however, is the disastrously low fertility of diapause resisting larvae in the fall. This should, at least in some years and probably in most, confine the developmental variation away from the fast-nondiapause extreme.

Variation in the development-time and diapause traits is selected against at both extremes. The pure slow-diapause extreme loses in direct reproductive competition with any mixed strategy. The pure fast-nondiapause extreme loses to some mixed strategies because its net annual reproductive rate, the geometric mean of R_0, will closely approach zero in some, perhaps most, years. The highest expressed fitness over the run of years falls to some best expression of the mixed strategy, which itself is the consequence of the mixing effects of genetic segregation, chromosomal reassortment, and recombination.

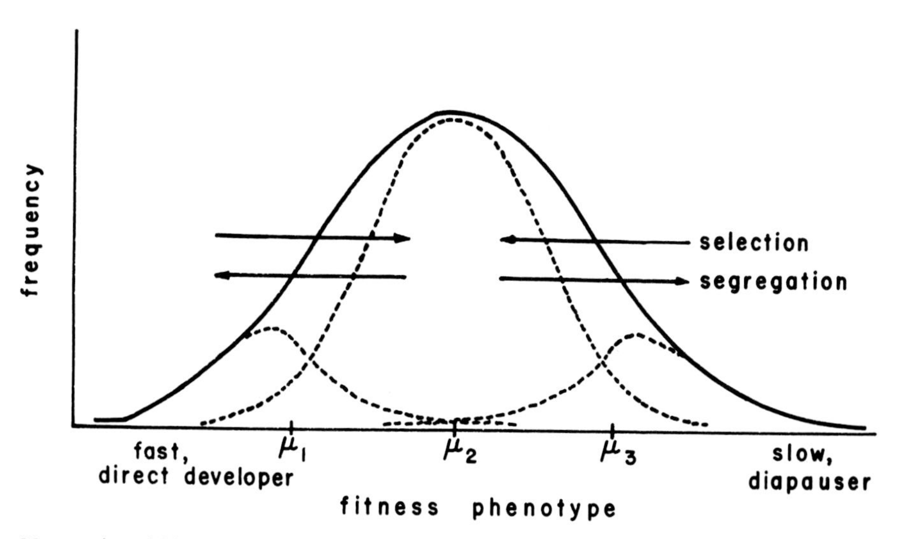

Fig. 13. A simplified model illustrating the balance of natural selection against genetic segregation (and recombination) for varying fitness phenotype scores along a spectrum from extreme fast-direct developers to extreme slow-diapausers. The μ represent means or level constants for three hypothetical, quantitatively varying, subpopulations within the full continuous variation in fitness of the entire population.

Discussion

The balance of segregation, chromosomal reassortment, and genetic recombination against selection for these developmental fitness characters will maintain a stable V_A for fitness greater than zero under the genetic and ecological conditions found for *W. smithii*. This contradicts the strict interpretation of Fisher's theorem. Such balancing selection is summarized in Figure 13. Solely for illustrative purposes, the figure depicts three quantitative modalities such that any phenotype is given by the following expression:

$$x_i = \mu_i + e$$

where x_i is the phenotypic score of any member of the ith (1, 2, or 3 in Fig. 13) modality, μ_i is the mean additive genetic score for the ith modality, and e is an error term containing environmental and nonadditive genetic effects. Obviously the model expands to any arbitrary number of modalities, the maximum being the number of genetically unique types in the population. All my data suggest that mixing through segregation, reassortment, and recombination in nature occurs along two-thirds or more of the full spectrum of variation in developmental time and diapause which can be exposed in the laboratory. This is close to a quantitative genetic analog of the segregation-selection balance suggested theoretically by Haldane and Jayakar (1963) for a simple genetic polymorphism. Although here, I do not require that selection fluctuate in direction as did Haldane and Jayakar. The direction of selection probably does fluctuate both within and between years, but this is not necessary to ensure persistent variation in expressed fitness.

This complex polymorphism in fitness cannot be looked at as a segregation or recombination load (Crow and Kimura, 1970). What we might call genetic load, turns out to be precisely the ecological adaptive mechanism through which selection favors the lineages of individuals closest to the best mixed strategy at any time. And the better mixed strategies can only be achieved through chromosomal segregation and reassortment and possibly also through intermediate levels of recombination. A role for intermediate levels of crossing-over as a generator of fitness variation seems plausible here since *W. smithii* has only three (haploid number) chromosomes. Organisms with higher chromosome numbers might require less recombination to generate comparable variation.

If a complex polymorphism in fitness traits can be "stabilized," we then understand one possible reason why "the genotype does not congeal," and how a great field of more superficial genetic variation tied to fitness variation can be held in a natural population. I suspect that an amplification effect could exist in such a situation; relatively modest amounts of additive genetic variance in fitness may support large fields of qualitative, additive, and nonadditive genetic variation in more superficial traits. We simply don't know this relationship now. We don't even know the extent of genetic variation in fitness characters for natural populations. Genetic variation for morphology, behavior, and ecology as well as for structural gene loci and chromosomal variation may be very differently coupled to fitness variation in different species and at different times in the history of the same evolutionary lineage. We know almost nothing of this coupling process.

A "stabilized" V_A for fitness gives persistent ecological and evolutionary flexibility to individuals and the population simultaneously. Selection for both individual and population properties become much the same thing. Individual mixed strategies are corrected, and continue to be selectively advantageous, as a consequence of genetic reassortment within the field of variation near the population mean. And the variation near the population mean is a record of recent success under selection.

From the findings presented here we cannot conclude that V_A for fitness in *W. smithii* is perfectly stabilized (i.e., static) at Kennedy Bog, a slow directional process could still be underway. In fact, a quasi-stable V_A for fitness is required to account specifically for the known latitudinal shifts in development time and diapause tendency in *Wyeomyia smithii*.

It is important that variation in development time and diapause tendency turn out to make such large contributions to fitness variation, for we can see ecological meaning in the control these characters exert over variation in the timing and number of generations each year and in the various relations to seasonal changes in resource availability. If the mechanism uncovered is at all general, the ecologically meaningful fitness components may nevertheless be quite different ones for other species (e.g., fertility or survival). It will be interesting to know if the juxtaposition of extreme fitness variants is a common property of population genetic structures.

A direct role for sexuality in fitness has emerged here. There is no other way to achieve the required continuing genetic reassortment and intermediate levels of recombination except through sexuality. Over a short time span a parthenogen or inbred line with the genome for the best mixed strategy might seem a possible alternative. But a fixed "best mixed strategy" is undoubtedly an abstraction even in the short run. The best mixed strategy is one always changing, always subject to cautious revision as the flow of environments and generations unfolds.

Summary

It is suggested that the directional aspect of Fisher's fundamental theorem of natural selection may raise several severe problems as a general mechanism of microevolution. A mechanism which incorporates a quasi-stable, additive, genetic variance in fitness may be more appropriate for both ecological and evolutionary reasons.

Pertinent results from past studies of the genetics and ecology of the pitcher-plant mosquito *Wyeomyia smithii* are reviewed. In these, two fitness characters, development time and the tendency to diapause under warm season conditions are central. It appears that *W. smithii* populations do have a quasi-stable, genetic variance in fitness, contrary to Fisher's theorem, and that this stabilization is achieved through a mixed life-history strategy dependent on chromosomal reassortment and possibly on intermediate levels of genetic recombination. The ecological adaptation thus achieved becomes clear through the control the developmental characters exert over the number and timing of the generations each year. This mechanism makes selection for individual and population properties the same process. A direct role for, and advantage to, sexual reproduction becomes clear.

Acknowledgments I am deeply grateful to Nancy Istock, Joyce Horstman, Robert Spry, Harold Zimmer, Steve Wasserman, David Sloan Wilson, James Zisfein, William Weisburg, and Karen Vavra for their help with the studies reviewed here. I thank Ernst Caspari, Julia Graham, Hugh Dingle, Uzi Nur, Dave Parker, and John Jaenike for many helpful comments, and Elaine Duren for typing the manuscript. This work was supported by National Science Foundation (USA) grants over the past seven years.

References

Ayala, F.J.: Molecular Genetics and Evolution. In: Molecular Evolution. Ayala, F.J. (ed.). Sunderland: Sinauer Associates 1976.

Bradshaw, W.E., Lounibos, T.P.: Photoperiodic control of development in the pitcher-plant mosquito, *Wyeomyia smithii*. Can J. Zool. *50*, 713-719 (1972).

Crow, J.F., Kimura, M.: An Introduction to Population Genetics Theory. New York: Harper and Row 1970.

Dobzhansky, T.: Genetics of the Evolutionary Process: New York Columbia Univ. Press 1970.

Evans, K.W., Brust, R.A.: Induction and termination of diapause in *Wyeomyia smithii* (Diptera:Culicidae) and larval survival studies at subzero temperature. Can. Ent. *104*, 1937-1950 (1972).

Falconer, D.A.: Quantitative Genetics. New York: Ronald Press 1960.

Feldman, M.W., Lewontin, R.C.: The heritability hangup. Science *190*, 1163-1168 (1975).

Felsenstein, J.: The theoretical population genetics of variable selection and migration. Ann. Rev. Genet. *10*, 253-280 (1976).

Fisher, R.A.: The Genetical Theory of Natural Selection. New York: Dover 1958.

Haldane, J.B.S., Jayakar, S.D.: Polymorphism due to selection of varying direction. J. Genet. *58*, 237-242 (1963).

Haldane, J.B.S.: Natural selection. In: Darwin's Biological Work, Bell, P.R. (ed.). New York: John Wiley 1964.

Istock, C.A., Wasserman, S.S., Zimmer, H.: Ecology and evolution of the pitcher-plant mosquito. 1. Population dynamics and laboratory responses to food and population density. Evolution *29*, 296-312 (1975).

Istock, C.A., Zisfein, J., Vavra, K.J.: Ecology and evolution of the pitcher-plant mosquito. 2. The substructure of fitness. Evolution *30*, 535-547 (1976a).

Istock, C.A., Vavra, K.J., Zimmer, H.: Ecology and evolution of the pitcher-plant mosquito. 3. Resource tracking in a natural population. Evolution *30*, 548-557 (1976b).

Kimura, M.: On the change of population fitness by natural selection. Heredity *12*, 145-167 (1958).
Krebs, C.J.: Genetic and behavioral studies on fluctuating vole populations. Proc. Adv. Study, Inst. Dynamics Numbers Popul. pp. 243-256 (1970).
Leon, J.A.: Life histories as adaptive strategies. J. Theor. Biol. *60*, 301-335 (1976).
Levins, R.: Evolution in Changing Environments. Princeton: Princeton University Press 1968.
Lewontin, R.C.: Selection for colonizing ability. In: The Genetics of Colonizing Species. Baker, H. (ed.). New York: Academic Press 1965, pp. 77-94.
Lewontin, R.C.: The effect of genetic linkage on the mean fitness of a population. PNAS *68*, 984-986 (1971).
Lewontin, R.C.: The Genetic Basis of Evolutionary Change. New York: Columbia University Press 1974.
Mather, K., Jinks, J.J.: Biometrical Genetics. Ithaca: Cornell University Press 1974.
Maynard Smith, J.: The theory of games and the evolution of animal conflicts. J. Theor. Biol. *47*, 209-221 (1974).
Mayr, E.: Animal Species and Evolution. Cambridge: Harvard University Press 1963.
Mertz, D.B.: Life histories and ecological genetics. In: Readings in Ecology and Ecological Genetics. Connell, J.H., Mertz, D.B., Murdock, W.W. (eds.). New York: Harper and Row 1970.
Smith, J.B.: Life history of *Aedes smithii* Coq. J. New York Ent. Soc. *10*, 10-15 (1902).
Smith, S.M., Brust, R.A.: Photoperiodic control of the maintenance and termination of larval diapause in *Wyeomyia smithii*. Coq. (Diptera: Culicidae) with notes on oogenesis in the adult female. Can. J. Zool. 49, 1065-1073 (1971).
Thoday J.M., Boam, T.B.: Effects of disruptive selection. II. Polymorphism and divergence without isolation. Heredity *13*, 205-218 (1959).
Turner, J.R.G.: Why does the genotype not congeal? Evolution *21*, 645-656 (1967).
Wright, S.: The roles of mutation, inbreeding, crossbreeding, and selection in evolution. Proc. VI Int'l Cong. Genetics *I*, 356-366 (1932).

3 Migration, Diapause, and Life Histories

The final three papers in this book deal with species in which both migration and diapause are integral parts of life-history strategies. In his paper C. Solbreck first makes the point that migration and diapause are as much a part of life histories as the more traditional schedules of mortality, development, and births. Indeed the latter are profoundly influenced by the former as Istock also showed with *Wyeomyia*. As a result, Solbreck suggests that there is a hierarchical organization of the ways environmental information is used to control phenotypic switches between direct development, migration, and diapause. In Solbreck's terms choices are made between reproducing here, elsewhere, now, or later. Environmental influences on these choices are examined in the seed bug *Neacoryphus bicrucis*, a facultative migrant and diapauser, in terms of various life-history traits and the potential for exploiting various habitats. He concludes that like Waldbauer's saturniids, migrants are essentially "bet-hedgers" spreading egg production, for example, in space and time, although much remains to be learned about environmental cues and their relation to phenotypic (and genotypic) variation in life histories.

In his paper, K. Vepsäläinen compares ten species of European waterstriders (*Gerris*) with respect to life histories involving migration and diapause. *Gerris* are interesting in this regard because different species and populations show varying degrees of polymorphism in both voltinism, like Masaki's crickets, and wing-length. Populations may contain either or both long-winged migrants or short-winged nonmigrants. Vepsäläinen examines both genetic and environmental determinants of these traits with respect to factors such as habitat isolation and population density. Extreme habitat isolation, for example, leads to the evolution of genetic dominance for short-wings while a model of density-dependence predicts that alary dimorphism will be optimal over a wide range of environmental circumstances. Like *Neacoryphus* and saturniids, *Gerris*, too, seem to be bet-hedgers.

The final paper by H. Dingle compares migration and diapause strategies in species and populations of milkweed bugs (*Oncopeltus*) occurring in both temperate and tropical areas and on islands. In general tropical forms do not show a photoperiodically induced diapause, but local conditions must also be considered since the strongest diapause was found in *O. fasciatus* populations from South Florida. Diapause promotes migration in *O. fasciatus*, as also indicated by M.A. Rankin, and in species and populations which do not diapause migration is absent or greatly reduced. As was also found with *Gerris*, isolation, in the case of *Oncopeltus* on Caribbean islands, significantly reduces migratory capacity, although in these species without influencing relative wing lengths.

In all the above species, an intimate relation between migration and diapause in the evolution of life-history strategies is evident. Also apparent is the considerable genetic and environmental variation in these traits as also noted in Section 2 concerning diapause and development. Explaining this variation is an important task for entomologists and evolutionary biologists.

Migration, Diapause, and Direct Development as Alternative Life Histories in a Seed Bug, *Neacoryphus bicrucis*

CHRISTER SOLBRECK

Migration and Diapause as Life-History Traits

Nature is variable in space and time, and insects as well as other organisms are presumably adapted to cope with various degrees of environmental change. Thus organisms in relatively stable environments are thought to be on the average more specialized, produce fewer but larger offspring, live longer and consequently exhibit smaller fluctuations in population size than do organisms in relatively temporary environments. Accordingly different environments seem to accumulate - through immigration and evolution - species with particular life-history characteristics, resulting in communities and ecosystems of different qualities, for example with regard to their stability properties (Orians, 1975). Studies of life-history adaptations are thus of considerable interest for the entire field of ecology.

Insects living in highly variable environments will repeatedly face the problems of covering areas and time periods that are unsuitable for breeding. To bridge such gaps in space and time insects have evolved elaborate behavioral and physiological responses like migration and dormancy (as here exemplified by diapause). These phenomena have been shown to be integrated parts of the life history as they are strongly coupled to reproduction and other demographic events (Dingle, 1972, 1974a,b). Migration and diapause are hence important life-history traits to be studied along with the more commonly used variables like egg number, egg size, generation time, survival, and reproductive rates.

Migration and diapause share certain common features (e.g., Kennedy, 1961). Both usually occur before reproduction, migration displacing reproduction to some place elsewhere, and diapause delaying reproduction to some time later, usually much later (Fig. 1). (Migration of necessity also involves a reproductive delay but this is generally much shorter. Effects of such short delays are discussed below.) Accordingly migration can be regarded as escape in space and diapause as escape in time. However, insect migratory flights are of different kinds, and for a discussion about the relationships between migration and diapause two types of migration have to be distinguished. The first type is migration by young adults soon followed by breeding. This type of migration was denoted "class I" migration by Johnson (1969).

A second type of migration called "class III"* migrations by Johnson (1969) are those undertaken to and from diapause sites. In this case migration is intimately linked to the general diapause development, and the pre- and postdiapause migrations can usually be regarded as marking the beginning and end of the dormancy period.

---

* Johnson (1969) also distinguishes a type of migration called "class II." These are flights by young adults to feeding habitats soon followed by a return flight for oviposition. This type of migration is in many respects similar to "class I" and there is little reason to treat it as a separate case in the context of this paper.

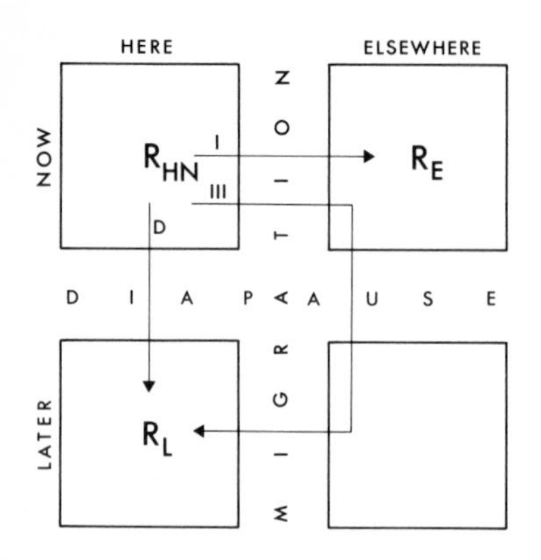

Fig. 1. Relations among reproduction, migration, and diapause. Arrows illustrate class I migrations (I), class III migrations (III), and diapause without migration (D). The three alternatives in *N. bicrucis* are reproduction here and now (R_{HN}), reproduction elsewhere (R_E) after migration, and reproduction much later (R_L) after diapause. For further explanations see text.

The two types of migration are illustrated in Figure 1. Other alternatives than those outlined in the figure can, of course, be visualized. There are, for example, many variations on the theme of class III migration. Thus postdiapause flights do not necessarily end up "here." However, this scheme focuses the attention on two basic types of migration: those soon followed by reproduction and those linked to a dormancy syndrome. In the following discussion diapause without migration will not be distinguished from diapause with migration, since their demographic consequences seem to be very similar, and the term migration will, unless otherwise stated, be used to denote only "class I" migrations.

Most life-history traits are intimately linked to each other. Thus the number of eggs and egg size are inversely related (e.g., Price, 1973) and maternal survival and egg production may be also (Murdoch, 1966). Migration and diapause will, for example, affect generation time and thus the intrinsic rate of increase. A change in one life-history trait will affect other traits, and a certain combination of traits that improve survival in a certain environment is called a life-history strategy or tactic (cf, Stearns, 1976).

Approaches to the Study of Life-History Strategies

A theory about how life histories are molded in different environments must consider the existing variability in life-history traits and environmental parameters. To answer the question what alternatives are favored in different environments, we have to decide in what terms to describe both life histories and environments. We also have to ascertain whether each animal is maximizing reproductive rates or minimizing risks of having no offspring at all (Stearns 1976, p. 41). There is not yet a generally adopted view on this subject, and there will probably not be any consensus until more empirical data are available.

In what ways can we study the adaptive significance of life-histories empirically? The most direct approach is to perform selection experiments as done by Dingle (1974b) on the bug *Oncopeltus fasciatus* and by Istock et al. (1976a; this volume) on the pitcher-plant mosquito, *Wyeomyia smithii*. Such studies will reveal the relative roles of genetic and phenotypic variation of life-history traits, and show to what extent these traits are amenable to selection.

A more common and less time-consuming approach is to correlate observed life-history patterns with particular environments. This can be done by comparing species living in different environments, preferably an assemblage of closely related species which share a basic physiological and morphological organization (Vepsäläinen, this volume). Organisms are always to some extent trapped in their evolutionary past. The more closely related two taxa are, the higher is the probability that traits are compared which are real evolutionary alternatives, and not merely the result of a ground plan that is not easily altered by natural selection. This criterion of organizational similarity is still more easily met by studing geographical populations or specific genotypes within a species.

Another way to correlate life histories and enviornments is to compare different phenotypic expressions possible for the individual. This can, for example, be done by comparing stationary individuals and migrants of a species with facultative migration, or nondiapause and diapause individuals in a species with facultative diapause.

Phenotypic plasticity, e.g., as expressed in facultative migration or diapause, is a likely attribute of organisms inhabiting rapidly changing environments (cf, Slobodkin, 1968). Phenotypically alternative life histories are interesting for studies of life-history strategies since they represent real alternatives. If it can be discerned under what sets of environmental conditions individuals switch from one life history to another, we can also gain some understanding about conditions promoting monomorphic life histories with regard to these traits. Shifts of response thresholds to environmental signals can be envisaged as a possible road to genetic fixation of monomorphic life histories. For example, the bug *Oncopeltus fasciatus* normally enters diapause under short-day conditions, but selection for nondiapause under short days results in a population with a shorter critical photoperiod for diapause induction (Dingle, 1974b).

Life-history strategies are specific compromises in the allocation of an individual's resources, which ensure survival in specific environments. I will mainly be concerned with the first aspect, namely the trade-offs between different life-history traits accompanying the switches between direct reproduction, migration, and diapause. Studies on such life-history consequences seem justified today in a field characterized by a flood of theoretical models and few empirical data.

I will also be concerned with the use of environmental information for tracking resources in space and time. It will be argued that simultaneous information in both time and space is essential for "decisions" whether to reproduce directly, migrate, or enter diapause, and that there exists a certain hierarchical organization in the use of environmental information for this purpose.

The Interface of Environmental Change in Space and Time

Imagine an archipelago of habitat patches variable in space and time, and an insect which may respond with either immediate reproduction, migration, or diapause. In order to develop an optimal strategy this insect needs information about environmental variability in *both* dimensions. Information about future changes "here" alone, or instantaneous differences between "here" and "elsewhere" alone are not sufficient but must be combined. Figure 2 illustrates the importance of considering habitat changes simultaneously in both space and time. In A and B

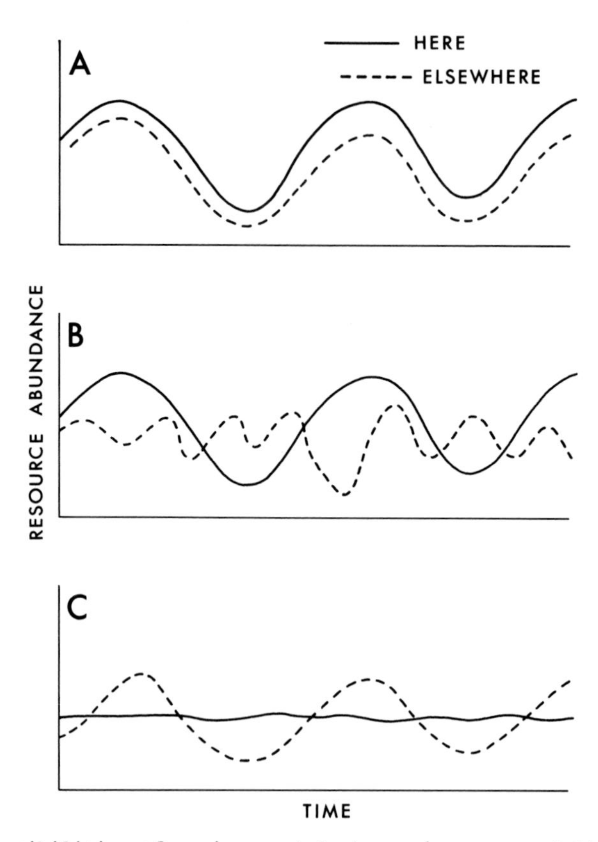

Fig. 2. Three possibilities of environmental change in space and time. (A) may represent the seasonal habitat changes for a diapausing insect, (B) the short-term variations in food resources for a migrant aphid, for example, and (C) the conditions for a long-range migrant with a source area allowing continuous reproduction and distant habitats being exploited during summer.

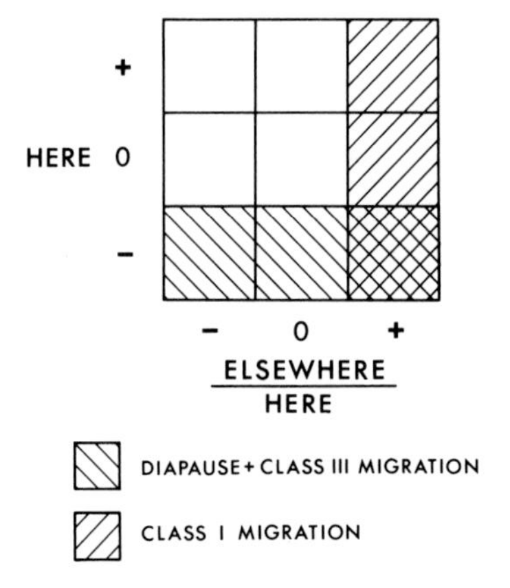

Fig. 3. The relation between habitat change in space and time and possibilities of evolving migration and diapause. (+) increase, (0) no change, and (-) decrease in resource abundance. Class I migrations are those soon followed by reproduction, and Class III migrations are those undertaken to and from diapause sites.

habitat changes "here" are identical but the differences in changes "elsewhere" will tend to promote diapause in A and migration in B. In C the insect inhabits an almost constant environment, but changes in conditions "elsewhere" provide opportunities for the evolution of migration. This may represent the conditions for certain long-range migrants with a source area allowing continuous reproduction and distant habitats being exploited during favorable seasons.

In order to evaluate the more exact conditions promoting migration and diapause one must, of course, consider more detailed information about environmental conditions like differences in resource abundance, duration of resource cycles, distances between habitat patches, etc. Alternative responses of the organism, like increased body size, longer life, and improved homeostatic mechanisms must also be considered. Nevertheless, this simple reasoning focuses the attention on two basic questions. First, how will conditions change "here"? And, secondly, how will conditions "elsewhere" change *in relation to* those "here"? Both these questions can be answered qualitatively in the form: improving (+), no change (0), and deteriorating (-). Possibilities for the evolution of migration and diapause only exist under certain combinations of environmental conditions (Fig. 3). Diapause is favored (although not a necessary consequence) when conditions "here" deteriorate and changes "elsewhere" are approximately in phase, while migration is favored by conditions improving "elsewhere" *relative* to "here." Migration could thus be favored even though conditions "here" improve.

Neacoryphus bicrucis - A Seed Bug

Seed bugs (Heteroptera, Lygaeidae) as their name implies feed mainly on seeds, and many species are easy to rear in the laboratory. The usually large size and bright colors of most members of the subfamily Lygaeinae make them easy to observe in the field. As seed feeders lygaeids are dependent upon food resources that are often highly variable in space and time (cf, Janzen, 1971), making them interesting objects for studies on migration and diapause (Dingle, 1974b; this volume).

Neacoryphus (Lygaeus) bicrucis (Say) (Heteroptera, Lygaeidae, Lygaeinae) has a wide geographic range, occurring from Brazil to southern Canada (Slater, 1964). It hibernates in the adult stage, and unlike its relative the large milkweed bug *Oncopeltus fasciatus*, it seems able to hibernate also along the northern border of its range. In summer the bug is found in weedy fields, along roadsides and similar habitats of early successional stages, but also in more permanent habitats like sand prairie. It feeds on flowers, specifically on the green ovulae or developing seeds, of a variety of plant species, possibly with a preference for composites. Seeds on the ground are probably also important food since the bug is easily reared on dry dandelion or sunflower seeds. The bug seems to have at least two generations in Iowa.

Glick (1939, 1960) collected several specimens of *N. bicrucis* high up in the air, one even at 7000 ft. Catches in beach washup and at light traps further point to a species with strong powers of dispersal by flight. Flights occur both by day and night. A fuller treatment of the field biology of the bug is given by Solbreck and Bornfeldt (in preparation).

N. bicrucis thus is dependent upon a food resource (immature and mature seeds) which is evidently highly variable in space and time, and it regularly engages in dispersal flights. Laboratory experiments have shown that the bug migrates when starved, but histolyzes its wing muscles and oviposits very soon after adult ecdysis when given ample food (Solbreck and Bornfeldt, in preparation). *N. bicrucis* has a fairly high intrinsic rate of increase (*r*) (Table 2) higher than the rates recorded for other Lygaeinae (Dingle, 1968b; Landahl and Root, 1969), which are usually regarded as fairly *r*-selected species. *N. bicrucis* thus could be characterized as a rather typical *r*-strategist.

Under conditions of *shortening* photoperiods *N. bicrucis* will enter a reproductive diapause (Solbreck, in preparation). Thus each adult female of *N. bicrucis* seems to have three alternative lines of development. It can (1) breed soon after adult eclosion without migrating, (2) migrate before reproduction, or (3) enter diapause resulting in a long reproductive delay (Fig. 1). In other words *N. bicrucis* is a facultative migrant and it has a facultative diapause.

N. bicrucis was reared on peeled sunflower seeds (unless otherwise stated) and distilled water from a cotton wick. Larvae were kept in plastic boxes 6 X 7 X 10 cm with a mesh lid. Cultures were started with 25 larvae. Adults were collected on the day of eclosion and maintained as pairs in petri dishes (diameter 9 cm, height 2.5-3 cm). Data on survival, development time, and egg production were collected daily, except for the later part of the egg-laying period when sometimes three-day intervals were used and egg batches were aged according to egg coloration. Flight tests were performed at 25-26° C and an illumination of 5000 lux. The bugs were glued to an applicator stick and flight started by swinging them in the air. The total time of five consecutive flights was recorded. Further details of rearing and flight testing are given by Solbreck and Bornfeldt (in preparation).

Evaluating Life-History Data

Life-history strategies should ultimately be evaluated against particular environmental conditions. Data on the natural environments of insects are, however, scarce. Environmental measures when available are usually crude, and most often they exist in the form of indirect indices. Measures of spatial and temporal variation of the environment - so important for the analysis of migration and diapause strategies - are rarely better than good guesses. The study of *N. bicrucis* is no exception to this, since it has been confined to laboratory conditions. It will be assumed that the migration and diapause strategy of *N. bicrucis* is a problem of resource (= food) tracking, thus regarding predation and parasitism to be unimportant. The problem of resource tracking seems important for most seed-feeding insects. Predator and parasite pressure, however, seems highly variable (Janzen, 1971). *N. bicrucis* and other members of the brightly colored Lygaeinae are regarded as aposematic insects since they contain cardiac glycosides (Scudder and Duffey, 1972), indicating low predation pressures. Nevertheless high rates of parasitism have sometimes been recorded, e.g., on eggs of two tropical *Oncopeltus* species by a scelionid wasp (Root and Chaplin, 1976) and by tachinid flies on *Lygaeus kalmii* (Dingle, personal communication).

A specific problem in dealing with life-history strategies is that many traits like birth and death rates can only be measured on populations. Consequently one has to approximate population averages as measures of individuals (cf, Stearns, 1976). This has also been done throughout this study.

Table 1. Symbols and formulae used

Symbol	Definition
l_x	Probability of surviving to age x.
m_x	Number of daughters per female in the age interval $x \pm 0.5$.
α	Age at first reproduction.
$R_0 = \sum_{x=\alpha}^{\infty} x l_x m_x$	Net reproductive rate.
r	Intrinsic rate of increase. Solved by trial and error substitution in the equation $1 = \sum_{x=\alpha}^{\infty} l_x m_x e^{-rx}$
$T = \frac{\ln R_0}{r}$	Mean generation time.
$r_c = \frac{\ln R_0}{T_c}$	Capacity for increase. $T_c = \frac{1}{R_0} \sum_{x=\alpha}^{\infty} x l_x m_x$

An often-overlooked problem when using life-history data is the risk of large systematic errors. Life-history traits like survival and egg production are often very sensitive to small variations in experimental conditions. Thus the particular food provided for laboratory-reared insects will greatly affect the life table data obtained, as exemplified for the lygaeids *Nysius vinitor* (Kehat and Wyndham, 1972b), *Lygaeus equestris* (Kugelberg, 1973a,b), and *Oncopeltus* spp (Root and Chaplin, 1976).

Even between-year variation in the quality of one kind of seed (*Cynanchum vincetoxicum*) greatly affected development time and adult weight in *Lygaeus equestris* (Kugelberg, 1977). Perhaps even more dramatic are the effects of temperature (e.g., Kehat and Wyndham, 1972a). In *N. bicrucis* a 4°C temperature increase from 23°C causes a doubling of r (cf Table 1 for definitions of symbols), while a corresponding temperature decrease results in a negative value of r. This is the result of dramatic changes in survival, timing of oviposition, and the number of eggs laid (Table 2, Fig. 4).

Another type of problem concerns the assessment of fitness characters and the definition of optimal environmental conditions. Different life-history traits often have different optima. In *N. bicrucis* survival is highest at 23°C, egg production and R_0 at 27°C, r at 31°C, and generation time reaches a minimum at 35°C (Table 2, Fig. 4.). A further illustration of this complexity is added by looking at the effects of varying food quality and quantity (Fig. 8). Food conditions have strong effects on egg production while maternal survival is much less affected.

Thus effects of environmental conditions on life-history traits are often profound and complex. When comparing different species we ought to proceed carefully by surveying a fairly large spectrum of environmental conditions, and by looking at the responses of different life-history traits. This should help us to avoid possible large systematic

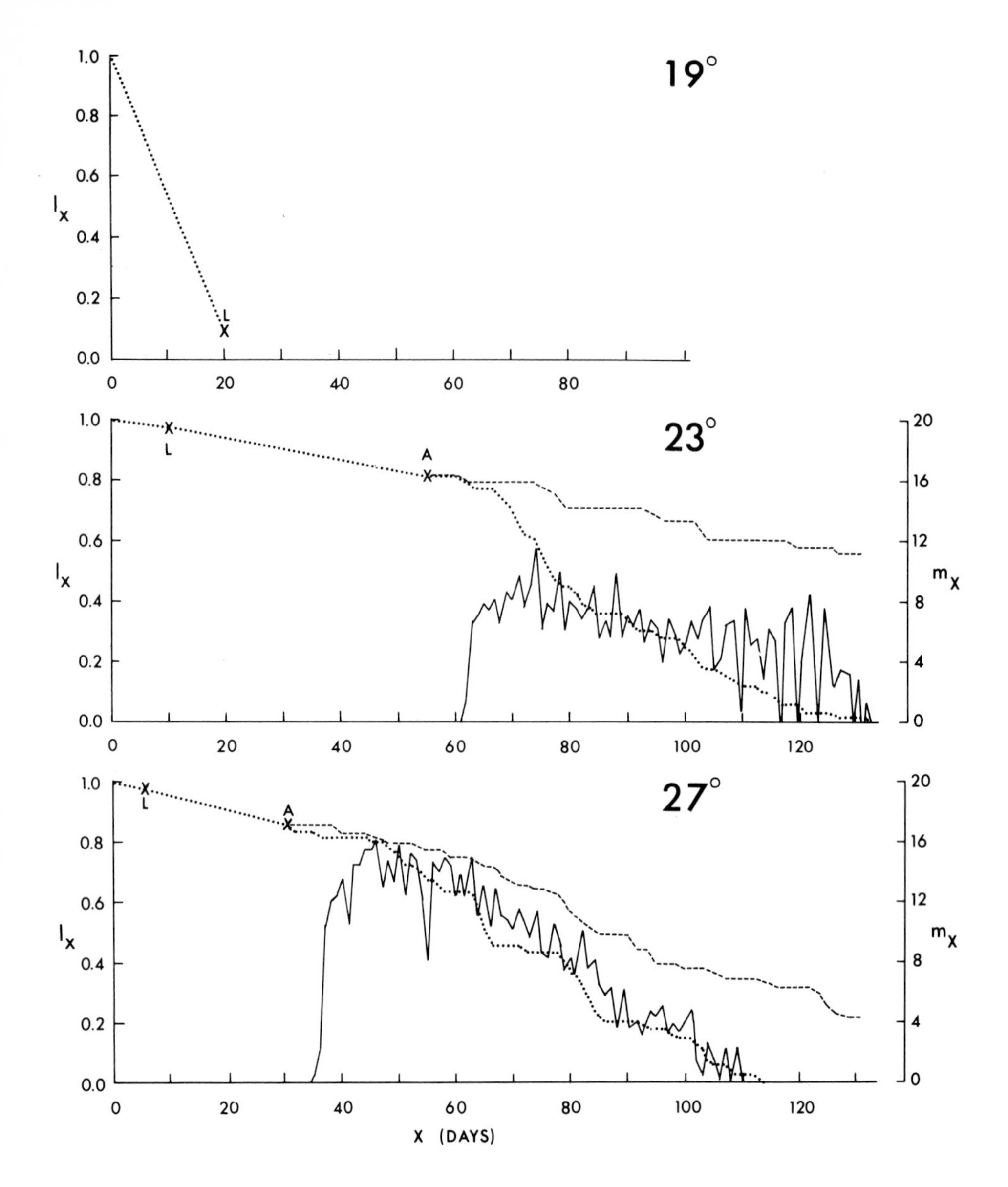

errors in our data collecting. In particular when natural feeding habits are poorly known - which is commonly the case - we should proceed with extra care when comparing different species.

Demographic Consequences of Diapause

Diapause has several important life-history consequences. One is the frequent synchronization of individual development resulting in an even-aged cohort emerging after diapause. This is brought about by two processes. First, individuals which have not yet reached the developmental stage adapted for diapause when conditions deteriorate

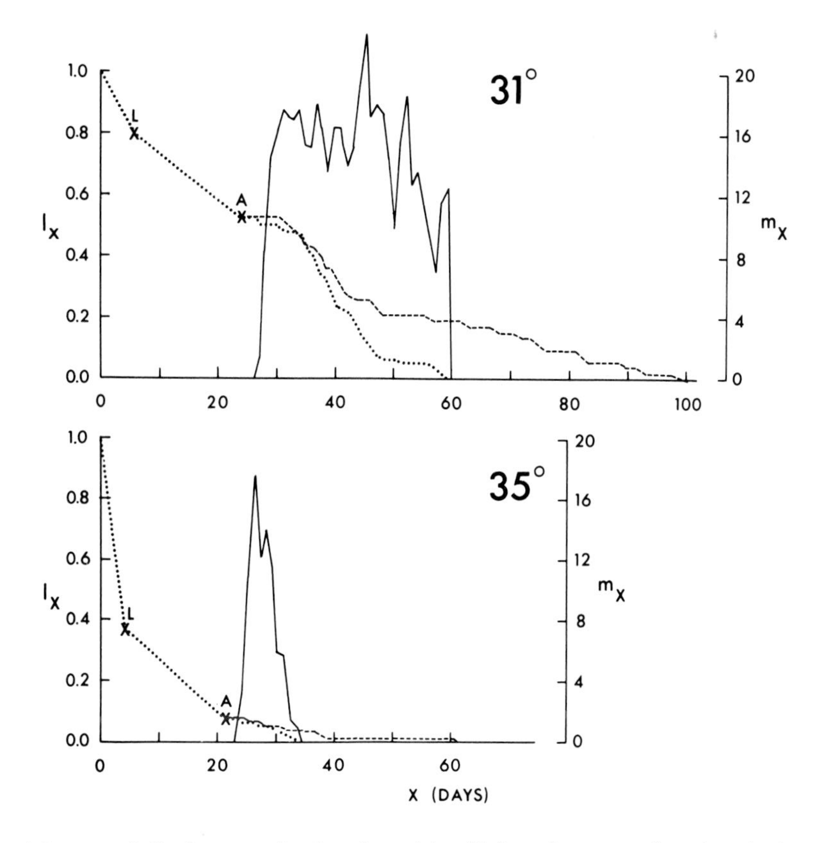

Fig. 4. (above and facing page). Survivorship (l_x) and egg production (m_x) curves for *N. bicrucis* at various temperatures under long-day (16L:8D) conditions and high food abundance (sunflower seeds). *L* denotes mean day for egg hatching and *A* that for adult eclosion. Sex ratio is 1:1 at adult eclosion. Survival of adult females is shown by dotted line and that of adult males by broken line. Continuous line shows m_x curve. For further explanations, see text and Tables 1 and 2.

Table 2. Demographic data for *N. bicrucis* reared at different temperatures at a 16L:8D light regimen and peeled sunflower seeds as food. Sample sizes for eggs and larvae exceed 500 except for eggs at 19°C (N = 161) and larvae at 31°C (N = 275) and 35°C (N = 100). Numbers of adult pairs are 40 at 23°C, 38 at 27°C, 30 at 31°C, and 9 at 35°C. For further explanations see text and Table 1.

Temperature (°C)	R_0	r (day^{-1})	r_c (day^{-1})	T (days)	T_c (days)
19	--	--	--	--	--
23	150.0	0.067	0.063	75.1	79.7
27	388.2	0.123	0.105	48.5	56.6
31	109.6	0.139	0.130	33.9	36.1
35	4.4	0.053	0.053	27.8	27.8

will usually not survive the diapause period. Both in *Lygaeus equestris* (Solbreck, 1976) and *Oncopeltus fasciatus* (Dingle, 1972) only those individuals that reach adulthood before a certain date can survive the winter. Second, maturation processes at the end of the dormancy period often seem to bring about further synchronization of individual development (Tauber and Tauber, 1976). This maturation often seems to be a temperature-dependent process. In autumn the coccinellid *Coleomegilla maculata* flies to hibernation sites along forest edges and similar places, where it spends the winter in diapause. In spring it flies back to breeding areas. The timing of this migration, which also signifies the end of the winter dormancy, is controlled by the long-term experience of temperature above 15°C (Fig. 5) (Solbreck, 1974).

Sometimes diapause may result in the population diverging into cohorts emerging in different seasons as in the moth *Cecropia* (Waldbauer and Sternburg, 1973; Waldbauer, this volume) or in different years as in the Colorado beetle, *Leptinotarsa decemlineata* (Ushatinskaya, 1972) and the moth *Eriogaster lanestris* (Nordström et al., 1941). This is evidently an adaptation to an uncertain environment by spreading out risks (Cohen, 1970). These responses will also lead to the populations appearing in rather distinct cohorts.

If diapause occurs in the adult stage it is directly antagonistic to reproduction. Actually the criterion for adult diapause is usually a reproductive delay. Diapause is thus normally an all-or-none response, either diapause and no reproduction or maximum reproduction and no diapause, although the duration of the reproductive delay may vary between individuals (cf, Dingle, this volume). Sometimes, however, insects that have started to reproduce may switch to diapause. Thus egg-laying pine weevils, *Hylobius abietis* (Eidmann and Klingström, 1976) and carabid beetles (Vlijm and van Dijk, 1967) may stop ovipositing, enter diapause, and then start reproducing the next year again.

An important life-history aspect of diapause is the relationship between maternal survival rate and egg production (Murdoch, 1966). In laboratory experiments with *N. bicrucis* diapausing females have a prereproductive period that is five times longer than that of nondiapausing ones. Survival from adult eclosion to the start of reproduction is very high. Even among diapausing females survival is close to 100% for the entire seven-week prereproductive period. Once reproduction starts, mortality greatly increases and follows the same path in both groups (Fig. 6). It is also evident that as soon as they start reproducing both groups do it at the same maximum rate (Fig. 6), as has also been shown for *Oncopeltus fasciatus* (Dingle, 1974a). Prereproductive mortality will certainly be higher under natural conditions, but this will not change the general picture of a strong trade-off between reproduction and maternal survival. The start of oviposition marks the start of a period with much increased female mortality. This is, however, not the case in males, since there is evidently no difference in survival between diapausing and nondiapausing individuals (Fig. 7).

How does the trade-off between maternal survival and reproduction relate to female reproductive effort, e.g., the number of eggs laid? Different reproductive rates were obtained by rearing *N. bicrucis* (under nondiapause conditions) on different qualities and quantities of seeds. The food regimen has a strong effect on oviposition rate but only little effect on maternal survival, at least during the first three weeks of adult life (Fig. 8). Although significant differences in survival for some groups of older females can be seen (e.g., between 4T and ∞T) there are no conclusive trends in the present material. There is thus no simple relationship with mortality increasing with increasing egg production. At least during early reproductive life

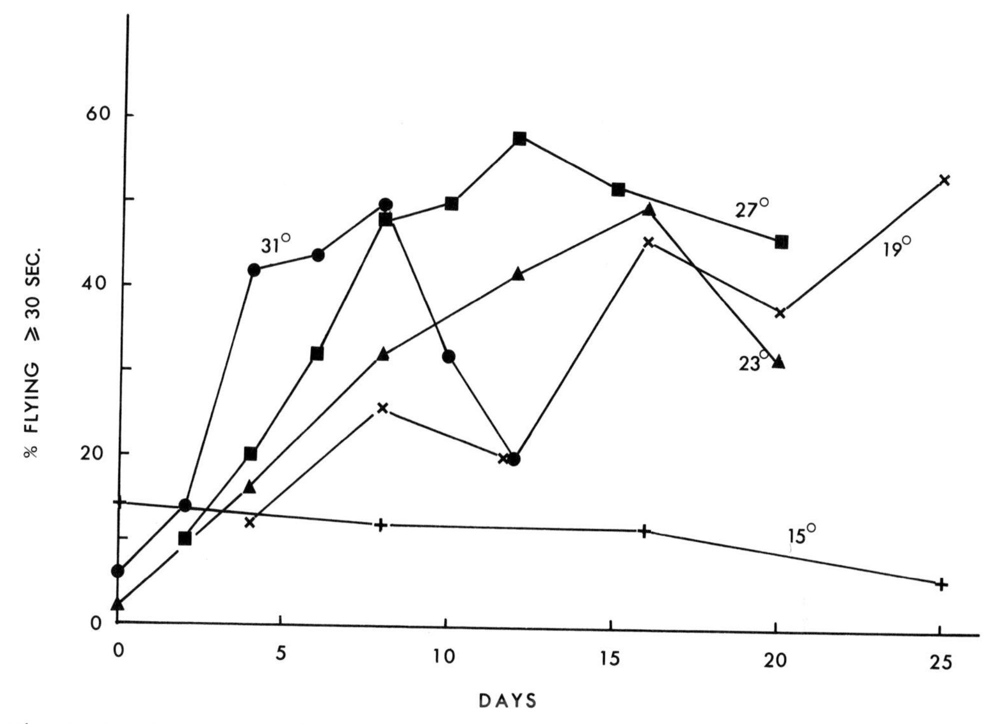

Fig. 5. Development of flight behavior of *Coleomegilla maculata* (Coccinellidae) collected at a hibernation site in late March, approximately one month before the spring migration, and incubated at 16L:8D and five different temperatures (from Solbreck, 1974).

maternal survival seems to be mainly a function of the time spent in a reproductive state, regardless of the number of eggs laid.

The close connection between diapause and voltinism also leads to important life-history consequences. Thus small changes in climatic conditions may lead to abrupt changes in the number of generations per year (Cohen, 1970; Masaki, this volume). Such changes in voltinism may be coupled to changes in body size and development time as shown for the cricket *Pteronemobius fascipes* by Masaki (1973), thus strongly affecting demographic traits (cf also Masaki, this volume).

Three conclusions emerge from this. Diapause often has the effect of synchronizing population age distribution and reproduction so as to occur in (one or more) even-aged cohorts. This indicates that diapause strategies should preferably be analyzed using discrete generation models. Secondly, there is often an important trade-off between maternal survival and reproduction, the exact nature of which should be investigated in more species. Thirdly, life-history traits like body size and generation time may sometimes be regarded as adaptations for ensuring proper seasonal synchronization.

Demographic Consequences of Migration

It is often claimed that migration is a characteristic of so-called *r*-strategists (Dingle, 1974a; Pianka, 1974; Southwood, 1976). This

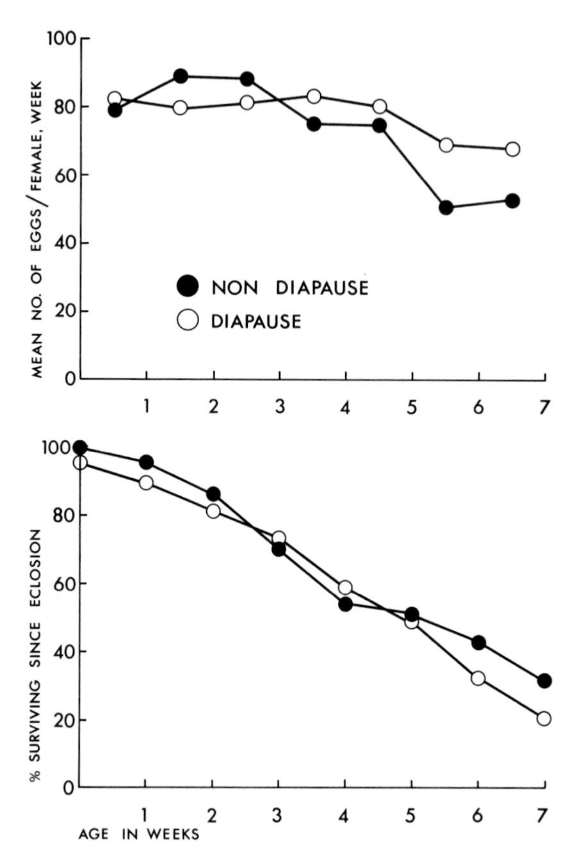

Fig. 6. Egg production and survival since adult eclosion of diapause and nondiapause females of *N. bicrucis*. All individuals were aged according to the day of first oviposition (α). Larvae were transfered between third and fifth instar from long-day (16L:8D) to short-day (12L:12D) conditions. The bugs were provided a constant ample supply of sunflower seeds at 23°C. Adults were reared as pairs. Those ovipositing 26 days posteclosion or later were classified as diapausing. Mean α for nondiapause individuals was 10.4 days and for diapause individuals 49.5 days posteclosion. Sample size at the start was 46 for diapause and 42 for nondiapause individuals.

easily leads to the assumption that migrants are selected for maximizing r, and thus to the view that migration should be analyzed by using models maximizing r. There is an obvious paradox in this reasoning (cf, Kennedy, 1975) since what is typical of migration is a reproductive delay often called the oogenesis flight syndrome (Johnson, 1969). Thus migrants in wing polymorphic water striders of the genus *Gerris* (Vepsäläinen, this volume) and leafhoppers (Waloff, 1973) oviposit later than wingless individuals. Mochida (1973) also found a decrease in egg production among the winged forms of the leafhopper *Javesella pellucida*.

Since $r \approx \ln R_0/T$ and since migration often seems to result in an increase in T and a decrease in R_0 (e.g., by affecting egg production), migrants should be characterized by a lowered r. But how much is r decreased by allowing a period of migration before reproduction?

The early studies on *N. bicrucis* had shown it to be a facultative migrant, either being unable to fly when reproduction could be maximized or

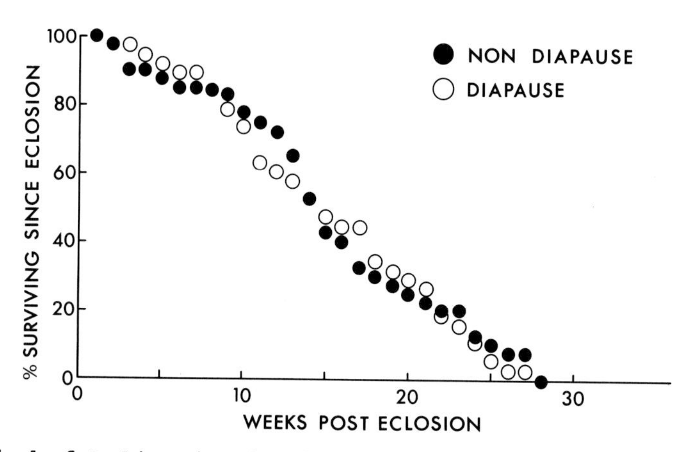

Fig. 7. Survival of *N. bicrucis* males from the day of adult eclosion. Males were classified according to the response of their female partner. For further explanations, see Figure 6.

flying when conditions for egg production were bad. At 27° C and a 16L:8D light regimen, *N. bicrucis* females exhibit very little flight if given abundant food. Their flight muscles are histolyzed and oviposition starts within 4-6 days postadult eclosion. But if they are starved (or unmated) they fly, reaching maximum flight capacity in about 4 days. The bugs survive several weeks of total starvation during which they also retain their flight capacity (Solbreck and Bornfeldt, in preparation).

On the basis of its natural history and demographic characteristics (Table 2) it was found above that *N. bicrucis* seemed to be a fairly typical r-strategist. This was the basis for studying the effects of migration on demographic traits in this bug, assuming r to be a fitness character.

It could be expected that bugs in the field would fly if subjected to approximately 4 to 6 days of starvation after adult eclosion (under otherwise comparable conditions). Accordingly females were starved for 4 or 6 days, flight-tested, and then given abundant food, thus simulating migration and successful colonization. These females responded by displacing their m_x curves for periods approximating the starvation periods. The form and area under these curves were not much affected by this experience (Fig. 9). Judging from Figure 8 maternal survival is not much affected by such a short starvation period. Assuming optimal conditions for migration, i.e., that all migrants are able to colonize optimal habitats, there would only be small effects of migration on R_0, and particularly on ln R_0. Migration will, however, affect mean generation time (T) by delaying the onset of oviposition. It seems a reasonable estimate that the bugs are able to migrate and colonize within 4-6 days after eclosion (cf Fig. 9). This would cause an increase in generation time of approximately 3 to 6 days as estimated from Figure 9. Since T at 27° C is 48.5 days (Table 2) this represents an increase of roughly 6-12% of T. This would mean that r decreased from 0.123 day^{-1} to 0.116 or 0.109 day^{-1} because of the migratory flight.

Even under most favorable conditions there would thus be a certain cost of migrating as manifested by a lowered r. Assuming that r is a fitness character we would expect migration to occur when r possible

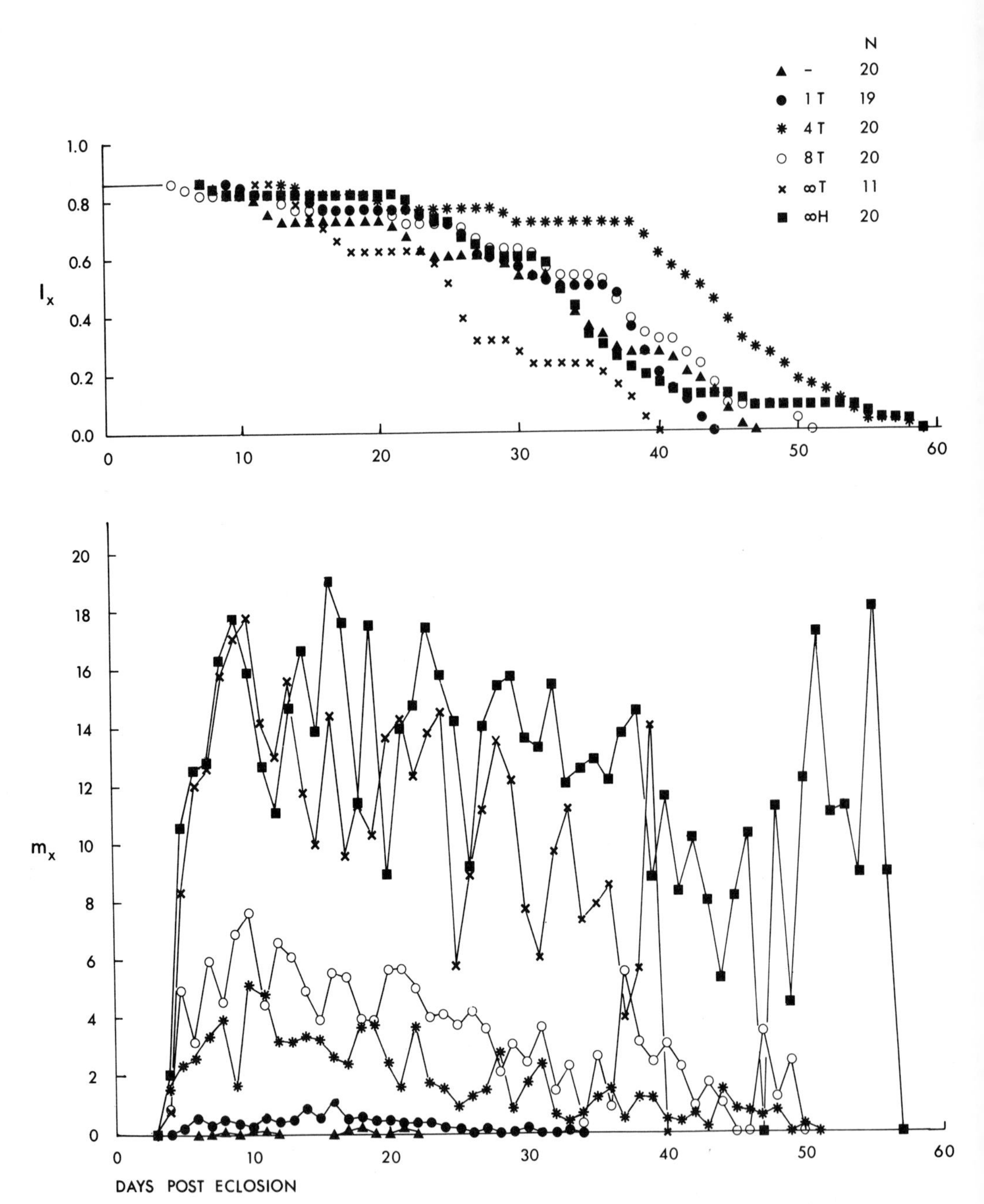

Fig. 8. Survival (l_x) and egg production (m_x) of *N. bicrucis* females provided different food quantities and qualities from adult eclosion onwards. Bugs reared as pairs at 27° C and 16L:8D. - = no food; 1T, 4T, 8T = 1, 4, or 8 dandelion (*Taraxacum*) seeds every second day. ∞T, ∞H = constant superabundance of dandelion or sunflower (*Helianthus*) seeds. *N* shows sample sizes.

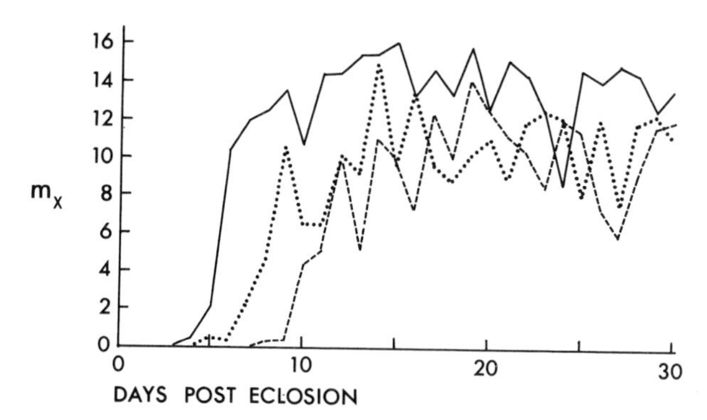

Fig. 9. Effects of short periods of starvation on egg production (m_x) of *N. bicrucis* females at 27°C and 16L:8D. ——— = constant supply of sunflower seeds, ······ = starved 4 days following adult eclosion, ----- = starved 6 days following adult eclosion. The latter two groups were flight-tested on day 4 and 6, respectively, and then allowed a constant supply of sunflower seeds.

to realize "here and now" (r_{HN}) is less than r that could be realized "elsewhere" (r_E) (cf Fig. 1). If $r_{HN} < r_E$ then $\ln R_{HN}/T < \ln R_E/(T + \Delta T)$ where R_E (net reproductive rate elsewhere) includes the hazards to the mother of the migratory flight as well as her reproductive success "elsewhere," and ΔT is the increase in T caused by migration.

How do *N. bicrucis* females respond when conditions "here and now" gradually deteriorate, thus causing R_{HN} to decrease? It seems reasonable to assume that present food resources are correlated with those in the near future and hence with an expected $R_{HN}/$. It also seems likely that as food resources and expected R_{HN} decreased, a limit will be reached when conditions elsewhere are likely to be better. If this is the case, there will be a threshold for R_{HN} when the bug switches from no migration to migration. An estimate of $\Delta T/T$ = 0.06-0.12, i.e., 6-12% of T, was calculated above. Estimates of R_{HN} at the migratory switch can be obtained by flight-testing bugs reared under conditions resulting in different net reproductive rates. (We thus assume that the insect is adapted to a certain environment and that by studying its responses we may get an idea of this environment.)

By subjecting bugs to different food regimens a spectrum of different net reproductive rates were obtained. The bugs were also flight-tested 10 days posteclosion. When the bug can realize a high net reproductive rate, flight is inhibited; but as R_0 drops to an interval roughly between 150 and 200, the bugs switch to migration. Below this threshold zone the migration rate is constant (Fig. 10).

We thus have estimates of $\Delta T/T$ and of R_{HN} for the migratory switch. The formula $\ln R_{HN}/T = \ln R_E / (T + \Delta T)$ can be rewritten $\ln R_E = \ln R_{HN} (1 + \Delta T/T)$. Using the lower values R_{HN} = 150 and $\Delta T/T$ = 0.06, we arrive at a value of 203 for R_E; and for the upper values R_{HN} = 200 and $\Delta T/T$ = 0.12, R_E becomes 378. If this model is valid it would mean that the insect is adapted to conditions where prospects for successful colonization are on the average very high.

Figure 11 shows the relationship among R_{HN}, R_E, $\Delta T/T$ and the migratory switch. The isoclines connect points where $r_{HN} = r_E$. Points below these curves represent conditions favoring migration. R_E = 388 represents

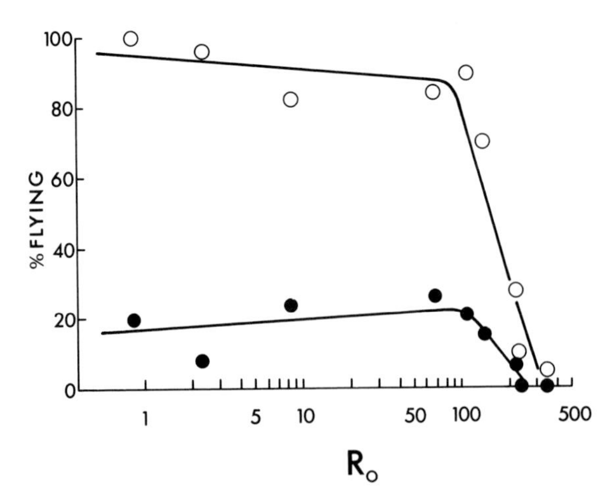

Fig. 10. Relations between net reproductive rate (R_0) and flight response on day 10 posteclosion. ○ = % flying, ● = % flying ⩾ 1 min. Different reproductive rates were obtained by varying the quantity of dandelion or sunflower seeds. Sample sizes are left to right 26, 10, 17, 19, 19, 20, 18, 10, and 19. Data from Solbreck and Bornfeldt (in preparation).

maximum attainable net reproductive rate in *N. bicrucis* (cf Table 2). Supposing that the migratory switch occurs at R_{HN} = 180 and $\Delta T/T$ = 0.1 (which are reasonable values for *N. bicrucis*), a point is defined which intersects the isocline for R_E = 303. We can see that for high values of R_E the migratory threshold (expressed as R_{HN}) is very sensitive to small changes in $\Delta T/T$. In this case, a 10% increase in T because of migration corresponds to 40.6% [(303 - 180)/303 = 0.406] decrease, e.g., in egg production. (As seen in Fig. 8, food shortage mainly affects egg production and not maternal survival in *N. bicrucis*.)

To what extent does *N. bicrucis* follow the model outlined above? Actually it violates some of its implicit assumptions. There is not complete antagonism between migration and reproduction. The bugs can have a fairly high egg production and still retain maximum flight potential (Fig. 9). Under conditions of moderate to absolute food shortage the bug obviously flies *and* oviposits in many places. It was also found that females engaged in reproduction and being unable to fly could sometimes regain their flight capacity later in life if subjected to prolonged starvation (Solbreck and Bornfeldt, in preparation). We thus have to reconsider what the alternatives for *N. bicrucis* really are. First, there is not just one point of decision about migrating or not. Corrections can be made, e.g., by switching to migration after a period of reproduction. Second, the migratory alternative does not mean a total suppression of reproduction, but rather a life history with both flight and oviposition although at a reduced rate. These life-history characteristics as well as the extended oviposition period, the long-lived males, and the repeated matings are not to be expected of a species maximizing r. They rather suggest a species primarily adapted to spread out risks in time and space (den Boer, 1968). Thus *N. bicrucis* evidently is not selected for maximizing r, making the above-used model an unsatisfactory description of reality.

What has been said above refers to the female strategy. Males fly both when starved and given abundant food, and their survival is obviously not affected by mating (Fig. 7). Their strategy seems to be to mate as often as possible over an extended period and in many places.

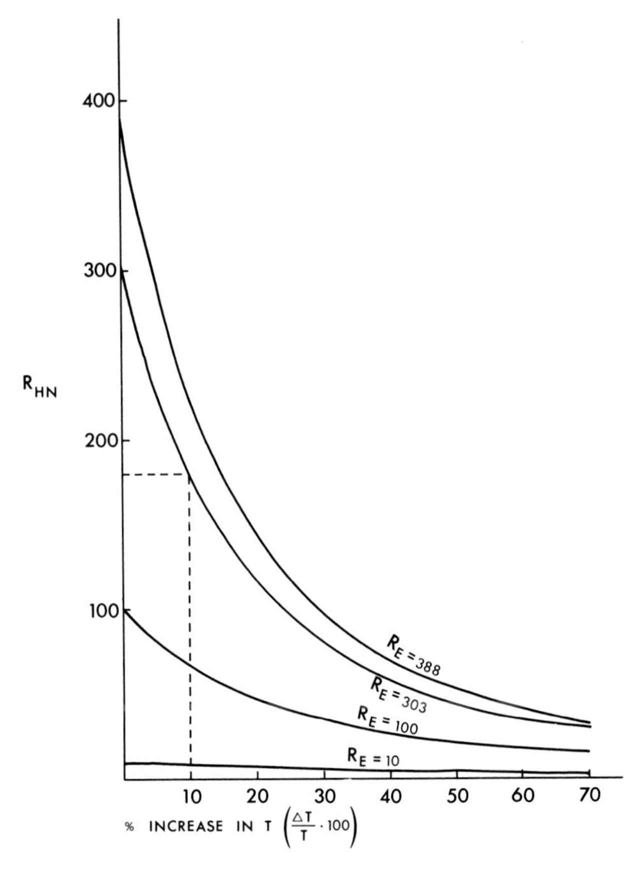

Fig. 11. Relations between net reproductive rates of residents (R_{HN}) and migrants (R_E) in relation to changes in mean generation time (T). The isoclines connect points with equal r:s. For further explanations, see text.

Relations Between Diapause and Migration

N. bicrucis responds to several environmental signals when switching among its three alternative life histories. These responses are arranged in a hierarchical system. When experiencing *shortening* daylengths the insect responds with diapause (Solbreck, unpublished). This response makes the insect insensitive to information about mates and food resources. However, under conditions of constant (and probably also of increasing) photoperiods females respond to lack of mates or food shortage by migrating. Only fed *and* mated females respond with direct reproduction without migration (Solbreck & Bornfeldt, in preparation) (Fig. 12).

There is thus a hierarchy of responses with regard to migration and diapause conditions. Going back to Figures 2 and 3, one would expect the insect to extract information about future changes both in time and space. Photoperiod, often coupled with temperature, usually seems

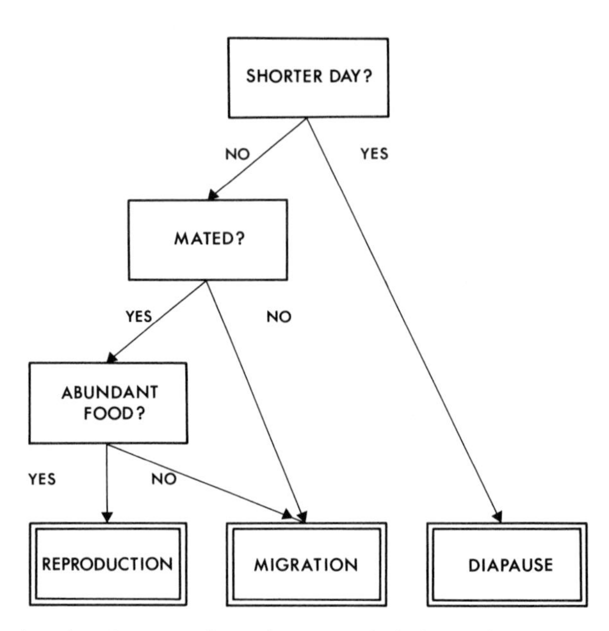

Fig. 12. Organization in the use of environmental information by *N. bicrucis* females. The location of the boxes "mated?" and "abundant food?" are interchangeable. For further explanations, see text.

to be the best kind of information signaling synchronous deterioration over large areas. The proper response in such a situation would be diapause, sometimes accompanied by class III migrations. It is also evident that signals which otherwise lead to migration (class I) should be blocked, hence the hierarchical organization where diapause responses have priority.

According to Figure 3 we should expect migration when conditions "elsewhere" improve relative to "here." (This difference must of course be large enough to compensate for losses during migration, etc.) Information about conditions favoring migration often stems from deteriorating conditions "here" providing there are no photoperiodic or temperature signals about synchrony in this deterioration. Decreasing food quality or quantity, crowding, and lack of mate seem the most appropriate signals for deteriorating conditions "here." These kinds of signals also seem to be the most often used for initiating class I migrations. Thirst, starvation, or low quality of food affect migration in the lygaeid bug *Nysius vinitor* in Australia (Kehat and Wyndham, 1973). Starvation causes migration in otherwise flightless cotton stainer bugs, *Dysdercus fasciatus* (Pyrrhocoridae), and it enhances migration in *D. nigrofasciatus* and *D. superstitiosus* (Dingle and Arora, 1973). Lack of mate has also been shown to affect flight in the latter species (Gatehouse and Hall, 1976). In *D. intermedius* migration is stimulated by starvation, lack of mate and number of conspecifics, but not of photoperiod or temperature (Edwards, 1969). In *Lygaeus kalmii* migration is induced by lack of mate and food shortage, while winter dormancy is induced by lowered temperatures (Caldwell, 1974). Food shortage, crowding, and lack of mate cause moderate delays in the onset of oviposition and enhance flight (obviously class I migration) in *Oncopeltus fasciatus*, but photoperiod and temperature have overriding effects inducing diapause (which is a much longer reproductive delay) and the accompanying long-range migrations (Caldwell, 1974; Dingle, 1968a,b, 1974a,b). In water striders (*Gerris*) photoperiodic conditions affect wing development, but this is linked to diapause induction. During nondiapause conditions long-wingedness and migration are enhanced by high temperatures, crowding, and food shortage (Vepsäläinen, this volume). In the predatory stilt bug *Jalysus spinosus* (Berytidae) diapause and associated migrations are influenced by photoperiod regardless of food conditions, while summer migrations are affected by host plant and food availability (Elsey, 1974).

In cases where the individual insect faces the three alternative life histories, direct development, migration, and diapause, photoperiod and temperature usually seem to inform about diapause conditions, and further have priority over information about the other alternatives. Although cases are known of diapause induced by food conditions alone, this seems to be uncommon (Tauber and Tauber, 1973). Information about the other two alternatives generally seem to come from local conditions concerning food and conspecifics.

Conclusions

Migration and diapause are adaptations for insects exploiting temporary habitats, temporariness being a function of both space and time. Organisms adapted to such environments have been given names like vagrants and opportunists, referring to their abilities for rapid population growth following colonization. What are the properties of insects thus categorized? I would like to dwell on two aspects of their biology. First, tracking variable resources obviously puts a high premium on abilities to extract information about future and distant environments. Can we discern any general trends with regard to the information channels used? Secondly, certain demographic characteristics are favored in such environments. What are these and what are the trade-offs involved?

Insects have evolved abilities to respond to a variety of environmental signals, enabling them to migrate or diapause under appropriate conditions. Insects perceive information from their immediate environment and from larger regions. Both types of information must be weighed against each other before proper "decisions" can be made. We should expect insects in the first place to respond to signals correlated with the synchronous deterioration of resources over large areas. The proper response in this case is diapause, sometimes coupled with migrations to and from diapause sites. Photoperiodic conditions are often well-correlated with such synchronous habitat changes and seem to provide the most commonly used signals for induction of diapause (cf, Cohen, 1967).

Improvements of habitats "elsewhere" in relation to "here" provide opportunities for migration (cf, Gadgil, 1971). Direct information about improving conditions "elsewhere" is rarely available. Decisions about migrating or not thus usually have to rest upon the assessment of deteriorating conditions "here," e.g., from crowding and decreasing food quality and quantity, coupled to information that other habitat patches are *not* synchronously deteriorating. This latter type of information is usually obtainable from photoperiodic and temperature conditions.

What are the demographic characteristics needed by an insect exploiting variable resources? Temporariness implies frequent outbursts in time and space of resource abundance. During such outbursts competition is evidently low, implying that many and small eggs should be an advantage. Temporariness, however, also means frequent catastrophes with high risks that all offspring in one place may die. Such conditions evidently favor a strategy where risks are spread out in space and time. This could be accomplished by an extended egg-laying period and by deposition of eggs in many places as done by *N. bicrucis*. Only when conditions are very favorable does *N. bicrucis* abandon flight. A similar strategy is evidently employed by males who mate repeatedly and who retain their flight capacity for most of their adult life.

How does such a strategy relate to r-selection? High egg production is compatible with r-strategy but the spreading out of eggs in space and time obviously is not. Particularly among insects with high egg-production generation time (T) will greatly affect the value of r (cf Fig. 10). Opportunistic species have efficient means of extracting information about the environment and concomitant abilities to synchronize their breeding in space and time with the onset of favorable conditions. Synchronization per se seems a very important characteristic of opportunistic species. This probably often means reproductive delays, increased T , and, thus, lowered r. Since r is generally so sensitive to small changes in T, such losses in reproductive rates seem difficult to compensate for by increased egg production. The intrinsic rate of increase is a complex variable which has to be measured over an entire generation. Several important steps - like migration and diapause - have to be taken before successful timing and colonization is a fact and high reproductive rates per se might be favored. A further consequence of synchronization in time is that populations will frequently appear in fairly discrete cohorts, making the use of models with continuous population growth (like r of the logistic equation) questionable.

It seems that spreading out the risks in space and time (den Boer, 1968) of total reproductive failure is a common strategy of species inhabiting temporary habitats. This as well as the importance of synchronization per se makes it highly questionable if r generally is a fitness character for opportunistic species. Such a minimization of risk or "bet-hedging" strategy is not a strategy for maximizing r (Stearns, 1976). r is determined as the outcome of selection on several partly independent or even antagonistic life-history traits. Future theories will need to consider more specific selection pressures on more specific traits (Wilbur et al., 1974).

Theories of life-history strategies - in which migration and diapause form important parts - often suffer from poor connection with empirical data. First we have problems in making reasonable assumptions because of scarce data, and second we have few data to test the outcomes of the models. More empirical data are badly needed to improve the realism of both assumptions and predictions in life-history models.

An optimal life-history strategy is the best-available alternative life history in a specific environmental context. Life histories are systems of interacting coadapted traits with the trade-offs among them (Stearns, 1976, p. 17). The complex interplay between more-detailed environmental conditions and insect life histories remains a largely unexplored area for empirical studies. Surprisingly little is known about actual interactions and trade-offs among life-history traits in insects. Phenotypical alternatives seem easily amenable to experimentation (e.g., Dixon, 1976), and will be important for the further exploration of life-history traits as fitness characters. Perhaps the weakest point lies in the measurement of the environments of insects. How do food resources and other environmental factors actually vary? How can they be exploited by insects? Important steps in this area have been taken lately by extending experimental studies to field conditions (e.g., Istock et al., 1976b).

Acknowledgments I am much indebted to Dr. Hugh Dingle for many valuable discussions and for much encouragement. He, Dr. Kari Vepsäläinen, and Dr. Christer Wiklund made valuable comments on the manuscript for which I am very thankful. Thanks are also due to Nigel Blakley, Ingela Bornfeldt, and Birgitta Koski for technical assistance during various parts of this work.

References

Boer, P.J. den: Spreading of risk and stabilization of animal numbers. Acta Biotheor. *18*, 165-194 (1968).

Caldwell, R.L.: A comparison of the migratory strategies of two milkweed bugs. In: Experimental Analysis of Insect Behaviour. Barton Browne, L. (ed.). Berlin-Heidelberg-New York: Springer 1974, pp. 304-316.

Cohen, D.: Optimization of seasonal migratory behavior. Amer. Natur. *101*, 5-17 (1967).

Cohen, D.: A theoretical model for the optimal timing of diapause. Amer. Natur. *104*, 389-400 (1970).

Dingle, H.: The influence of environment and heredity on flight activity in the milkweed bug *Oncopeltus*. J. Exp. Biol. *48*, 175-184 (1968a).

Dingle, H.: Life history and population consequences of density, photoperiod, and temperature in a migrant insect, the milkweed bug *Oncopeltus*. Amer. Natur. *102*, 149-163 (1968b).

Dingle, H.: Migration strategies of insects. Science *175*, 1327-1335 (1972).

Dingle, H.: Diapause in a migrant insect, the milkweed bug *Oncopeltus fasciatus* (Dallas) (Hemiptera: Lygaeidae). Oecologia (Berl.) *17*, 1-10 (1974a).

Dingle, H.: The experimental analysis of migration and life history strategies in insects. In: Experimental Analysis of Insect Behaviour. Barton Browne, L. (ed.). Berlin-Heidelberg_New York: Springer 1974b, pp. 329-342.

Dingle, H., Arora, G.: Experimental studies of migration in bugs of the genus *Dysdercus*. Oecologia (Berl.) *12*, 119-140 (1973).

Dixon, A.F.G.: Reproductive strategies of the alate morphs of the bird cherry-oat aphid *Rhopalosiphum padi* L. J. Anim. Ecol. *45*, 817-830 (1976).

Edwards, F.J.: Environmental control of flight muscle histolysis in the bug *Dysdercus intermedius*. J. Insect Physiol. *15*, 2013-2020 (1969).

Eidmann, H.H., Klingström, A.: Skadegörare i skogen Borås: LTs förlag 1976.

Elsey, K.D.: *Jalysus spinosus*: Effect of age, starvation, host plant, and photoperiod on flight activity. Environ. Entomol. *3*, 653-655 (1974).

Gadgil, M.: Dispersal: population consequences and evolution. Ecology *52*, 253-261 (1971).

Gatehouse, A.H., Hall, M.J.R.: The effect of isolation on flight and on the pre-oviposition period in unmated *Dysdercus superstitiosus*. Physiol. Entomol. *1*, 15-19 (1976).

Glick, P.A.: The distribution of insects, spiders, and mites in the air. U.S. Dept. Agric. Tech. Bull. *673*, 1-150 (1939).

Glick, P.A.: Collecting insects by airplane, with special reference to dispersal of the potato leafhopper. U.S. Dept. Agric. Tech. Bull. *1222*, 1-16 (1960).

Istock, C.A., Zisfein, J., Vavra, K.J.: Ecology and evolution of the pitcher-plant mosquito. 2. The substructure of fitness. Evolution *30*, 535-547 (1976a).

Istock, C.A., Vavra, K.J., Zimmer, H.: Ecology and evolution of the pitcher-plant mosquito. 3. Resource tracking by a natural population. Evolution *30*, 548-557 (1976b).

Janzen, D.H.: Seed predation by animals. Annu. Rev. Ecol. Syst. *2*, 465-492 (1971).

Johnson, C.G.: Migration and Dispersal of Insects by Flight. London: Methuen 1969.

Kehat, M., Wyndham, M.: The influence of temperature on development, longevity, and fecundity in the Rutherglen bug, *Nysius vinitor* (Hemiptera: Lygaeidae). Austral. J. Zool. *20*, 67-78 (1972a).

Kehat, M., Wyndham, M.: The effect of food and water on development, longevity, and fecundity in the Rutherglen bug, *Nysius vinitor* (Hemiptera: Lygaeidae). Austral. J. Zool. *20*, 119-130 (1972b).

Kehat, M., Wyndham, M.: The relation between food, age, and flight in the Rutherglen bug, *Nysius vinitor* (Hemiptera: Lygaeidae). Austral. J. Zool. *21*, 427-434 (1973).

Kennedy, J.S.: A turning point in the study of insect migration. Nature, Lond. *189*, 785-791 (1961).

Kennedy, J.S.: Insect dispersal. In: Insects, Science and Society. Pimentel, D. (ed.). New York-San Francisco-London: Academic Press 1975, pp. 103-119.

Kugelberg, O.: Larval development of *Lygaeus equestris* (Heteroptera, Lygaeidae) on different natural foods. Entomol. Exp. Appl. *16*, 165-177 (1973a).

Kugelberg, O.: Laboratory studies on the effects of different natural foods on the reproductive biology of *Lygaeus equestris* (L.) (Het. Lygaeidae). Entomol. Scand. *4*, 181-190 (1973b).

Kugelberg, O.: Food relations of a seed feeding insect, *Lygaeus equestris* (L.) (Heteroptera, Lygaeidae). PhD thesis, Univ. of Stockholm 1977.

Landahl, J.T., Root, R.B.: Differences in the life tables of tropical and temperate milkweed bugs, genus *Oncopeltus* (Hemiptera: Lygaeidae). Ecology *50*, 734-737 (1969).

Masaki, S.: Climatic adaptation and photoperiodic response in the band-legged ground cricket. Evolution *26*, 587-600 (1973).

Mochida, O.: The characters of the two wing-forms of *Javesella pellucida* (F.) (Homotera: Delphacidae), with special reference to reproduction. Trans. R. Entomol. Soc. Lond. *125*, 177-225 (1973).

Murdoch, W.W.: Population stability and life history phenomena. Amer. Natur. *100*, 5-11 (1966).

Nordström, F., Wahlgren, E., Tullgren, A.: Svenska fjärilar. Stockholm: Nordisk Familjeboks Förlag 1941.

Orians, G.H.: Diversity, stability and maturity in natural ecosystems. In: Unifying Concepts in Ecology. van Dobben, W.H., and Lowe-McConnell, R.H. (eds.). den Hague: Junk 1975.

Pianka, E.R.: Evolutionary Ecology. New York-Evanston-San Francisco-London: Harper & Row 1974.

Price, P.W.: Reproductive strategies in parasitoid wasps. Amer. Natur. *107*, 684-693 (1973).

Root, R.B., Chaplin, S.J.: The life-styles of tropical milkweed bugs, *Oncopeltus* (Hemiptera: Lygaeidae) utilizing the same hosts. Ecology *57*, 132-140 (1976).

Scudder, G.G.E., Duffey, S.S.: Cardiac glycosides in the Lygaeinae (Hemiptera: Lygaeidae). Can. J. Zool. *50*, 35-42 (1972).

Slater, J.A.: A catalogue of the Lygaeidae of the world. Storrs, Conn.: University of Connecticut (1964).

Slobodkin, L.B.: Toward a predictive theory of evolution. In: Population Biology and Evolution. Lewontin, R.C. (ed.). Syracuse, N.Y.: Syracuse University Press 1968, pp. 187-205.

Solbreck, C.: Maturation of post-hibernation flight behaviour in the coccinellid *Coleomegilla maculata* (DeGeer). Oecologia (Berl.) *17*, 265-275 (1974).

Solbreck, C.: Flight patterns of *Lygaeus equestris* (Heteroptera) in spring and autumn with special reference to the influence of weather. Oikos *27*, 134-143 (1976).

Solbreck, C., Bornfeldt, I.: Relations between environment, age, and flight in a seed bug, *Neacoryphus bicrucis* (Say) (Heteroptera, Lygaeidae). (In preparation.)

Southwood, T.R.E.: Bionomic strategies and population parameters. In: Theoretical Ecology. Principles and Applications. May, R.M. (ed.). Oxford-London-Edinburgh-Melbourne: Blackwell 1976, pp. 26-48.

Stearns, S.C.: Life-history tactics: A review of the ideas. Quart. Rev. Biol. *51*, 3-47 (1976).

Tauber, M.J., Tauber, C.A.: Insect phenology: Criteria for analyzing dormancy and for forecasting postdiapause development and reproduction in the field. Search (Agriculture), Cornell Univ. Agr. Exp. Sta., Ithaca. *3:12*, 1-16 (1973).

Tauber, M.J., Tauber, C.A.: Insect seasonality: Diapause maintenance, termination, and postdiapause development. Annu. Rev. Entomol. *21*, 81-107 (1976).

Ushatinskaya, R.S.: Perennial diapause of the Colorado potato beetle (*Leptinotarsa decemlineata* Say) and factors of its induction. Proc. K. Ned. Akad. Wet. Ser. C. *75*, 144-164 (1972).

Vlijm, L., van Dijk, T.S.: Ecological studies on carabid beetles. II. General pattern of population structure in *Calathus melanocephalus* (Linn.) at Schiermonnikoog. Z. Morph. Ökol. Tiere *58*, 396-404 (1967).

Waldbauer, G.P., Sternburg, J.G.: Polymorphic termination of diapause by *Cecropia*: Genetic and geographical aspects. Biol. Bull. *145*, 627-641 (1973).

Waloff, N.: Dispersal by flight of leafhoppers (Auchenorrhyncha: Homoptera). J. Appl. Ecol. *10*, 705-730 (1973).

Wilbur, H.M., Tinkle, D.W., Collins, J.P.: Environmental certainty, trophic level, and resource availability in life history evolution. Amer. Natur. *108*, 805-817 (1974).

Wing Dimorphism and Diapause in *Gerris*: Determination and Adaptive Significance

KARI VEPSÄLÄINEN

Introduction

Water-striders of the genus *Gerris* Fabr. consist of a small group of predatory bugs living on the water surface. In temperate climates the habitats often freeze in the winter, and gerrids overwinter on dry land as imagos. The morphology of the species restricts them to habitats protected from wave action. Many of such sites dry up temporarily, but permanent sites also exist.

The mosaic environment of water-striders includes both highly predictable (permanent waters; seasonal cycles) and unpredictable features (temporary waters). The extensive variation in space and time of water-strider environments has facilitated the evolution of a whole array of adaptive strategies in the genus *Gerris*. The strategies are variations on the central theme of wing dimorphism. Seasonal variation of wing length and diapause behavior are intimately connected.

Wing length variation in *Gerris* is extensive - the pattern varies with time, region, and species. In the following I shall describe the wing length variation of ten European *Gerris* species, and also explain the mechanisms that determine wing length and the adaptive significance of the various wing length strategies. The determination and role of diapause are also elucidated.

Polymorphism Vs. Polyphenism

In my earlier papers (reviewed in Vepsäläinen, 1974e) I have followed the definition of polymorphism by Ford (1965). He defined genetic polymorphism in a way which regrettably does not exclude environmentally switched polymorphism. In water-striders the application of such a definition cannot but confuse (Vepsäläinen, 1971b). Therefore I have replaced Ford's (1965) genetic polymorphism by *polymorphism* (Vepsäläinen, 1971a), which I define here as the occurrence together in the same population of two *(dimorphism)* or more discontinuous phenotypes (called morphs) belonging to the same stage in the life cycle of a species, in such proportions that the frequency of the rarest of them cannot be maintained merely by recurrent mutation. To make the term operational I specify that the frequency of the rarest phenotype must be at least 1% to allow it the status of a morph.

As the evolutionary constraints on polymorphism will differ depending on the mechanism of determination of the phenotypes (e.g., Vepsäläinen, 1974c; Shapiro, 1976), I define *genetic polymorphism* as a special case of polymorphism, in which the morphs are genetically differentiated. To indicate polymorphism in which the morphs arise by an environmental switch mechanism, I use *polyphenism* (cf. Mayr, 1963). In *seasonal polymorphism*, polymorphism is restricted to part of the year; the mechanism of *seasonal polyphenism* is implicit here (cf. Shapiro, 1976). Thus genetic polymorphism and polyphenism are subsets of the upper-level definition of polymorphism, and each case classified as polymorphism can be characterized by either one of the lower-level terms if the mechanism of morph determination is known.

Distribution and Voltinism

Most detailed data are available from Finland and Hungary, and this summary is based on these (Vepsäläinen, 1973a, 1974a,c). *Gerris odontogaster* and *G. lacustris* are the commonest and most frequent European water-strider species. The former is distributed to the Arctic Ocean in the north, and the northern border of *G. lacustris* is approximately at the Arctic Circle. Northernmost populations are always univoltine, and the border line between univoltine and partially bivoltine populations runs through Central Finland at about 63°N lat. The nondiapause proportion which produces the partial second generation is very small in Finnish *G. lacustris* populations, and does not exist at all in most populations and years. Central European populations are usually totally bivoltine, and in Hungary *G. odontogaster* may produce a partial third generation.

The voltinism of *G. argentatus* and *G. paludum* is similar to that of the above species pair, and the northern limit of partial bivoltinism is also the northern limit of distribution. *G. rufoscutellatus* is distributed to the Arctic Circle in the north, and is virtually univoltine in Finland; a minute partial second generation may occasionally exist. In Hungary the species is rare, and Mitis (1937) claimed univoltinism in Austria, but this should be checked. *G. najas* is a univoltine, rare species in Finland, as it seems to be in most parts of Europe. *G. lateralis* is commonest in northern and Central Finland, and is replaced by *G. asper* in Hungary. *G. lateralis* is usually univoltine in Finland, but partial voltinism, even if not known, may occur locally in Central Finland.

The only generation of univoltine populations is called the *winter* or *diapause generation,* as is the second generation of bivoltine populations. In partially bivoltine populations the overwintering group consists of the latter part of the first generation and all imagos of the partial second generation. That part of the first generation which does not overwinter but reproduces without diapause is called the *midsummer* or *nondiapause generation,* i.e., in completely bivoltine populations this is the first generation (all imagos).

Wing Length Patterns

Wing lengths of populations were studied in connection with the life cycles. Thus the age of individuals, and the state of flight muscles and ovaries, among others, were inspected in many species and regions (Andersen, 1973; Vepsäläinen 1974a,c; Vepsäläinen and Krajewski, 1974; Vepsäläinen and Nieser, 1977). Sometimes population studies were undertaken (e.g., on *G. odontogaster;* Vepsäläinen, 1971a). In some species the reproductive stage of females and the generation to which they belong can be asserted with high confidence merely by noting the coloration of the venter (Table 1; see further: Andersen 1973; Vepsäläinen, 1974a). The following summary of wing length patterns is taken mainly from Vepsäläinen (1974a,c).

Table 1. Color pattern of the completely pigmented female venter of various *Gerris* species in individuals which reproduce without preceding diapause (nondiapause) and those which overwinter before reproduction (diapause). As a rule, males have dark venter. ♀ = females; + = part of the females (not obligate); - = not any female; ? = usually as in *G. argentatus* but deviations may occur

	Thoracic venter		Sternites	
	(partially) light	dark	(partially) light	dark
G. odontogaster				
nondiapause	+	+	♀	-
diapause	-	♀	-	♀
G. argentatus				
nondiapause	+	+	♀	-
diapause	-	♀	-	♀
G. paludum				
nondiapause	+	?	+	?
diapause	-	♀	-	♀
G. lacustris				
nondiapause	♀	-	♀	-
diapause	-	♀	+	+
G. thoracicus				
nondiapause	-	♀	-	♀
diapause	-	♀	-	♀
G. rufoscutellatus				
diapause	-	♀	-	♀
G. lateralis				
diapause	-	♀	-	♀
G. sphagnetorum				
diapause	-	♀	-	♀
G. najas				
diapause	-	♀	-	♀

There is considerable diversity among the different species with respect to wing length variation (Table 2). With a few exceptional populations, three of the species are monomorphic with respect to wing length: *G. rufoscutellatus* is longwinged, and *G. najas* and *G. sphagnetorum* are usually wingless. Four species, *G. thoracicus*, *G. odontogaster*, *G. argentatus*, and *G. paludum*, have seasonally dimorphic (i.e., seasonally polyphenic) populations. Their overwintering population is longwinged, but the midsummer nondiapause generation (i.e., the midsummer offspring which reproduce without a preceding diapause and which do not overwinter) is dimorphic or shortwinged. Moreover, certain multivoltine populations are always longwinged, such as *G. thoracicus* in Hungary. Finally, most *G. lacustris*, *G. lateralis*, and *G. asper* populations are permanently dimorphic. In these species dimorphism is not restricted to midsummer but is extended as well into the overwintering generation; univoltine populations are also dimorphic. In addition, pure shortwinged or wingless populations exist in these species. The picture given in Table 2 is not re-

Table 2. Wing-length patterns of the *Gerris* Fabr. species studied in southern Finland (about 60° N lat.) and Hungary (about 47° N lat.). *LW* = long-winged, *SW* = short-winged, ? = not known, - = does not exist

Wing-Length Pattern	Winter = Diapause Generation (Univoltine Populations)	Summer = Nondiapause Generation of Multivoltine Populations		Additional Notes
		Finland	Hungary	
I. *LW* monomorphism				
G. rufoscutellatus Lt.	*LW*	*LW*	*LW*	*SW* known
II. Seasonal dimorphism (seasonal polyphenism)				
G. thoracicus Schumm.	*LW*	*SW*	*LW*	Type I in Hungary
G. odontogaster (Zett.)	*LW*	*SW*	Dimorphic	
G. argentatus Schumm.	*LW*	*SW*	Dimorphic	
G. paludum Fabr.	*LW*	*SW*	Dimorphic	
III. Permanent dimorphism (genetic polymorphism)				
G. lacustris (L.)	Dimorphic	*SW*	Dimorphic	Partially type II
G. lateralis Schumm.	Dimorphic	?	-	
G. asper Fieb.	Dimorphic	-	?	
IV. *SW* monomorphism				
G. najas DeG.	*SW*	-	?	Dimorphic in Poland
G. sphagnetorum Gaun.	*SW*	?	-	*LW* known

stricted to Finland and Hungary, but can be extended over all of Europe (Vepsäläinen, 1974c, pp. 32-34).

In the following I give an example of a most diversified system of wing lengths in water-striders. Figure 1 gives the seasonal changes of different female forms of *G. lacustris* in a metapopulation in the River Grabia and its inundation area, Poland (from Vepsäläinen and Krajewski, 1974). In spring overwintered imagos colonize the waters; shortwinged females (b in the figure) have been born at the site, but part of the longwinged ones (a) may have flown from elsewhere. The indirect flight muscles of most longwinged females are histolyzed after the commencement of egg-laying, which is extended over several weeks. The population numbers decrease as a function of time, and the number of developing larvae increases; the major part of the first-generation larvae have reached the fourth instar before summer solstice. Hence (see the section on wing length and diapause determination) the major part of the larvae develops to nondiapausing imagos (c-f). In some of the newly emerged

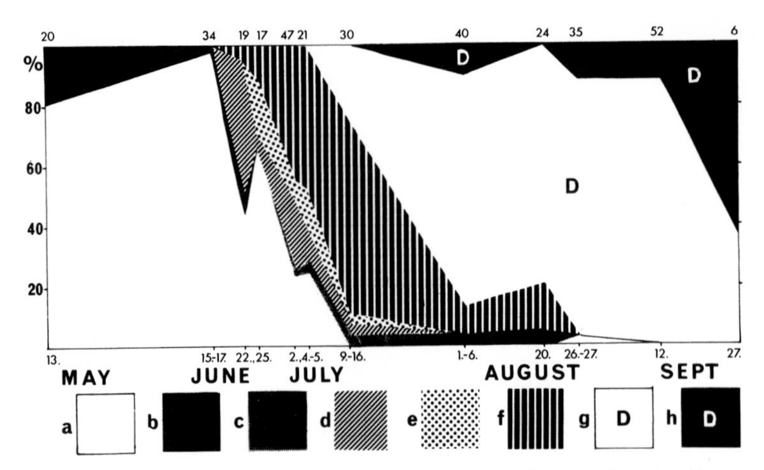

Fig. 1. Seasonal changes in the proportions of the different female forms of *G. lacustris* in the River Grabia and its inundation area, Poland. The number of dissected specimens per time period indicated above. (a) Longwinged, hard chitin, dark thoracic venter, eggs; (b) as category a but shortwinged; (c) longwinged, soft chitin, pale thoracic venter, maturing or mature eggs; (d) as category c but shortwinged; (e) longwinged, hard chitin, pale thoracic venter, eggs; (f) as category e but shortwinged; (g) longwinged, either soft or hard chitin, dark thoracic venter, immature; (h) as g but shortwinged. (From Vepsäläinen and Krajewski, 1974.)

females the ovaries are still poorly developed (c-d), but in most individuals positive signs for nondiapause ovaries are seen. When the chitin has had time to harden, all the females (e-f) have mature or maturing eggs. Longwinged individuals (c, e) develop functional flight muscles, and hence their teneral period before ovarian maturation is longer than that of shortwinged females (d, f), which do not build flight musculature. Some of the longwinged individuals may perform prereproductive flights, but it is not known whether they maintain flight ability for possible interreproductive flights. After an egg-laying period of some to several weeks the nondiapause individuals die, and the second generation (g, h) develops diapause and moves after teneral development to overwintering sites on land. Part of the longwinged individuals (g) may fly even long distances to wintering sites.

Wing dimorphism of *G. lacustris* is basically genetic polymorphism, and most overwintering shortwinged individuals differ genetically from longwinged ones. However, in late summer part of the recessive genotype (which has earlier developed long wings, g) grow to shortwinged imagos (h). In midsummer, part of the shortwinged nondiapause imagos (d, f) have developed by genetic switch, but a large proportion of them belongs to the same genotype (the recessive homozygote) as the longwinged individuals (c, e); this is due to temperature differences in the environment of the developing larvae. Accordingly, in addition to genetic polymorphism, *G. lacustris* may also show seasonal polyphenism for wing length. The differences between populations may be striking: The life cycle of an isolated *G. lacustris* population in southernmost Finland includes only group h (the shortwinged individuals before winter) and b (the same individuals after winter).

The shortwinged individuals of various species differ with respect to the degree of reduction of the wings (see Vepsäläinen, 1974c), but functionally all shortwinged imagos are unable to fly. The longwinged individuals are able to fly at least during part of their lives. During development shortwinged individuals do not build indirect flight muscles, as do longwinged individuals. There are many minor associated changes in the morphology of the morphs, but for these dissimilarities no selective differences have been demonstrated (cf. Steffan, 1973, on the sciarid fly *Plastosciara perniciosa* Edwards).

When rare or exceptional cases of wing length are considered (see last column in Table 2; for references see Vepsäläinen, 1974c, pp. 32-34), the following conclusion emerges: All *Gerris* species have the potential to produce both shortwinged and longwinged individuals. I have worked on the basis of the idea that the adaptive significance of wing length variation and dimorphism can best be explained by studying the wing length pattern and its determination in different, phylogenetically related species in various environments. Because the environment tends to change all the time, we expect that natural populations differ from theoretical optima due to history (Lewontin, 1966; see also Haldane, 1932, pp. 117-118; Haldane and Jayakar, 1965; Vepsäläinen, 1974c; Järvinen and Vepsäläinen, 1976), but it is, however, conceivable that natural populations tend to differ in the direction of their theoretical optima (Levins, 1968).

Determination of Wing Length and Diapause

The determination of wing length of water-striders has been illuminated by population studies in the field, and by crossing and culturing experiments in photoperiod-temperature chambers (Vepsäläinen, 1971a, 1974a,b,c,d). The ensuing model agrees with all data available from natural populations and laboratory experiments.

The reaction norm of each *Gerris* individual during larval development can be illustrated with a temperature-photoperiod coordinate system (Fig. 2). Note that the photoperiod axis also gives the direction of the change in the day length as well as absolute day lengths. One curve has been drawn for each genotype in this coordinate plane. This curve indicates the critical environmental conditions of the genotype during its stages of developmental option. The environmental conditions prevailing during a given moment of larval development can be represented with a dot lying either above or below the curve. For example, in *G. odontogaster*, which has been studied best, the fourth larval instar is the last critical developmental stage. Somewhat simplifying, if the environmental conditions prevailing during the fourth larval instar (the time of option) can be represented with a dot lying below the curve of the individual, a shortwinged imago without maturation of indirect flight muscles is produced. If the dot representing the environmental conditions lies above the curve, the eclosing imago is longwinged and will develop functional flight muscles. The fifth larval instar is not sensitive to environmental stimuli determining wing length. Note that the sensitivity of *Gerris* genotypes to temperature is very different in increasing and decreasing day-lengths (Vepsäläinen, 1974b).

The model presented is relatively schematic, and little is known about the shapes and exact position of the curves. The two curves represent a genetic switch, that is, two distinct genotypes which differ with respect to one allele or a supergene. In natural populations the curves are likely to be more or less blurred - but distinct - due to differences between the reaction norms of different individuals. In many species the genetic switch is apparently lacking, for temperature and photoperiod seem to have a similar effect on all individuals. This is the case in the seasonally dimorphic (seasonally polyphenic) and the

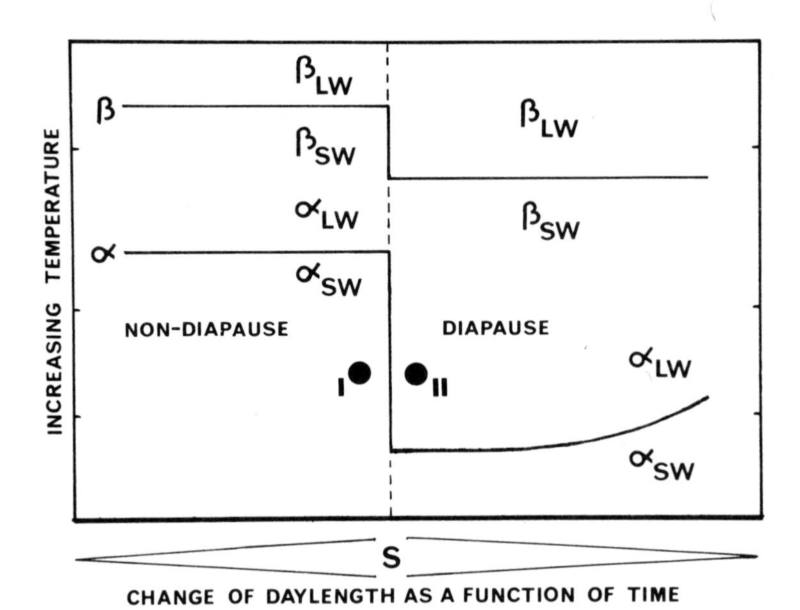

Fig. 2. The model of wing-length and diapause determination in *Gerris*. Along the *x*-axis change of daylength is incremental to point *S* (summer solstice) and decremental after it. Increasing temperature is given along the *y*-axis. Curves α and β stand for two different genotypes of larvae with different reaction norms in the temperature-illumination coordinate plane. The lower curve represents the recessive homozygote; the upper one the heterozygote and the dominant homozygote. The future morph of the developing larva depends on whether the ambient illumination-temperature conditions during its fourth instar can be presented with a dot below or above the curve of the individual. If below, for example, the resultant phenotype is a shortwinged imago (e.g., α_{SW} for genotype "α" at point I; compare to the phenotype α_{LW} (longwinged imago) of the same genotype α, which spends its fourth larval instar in conditions given by point II, and phenotype β_{SW} (shortwinged) of the genotype β at that point). Diapause develops if the "environmental dot" falls to the right of the vertical line at *S* (summer solstice). *LW* = longwinged, *SW* = shortwinged. (Modified from Vepsäläinen, 1974b.)

permanently monomorphic species. Only in permanently dimorphic species (see Table 2) does a genetic switch clearly operate (genetic polymorphism). The lower curve in Figure 2 represents the recessive homozygote, while the upper one represents the heterozygote and the dominant homozygote. It is not known whether the heterozygotes have reaction norms intermediate between the homozygotes or not. If they have, the upper curve must be split into two. Also, it is expected that not only the positions of the curves (i.e., sensitivity to temperature) of various species are different, but also those of various populations of the same species living in distinct climatic regions.

In addition to photoperiod and temperature, crowding and the amount and quality of food may function as developmental switches affecting the morphology (e.g., the wing length) of insects. This has lucidly been demonstrated by Lees (e.g., 1966) for aphids. Recently Järvinen and Vepsäläinen (1976) suggested that the amount of food is not expected to affect wing length determination in *Gerris*. The argument is based on the fact that water-striders cannot deplete their future resources but are dependent on the influx of insects trapped on the water surface. The length of the wings is determined about two weeks before the final ecdysis, i.e., in a situation which predicts poorly the amount of food available at the time the longwinged imago attains flight ability. *Gerris* food is very heterogeneous arthropod material consisting of what happens to drop (or come up) to the water surface. Accordingly the quality of food has virtually no predictive value and does not make sense as a switch mechanism in wing length determination of *Gerris*. The same reasoning also renders crowding improbable as a switch. It is the case, however, that higher densities result in less food per capita, and intensified cannibalism (see Järvinen and Vepsäläinen, 1976), and hence we should *a priori* expect crowding to promote longwingedness if it had any effect on wing length. In *G. lacustris*, a most heavy crowding during late larval development was, however, observed in an almost totally shortwinged population (Vepsäläinen, 1973a). The situation may, of course, be different in other environments and other species.

Diapause determination is easily included in the model (Fig. 2). Long, lengthening illumination averts diapause, but shortening day lengths always induce diapause. The last developmental stage sensitive to photoperiod is, as in wing length determination, the fourth larval instar. In stationary long-day rhythms or in constant light the reaction of water-striders is ambivalent with respect to both diapause and wing length (Vepsäläinen, 1971a, 1974d).

Note that conditions (shortening illumination rhythms) which switch diapause development also have a radical effect on the temperature threshold of long- vs. shortwingedness. Hence, in seasonally polyphenic species the correlation between diapause and long wings is very high, but in low temperatures (in late summer and autumn in nature) we can expect shortwinged individuals from the α-curve (Fig. 2). Indeed, a small proportion of the *G. odontogaster* population develops short wings and diapauses in late summer, and in the permanently dimorphic species *G. lacustris* it has been shown that the frequency of shortwinged imagos becomes very high in individuals eclosing in late summer both in Denmark and Finland (Andersen, 1973; Vepsäläinen 1974c; see also Mielewczyk, 1964, pp. 51-52, for *G. lacustris* in Poland). On the basis of sporadic laboratory evidence (Vepsäläinen, 1971a, 1974b, for *G. odontogaster* and *G. lacustris*, respectively) I have speculated that the additional late summer shortwinged imagos are produced not only by temperatures which fall below the α-curve, but that the sensitivity of the developing larva toward temperature again changes in short, decreasing illumination (Fig. 2; the right end of the curve α is supposed to bend upwards). It is also possible that the range of temperature variation could in combination with photoperiod serve as switch for wing length, as suggested by Honěk (1976) for the bug *Pyrrhocoris apterus*. The significance of temperature variation for development rates of insects is well known (see; e.g., Odum, 1971, p. 117).

Testing the Model of Wing Length Determination

The model (Vepsäläinen, 1974b) was built to agree with all data available from laboratory experiments and natural populations of almost all European species (Vepsäläinen, 1974b) studied in almost the whole of Europe (data and references in Vepsäläinen, 1974c). Also later results of collecting agree (Vepsäläinen and Nieser, 1977).

Much effort was put into seeking "exceptional" situations in nature to test the model. It is conceivable, for example, that the level of the α-curve (see Fig. 2) for *G. rufoscutellatus*, which is usually monomorphic for long wings, is adjusted in each geographical region according to its macroclimate. For example, in Central Europe, the general macro-climatic conditions are those of the lowlands. However, there are some places where the *local* temperatures are much the same as those prevailing about 2000 km northwards, in northern Finland. Due to gene flow from lowland populations to mountain populations we should expect to find shortwinged *G. rufoscutellatus* in midsummer populations of the high mountains in Central Europe. Indeed, a small number of this morph has been reported from the High Jesenik mountains in Czechoslovakia in the latter part of July (Stehlik, 1952). Also, in Hungary *G. thoracicus* is generally monomorphic for long wings. It is, however, known to develop shortwinged nondiapause individuals in Finland. Hence it was no surprise that in the cold, rainy May-June 1974, one shortwinged imago was collected in Hungary (unpublished). All Central European populations of this species are usually longwinged, but again an exception comes from the High Jesenik mountains: five shortwinged were found among 21 imagos collected in July (Stehlik, 1952).

Recently Mielewczyk (1976) reported from Poland his observations on *G. paludum*, which at present cannot be explained by the temperature-photoperiod model. The population inhabited a reservoir which remained warm throughout the year. The frequency of shortwinged individuals was exceptionally high (about 90%), and their wings were exceptionally short. Shortwinged individuals predominated in all samples (first taken in March, last in November). In nearby waters with normal temperatures only longwinged individuals were found. Mielewczyk (1976), who concentrates his discussion on this special case and its significance to wing-length determination hypotheses, does not suggest that high temperatures as such give rise to shortwinged water-striders, but he suggests that superabundant food (an outcome of high temperatures) could be one factor for the high frequency of shortwingedness. As we have shown (Järvinen and Vepsäläinen, 1976), stability and high productivity together favor shortwingedness. However, we also showed that the amount of food is not likely to *determine* the wing length. Here we have a rewarding task: To reveal the determination mechanism by laboratory experiments and intensive life-cycle studies in the field. I propose two testable mechanisms: 1. The curve α in Figure 2 has been lifted upwards so that even high temperatures do not give longwinged imagos. This means that sufficient genetic variance has been available for selection to shift the population toward curve β and eliminate the temperature switch. 2. An occasional mutant, conforming with curve β would have been selected for in this "superoptimal" locality.

The Adaptive Significance of Diapause Determination

There is a fundamental difference between diapause and migration as "escape strategies." Diapause is a "sit-tight tactic" (Southwood, 1975, p. 184) when adverse conditions are *expected*, i.e., predictable, all over the dispersal area of the species (common macroclimatic, usually seasonal changes). Migration, "seeking pastures new" (Southwood, 1975), is an effective strategy in environments changing unpredictably due to varying weather affecting local sites differentially (see Fig. 2 in Solbreck, this volume). Winter affects all (overwintering) individuals, conditions met by migration need not. The often keen association between diapause and migration is partly due to the fact that migration is often needed in traveling between disconnected diapause and breeding sites (see Kennedy, 1975). The adaptive significance of migration immediately after diapause instead of later emigration is explained by the high reproductive potential of the migrants (often r-selected due to temporal heterogeneity of their environment) at the stage when developmental and overwintering mortalities have been suffered, but reproduction is still ahead (Dingle, 1972).

The adaptiveness of diapause in temperate-latitude *Gerris* is obvious, as winter is unfavorable for reproduction everywhere in the species' range. Even where the waters do not usually freeze in winter, low ambient temperatures and scarcity of food favor diapausing individuals over nondiapausing (e.g., in Portugal; Vepsäläinen and Nieser, 1977).

As a rule insect populations have adjusted the beginning of diapause to *absolute* day lengths, not to the change of day length as have water-striders. When absolute day lengths are employed, large regional differences are possible, and usually occur in the onset of diapause and in patterns of voltinism (e.g., Danilevskii, 1965; see, however, Tauber and Tauber, this volume). But in *Gerris* the diapause determination occurs at the same time in the univoltine populations of northern Finland and the multivoltine populations of Hungary (Vepsäläinen, 1974a,c).

Why then is diapause in *Gerris* determined in climatically very different regions at the same time, that is, in all individuals developing their fourth larval instar after the summer solstice? This efficient control of voltinism which, however, affords differences between populations of different regions depending on the length of the season *before* summer solstice, can be explained by resource-limitation of water-strider populations. As water-striders fail to overwinter after reproduction, it is not advantageous to produce more generations than are allowed by the carrying capacity of the environment, plus the individuals which emigrate (Järvinen and Vepsäläinen, 1976). Further, it is also possible to vary the length of the reproductive season after solstice, viz. by extending the life of nondiapausing individuals and their egg-laying, and so decreasing crowding. Finally, *G. rufoscutellatus* is said to delay the beginning of egg-laying in Austria to the extent that bivoltinism is averted (Mitis, 1937).

Adaptive Significance of Wing-Length Strategies

Recently the adaptive significance of wing-length strategies in *Gerris* have been analyzed by three different models. The first, applied by Vepsäläinen (1974c, pp. 39-51), uses the fitness set technique (Levins, 1968), which is a powerful means of sorting out qualitative results. The second model is that by Järvinen (1976), based on Levins's (1969) migration-extinction models. Järvinen and Vepsäläinen (1976) studied the effect of resource limitation in their third model. In the following discussion, I use the results of all three models. I shall begin with a local treatment and show that dimorphism is possible even if not the optimal strategy. At the metapopulation level where attention is focused on a set of local populations connected by dispersing individuals, dimorphism can be shown to be optimal over a wide range of circumstances. However, both shortwinged and longwinged monomorphic optima also exist.

Local Optima and Natural Populations

Temporarity, i.e., drying up, of *Gerris* habitats during the reproductive season (duration instability, applying the terminology by Southwood, 1976) implies extinction of populations. Even a short dry phase of some days destroys the developing larvae through desiccation and starvation. As John Spence (personal communication) has been able to show with the Nearctic *Gerris pingreensis*, imagos of at least some water-strider species can withstand dried-up ponds through considerable periods: Imagos were caught in pitfall traps on a dry pond bottom over a month after drying-up. As soon as a dried-up site is refilled with water, it is open for recolonization of water-striders.

It is not known how shortwinged water-striders colonize. For instance, longwinged females (the recessive homozygotes; for clarity I assume here a genetic switch, as, e.g., in *G. lacustris*) may have occasionally mated with shortwinged males (the carriers of the dominant allele for short wings) before the colonization flight. (As was shown from individuals captured during spring dispersal flights (Landin and Vepsäläinen, 1977), the flight-oogenesis syndrome (Johnson, 1969) is the rule in many water-strider species, that is, the migrant females have undeveloped ovaries.) Another possibility is "waiting for mutation." Assuming a mutation rate of μ from the allele for longwingedness to the allele for shortwingedness, and an effective population size of 500, with an average number of 100 zygotes per individual, gives one heterozygote per $10^{-5}\mu^{-1}$ generations. Assuming a probability of .01 that the zygote reaches maturity (stable population size, no selective differences between shortwinged and longwinged individuals during development), gives one mature shortwinged imago in $10^{-3}\mu^{-1}$ generations. If the genetic switch were operated by a supergene, the change from longwingedness to shortwingedness might be more frequent, but supposedly still occasional. So, preflight mating is certainly the most probable means of colonizing in the time scale used here.

In the colonization phase of a site, longwingedness is the optimal strategy. This does not, however, ensure the establishment of the population. For instance, if the population site dries up frequently, it may be unsuitable also for "optimum wing-length strategists" - the drying-ups occur more frequently than colonizations and the optimal longwinged strategy cannot establish the population at the site (Vepsäläinen, 1974c; Järvinen, 1976). Also, isolation together with an even lower degree of temporariness of the habitat restricts severely the founding of *Gerris* populations. The offspring of the occasional immigrant are longwinged (at least in the overwintering generation), which may leave the site, but - due to isolation - do not usually find it again. On the other hand, shortwinged water-striders are poor colonizers and are liable to arrive less often than the drought periods occur.

Once the shortwinged individuals have been introduced into the site the population evolves *toward* its local optimum, monomorphic shortwingedness. The advantage of shortwingedness is most pronounced in isolated sites, and also in multivoltine populations as the shortwinged imagos mature earlier. The teneral period is also shorter as shortwinged imagos do not develop flight muscles which would interfere with the maturation of the gonads (Andersen, 1973; Vepsäläinen, 1974c). Temporary drying up of the habitat begins the morphism cycle (Järvinen and Vepsäläinen, 1976) anew with the immigration of longwinged water-striders. So, summarizing, longwingedness is the strategy *between* sites, but shortwingedness is the strategy *within* a site (Vepsäläinen, 1974c; Järvinen, 1976; Järvinen and Vepsäläinen, 1976).

However, what we see in nature are not ideal manifestations of local optima; rather natural populations are expected to differ toward their theoretical optima. True, it has been documented, for instance in *G. lacustris* and *G. lateralis*, that permanency and isolation promote monomorphic shortwingedness (Vepsäläinen, 1974c). Within the range of a

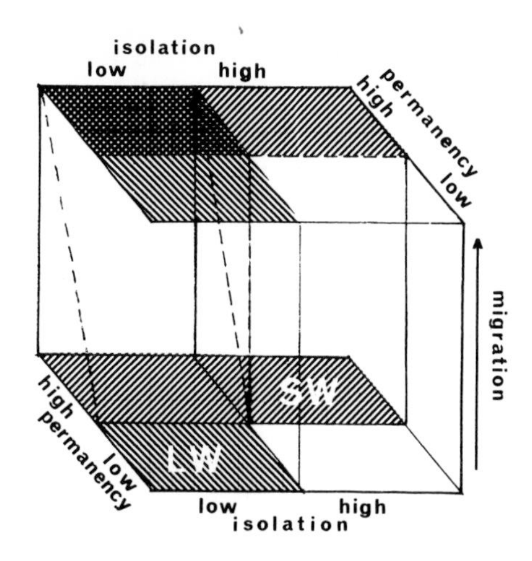

Fig. 3. The local optimal wing-length strategies in an isolation-permanency checkerboard (below). The mixing effect of migration is projected to the upper level. In places (low isolation and high permanency) dimorphism is produced, as indicated by the mixed cross-hatching at upper left. The visualization is simplified as permanence and isolation are not independent of each other, but the increase of isolation demands increased permanency to allow population existence. Accordingly, the white area picturing sites unsuitable for *Gerris* populations should be larger and its border should run obliquely from the lower-left to the upper-right corner over both checkerboards.

metapopulation where the isolation of *Gerris* sites ("islands") is weaker, local dimorphism is seen instead of the shortwinged populations predicted to be optimal by examination at the level of local populations if the area is heterogeneous with respect to the risks of drying up. The function of long wings (and functional flight muscles) is dispersal, and thus longwinged imagos will immigrate not only to newly arisen favorable sites but also into permanent, otherwise shortwinged populations. Understandably longwinged imagos do not avoid habitats favorable for shortwinged water-striders. Thus migration promotes dimorphism by mixing two local optima. This can be visualized as two checkerboards of habitat permanence and isolation, where the lower checkerboard gives the local optima and the upper one the realized situation due to migration (Fig. 3).

In the discussion above environmentally determined wing-length patterns were not treated specifically - the reason for this will be evident in the next sections. The general conclusions concerning local optima and the mixing effect of migration are valid independent of the mechanism of wing-length determination. Of course, when wing length is exclusively environmentally determined, regions of univoltine populations fall outside this generalization, as the species are there longwinged through the year due to the very mechanism of wing-length determination. I only explain here a special case of dimorphism, documented in *G. lacustris*, even in those populations where only the recessive homozygote is present. The overwintering diapause generation consists of both longwinged and shortwinged individuals; at least part of the latter develop by environmental switch (short, shortening days, and/or low temperatures) from recessive homozygotes in late summer. The adaptiveness of switching to shortwingedness is evident. Development toward longwingedness in late summer is selected against because longwinged individuals have a longer teneral period with flight muscle maturation. Due to low temperatures and often scarce food the delay in development and hence attainment of cold-hardiness is appreciable (Vepsäläinen, 1974c). Environmental switch mechanisms are used to track more or less predictably changing environments. The successive seasonal environments (the one earlier in summer, the other closer to winter) are here met by different monomorphic strategies (longwinged and shortwinged, respectively). Hence there will be in the same overwintering generation two morphs, each the adaptive one judged by its developmental history (that is, by the conditions prevailing during the time of developmental option of the larva).

In most *Gerris* species the indirect flight muscles histolyze in females and some of the males soon after the commencement of reproduction (Vepsäläinen, 1971a, 1974c). Hence the spring dispersal (see Landin and Vepsäläinen, 1977) takes place when the reproductive value of the migrants is high (after juvenile and winter mortalities but before reproduction has begun). This is in accordance with observations in insects in general (Johnson, 1966), and with theoretical predictions (MacArthur and Wilson, 1967). The Janus-faced aspect of longwingedness in these species is here revealed. When dispersal (between-sites strategy) has served its function, longwinged individuals adopt the role of shortwingedness (within-site strategy).

Regional Optima - Temporary Habitats

In a system where the function of one morph is dispersal, inspection of local strategies does not reveal optimal population strategies. The adaptive significance must be sought for by measuring the fitness of each morph by net harvest over all sites exploited. This treatment affects especially the longwinged morph in a metapopulation, an archipelago of *Gerris* sites, where the local populations are connected by dispersing individuals. I first describe the fitness set technique of Levins (1961, 1968), as this is central for the following.

By fitness sets strategies in heterogeneous environments are studied. In the simplest cases, fitnesses of individuals and populations are pictured in a right-angled coordinate system of two environments, where point zero (zero fitness in both environments) forms the lower left corner, and fitnesses increase up and toward right (e.g., Fig. 4a). The fitness of each individual is given with a coordinate point. Through a given fitness point two lines, one vertical and the other horizontal, can be imagined. These right-angled lines delimit four sectors. All individuals which have a point in the sector toward the origin have a lower fitness than the one in the intersection point of the imagined lines. Up and toward right, in the opposite sector, lie the points of higher fitness. The fitness relations of the points lying in the two other sectors depend on the ratio and so-called grain of the environments.

The set of all possible fitness points for the population is the fitness set. In the case of one point, all individuals in the population have equal fitnesses in both environments, and that is also the fitness of the population. If the population is composed of individuals with differing fitnesses the fitness set is formed of more than one point. The average fitness of the population depends on the ratio of the phenotypes and the grain and ratio of the environments. Natural selection is presumed to push the phenotype ratio toward the optimal one, i.e., the highest attainable one with the fitness set. To find the optimal strategy of the population, adaptive functions and fitness sets must be combined. Adaptive functions are families of either straight or curved lines, which connect points of equal fitness. Thus an adaptive function (a straight line) can never have a positive slope, and neither can the tails of a curved line bend up to the right, i.e., the tails must approach asymptotically the coordinate axes (cf., e.g., Levins, 1961, Fig. 1d; 1968, Fig. 4.3; Vepsäläinen, 1974c, Fig. 9). When adaptive functions are allowed to approach zero from $+\infty$, the

---

cold-hardiness of individuals which have not (yet) developed diapause (e.g., because of prolonged teneral development due to flight muscle maturation) is very poor (Vepsäläinen, 1974c).

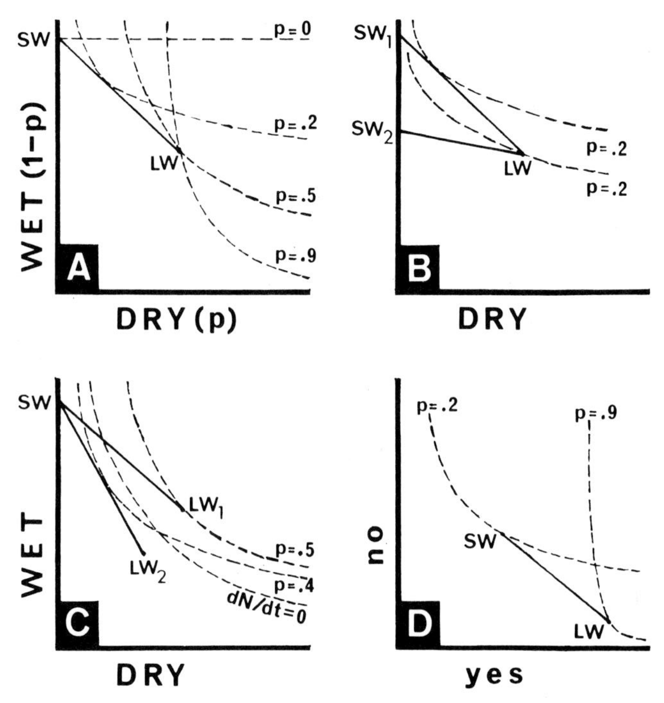

Fig. 4. Semirealistic examples of fitness sets and optimal strategies of *Gerris* populations living in temporary habitats. The relevant genotype here is the recessive homozygote, which has the potential for either wing morph, depending on the environmental switch signals. *LW* = fitness point for longwinged, *SW* = shortwinged imago; W_{SW} = fitness of shortwinged in wet, W_{LW} = fitness of longwinged in wet = for simplicity, the same in case of local drought. Fitness of shortwinged in dry is zero. Now the average fitness of the population is $\bar{W} = [W_{SW} - Q(W_{SW} - W_{LW})]^{1-p} \cdot (Q \cdot W_{LW})^{p}$, where Q = the proportion of longwinged imagos and p the probability that the habitat dries up during the development of the following generation. (a) $W_{SW} = 20$, $W_{LW} = 10$; no risk of drying, $p = 0$: $Q = 0$ (monomorphism for short wings), $\bar{W} = 20$; $p = 0.2$: $Q = 0.4$ (dimorphism), $\bar{W} = 12.2$; $p = 0.5$: $Q = 1.0$ (monomorphism for long wings), $\bar{W} = 10$; $p = 0.9$: $Q = 1.0$, $\bar{W} = 10$. When days are shortening read always "dry" or $p = 1.0$ (predicts winter): longwingedness is optimal. (b) The fitness set given by the line $SW_1 = LW$ is the same as in Fig. 4a. Crowding affects by decreasing the fitness of shortwinged individuals unable to emigrate to less crowded sites and now $W_{SW} = 12.5$, and the fitness set is $SW_2 = LW$. With $p = 0.2$ crowding alters the optimal strategy from dimorphism to monomorphic longwingedness, and decreases the population fitness $\bar{W}$ from 12.2 to 10. (c) The fitness set given by the line $SW = LW_1$ is the same as in Fig. 4a. Migration losses and unsuccessful colonization (due to isolation) move the fitness point of the longwinged morph to LW_2 ($W_{LW} = 7$). If the drought probability is $p = 0.5$, monomorphism for long wings is optimal with $W_{LW} = 10$ ($\bar{W} = 10$). Let us now suppose that a stable population is maintained by an average population fitness of $\bar{W} = 8$ (i.e., $dN/dt = 0$). Now crowding ($W_{LW} = 7$) leads to negative rate of population increase, when $p = 0.5$. This means that whatever the optimal strategy with $p = 0.5$ and the fitness set $SW = LW_2$, the population is not able to sustain itself (line $dN/dt = 0$ in the figure does not intersect the fitness set). By increasing habitat stability to $p = 0.4$ an optimal strategy of $Q = 0.6$ and $\bar{W} = 8$ will maintain equilibrium population. (d) In late summer and autumn the two habitat axes "yes" and "no" refer to answers to the question "Is there time for both maturation of flight apparatus and development of diapause?" The fitness point of longwinged in "yes" situation is higher than that of shortwinged imagos, as some of the populations of the recessive homozygote (relevant here) live in more or less temporary waters, which are perhaps not suitable for reproduction after winter. The

point of tangency with the fitness set is found graphically. This point shows the optimum strategy (or the best attainable one) of the population, which is composed of one phenotype or a mixture of different phenotypes.
The fitness set is either convex (i.e., bulges outwards from the origin) or concave (the upper-right-hand boundary bulges toward the origin). All fitness points of mixed-phenotype populations lie on the straight line connecting the phenotypes. Thus with convex fitness sets all points of mixed strategies have always a fitness point of an individual phenotype on the upper-right hand, and only monomorphic optima exist. With a concave fitness set, the point for optimal strategy lies on the straight line connecting the fitness points of the two extreme specialists, i.e., on the straight line forming the upper-right-hand boundary of the so-called extended fitness set. Now a straight-line adaptive function gives always monomorphic optima, but curved lines may give monomorphic or polymorphic optima (e.g., Fig. 4a).

The shape of the adaptive function depends on the so-called grain of the environment. In a fine-grained environment there is no uncertainty for the individual. The individual meets the different environments in those ratios they occur (temporally and/or spatially), and accordingly the adaptive function is the arithmetic mean of fitnesses in the two environments, that is, a straight line with a slope which depends on the weights of the two habitats. A coarse-grained environment presents itself to the individual as alternatives, which mean uncertainty for the individual, and the function (logarithmic in Levins's examples) is a curved line.

Above, the conclusion was reached for *Gerris* that shortwingedness is always the optimal within-site strategy, and longwingedness always the optimal between-sites strategy. Clearly, if the environment for the following generation is dry, the optimal strategy will be longwingedness. If the environment will continue to be wet, that is, favorable for reproduction, the optimal strategy is shortwingedness (due to earlier commencement of reproduction in shortwinged nondiapause individuals, and no losses through flight). (Note that resource limitation is not here taken into account; its effects have been studied with another model, referred to later on.) By using fitness sets I discuss now the possibilities and limitations of dimorphism as an optimal strategy in temporary habitats of *Gerris*. The two environments are wet (there is time for another generation at the site) and dry (the site will dry up during the development of the next generation). The phenotypes are the longwinged (long teneral development before maturation, ability to fly) and the shortwinged imago (short teneral period, undeveloped indirect flight muscles). It is easy to see that if any intermediate phenotypes existed (e.g., resorbtion of flight musculature during the late teneral development of longwinged, or development of flight musculature in shortwinged individuals), they would have a reduced fitness in both environments. Hence the fitness set is concave, and we need only consider the longwinged phenotype able to fly at least during part of its life, and the shortwinged phenotype without flight musculature. The relevant part of the fitness set is formed by the upper-right-hand boundary of the extended fitness set, i.e., the line joining the points of the two phenotypes. The fitness losses due to migration of longwinged individuals may be assumed to be about the same independent of the local environment to come (wet or dry - of course this is simplified, as drought is likely to affect also other sites of the metapopulation area). The fitness of shortwinged individuals during local drought is zero, but in wet habitat high (complications due to resource limitation are discussed below) - accordingly the fitness set is asymmetric. The environments wet and dry are alternatives to water-striders, and thus the environment is coarse-grained and the adaptive function is a curved line (Fig. 4a).

The optimum wing-length strategy is now monomorphism of either phenotype or dimorphism. Applying Levins's (1968) analysis of optimal diapause behavior, it can be deduced that the limits for dimorphism become wider when the fitness difference of the phenotypes in wet increases, i.e., when the advantage of fast teneral development of the shortwinged morph increases. If the advantage is small, the transition from one kind of monomorphism to the other is fairly abrupt. The point of tangency of the adaptive function with the extended fitness set divides (in the case of dimorphism) the line connecting the fitness points of the two morphs, longwinged and shortwinged, into two parts, which represent the ratios between the morphs in optimal strategy (Fig. 4a). The analysis is, of course, simplified. For instance, there is evidence that the flight motivation increases with temperature and in face of drought. Likewise, crowding is supposed to increase the emigration of longwinged imagos. Also, crowding favors emigration and thus longwingedness (resource limitation of many *Gerris* populations was suggested by Järvinen and Vepsäläinen, 1976). The effect of resource limitation can be analyzed with the fitness set technique by substituting "not crowded" (or plenty of food, or low degree of cannibalism) and "crowded" (or starvation, or heavy cannibalism) for "wet" and "dry," respectively. Of course the mutual fitness relations of the two morphs may also be changed (Fig. 4b). The effect of resource limitation is presumably less drastic than drying-up of the habitat for *Gerris*. In the latter case, it is a question of duration instability, in the former case of temporal variability in regard to carrying capacity (see the habitat classification by Southwood, 1976).

The idea of metapopulation was made explicit in the above treatment. Isolation of habitats for reproduction decreases the probability of longwinged imagos finding another site after migration. Hence increasing isolation moves the fitness point of longwinged individuals toward the origin (Fig. 4c). By drawing to the fitness set a line where the net increase of population numbers is zero, it is easily seen that a high degree of isolation in combination even with a moderate degree of temporarity will render the area unsuitable even for optimal wing-length strategists among *Gerris*. Isolation monomorphism for short wings may maintain the population when the probability of "wet" approaches unity (i.e., in stable habitats; Fig. 4a, $p = 0$). When drought periods occur with intervals of about the time necessary for offspring development, it is improbable that the site will support any kind of *Gerris* population (see also Vepsäläinen, 1974c; Järvinen, 1976; Järvinen and Vepsäläinen, 1976). The wet periods should allow not only offspring development but also give time for the preceding colonization phase. Isolation, of course, lengthens the necessary interval between successive drought periods required for successful reproduction. With fitness sets, the interplay of isolation and resource limitation with environmental heterogeneity are taken into account by decreasing and increasing the fitness of the longwinged morph, respectively.

Which ones of the realized wing-length strategies of *Gerris* can now be explained with the above treatment? Implicit in the whole concept of temporary uncertainty is that the two phases, wet and dry, alternate on a shorter time scale than an evolutionary one. Therefore, genetically determined shortwingedness is excluded from the examination by the choice of the environmental axes. When the probability of wet approaches unity, the environment no longer consists of alternatives; the habitat is permanent and allows for genetical changes of strategies. I propose that in *Gerris* environments, drought intervals of at least tens to hundreds of years are needed to call them permanent in the evolutionary sense. Of course, isolation of suitable breeding sites in the metapopulation area lengthens the uninterrupted wet period during reproductive seasons needed for maintenance of genetically determined shortwingedness. Accordingly, the above analysis explains only monomorphism and polyphenism, and genetic polymorphism is left to the next section.

The number and habitat spectrum of the sites suitable for water-striders in summer depends much on the rains (snow and water) in winter, and also on the weather conditions of spring and early summer. Hence the suitability of the site for reproduction must be evaluated anew every spring, and monomorphism for long wings is the overwintering strategy in all but most permanent habitats (Fig. 4a). Of course there are other factors which may increase the adaptiveness of longwingedness in winter. For example, in temporally heterogeneous environments fairly frequent population extinctions are part of the game (winter mortality takes also its tax), and longwinged individuals may gain an advantage in colonizing empty sites. Hence spring dispersal is a rule for all *Gerris* species of temporary or semipermanent habitats. The mechanism for determination of long wings in this colonizing group is by shortening day lengths (foretelling winter and also inducing diapause).

Analysis of midsummer generations which reproduce without preceding diapause (Fig. 4a) shows that when the risk of drying up of the habitat is negligible (p = 0) during the late summer, shortwingedness is favored by natural selection. When the probability of dry is over zero, the adaptive function approaches the origin from ∞ more from the upper-right hand, and the optimal strategy is dimorphic (moderate drought risk, p = .2). When the probability of drought increases, longwingedness will be optimal ($p \geqslant .5$). Naturally, with very high drought risk not even optimal strategies are able to maintain the population, as the rate of increase of the population will always be negative. The predictions obtained from the above analysis agree well with observations on natural populations (Table 2). All univoltine populations of species living in more or less temporary environments are longwinged (*G. odontogaster*, *G. thoracicus*). In Europe, the drought risk of small ponds and ditches increases from Finland to Hungary. So in Finland the critical period for *Gerris* habitats occurs in early summer, but in Hungary the rainy months are May and June and the rest of summer is remarkably dry. In Finland July and August are the rainiest months (for more detailed data and references, see Vepsäläinen, 1974c, p. 40). The overwintered individuals select suitable habitats for reproduction during dispersal flights. Of these some may, of course, dry anyway and the developing offspring (and also those parentals which have absorbed their flight muscles) go extinct. Those Finnish habitats which produce a first offspring generation are not likely to dry in the more rainy late summer when the (partial) second generation develops. Hence losses of pre- and interreproductive dispersal in midsummer may be greater than the advantages of flight. Moreover, due to the shortened teneral development of shortwinged imagos they are able to commence reproduction some weeks earlier than the longwinged individuals. This is an important advantage in regions of short summers (Vepsäläinen, 1974c) and hard winters. Hence the strategy of the midsummer generation is shortwingedness. Observations on all relevant Finnish species in Finland agree with the fitness set predictions: The reproducing part of the first generation of (partially) bivoltine populations of *G. odontogaster*, *G. argentatus*, *G. paludum*, and *G. thoracicus* is shortwinged (Vepsäläinen, 1971a, 1974c; see also Section V in the latter for references to the situation in Denmark and England). Further south in Poland, the populations are usually completely bivoltine. The drought risk is higher than in Finland, and the most severe droughts occur later in the summer. By prediction (cf. Fig. 4a) dimorphic and even longwinged midsummer populations evolve. In Poland and Hungary, the midsummer reproducing populations are dimorphic in the species of temporary habitats (*G. odontogaster, G. argentatus*, and *G. paludum*). In Hungary where the risk of drying up of the habitat is very high and increases as the summer grows older, the midsummer generation of *G. thoracicus*, an inhabitant of ponds, pits and ditches which face the most severe droughts in July-August, is totally longwinged and the individuals perform interreproductive flights (Vepsäläinen, 1974a). This is in accordance with the results of Levins (1964; see also Dingle, 1974) that the optimal amount of migration (and hence selection for longwingedness) is increased by the short-term temporal variance of the environment.

The mechanism of wing-length determination fits well in the above increase of longwinged frequency toward the south. In larvae which develop to a nondiapause imago in lengthening illumination (i.e., in the first part of summer), high temperatures (predicting drying up of the habitat) during the sensitive period of the larva (about the fourth instar in *G. odontogaster*) induce a longwinged imago capable of seeking new sites for reproduction; lower temperatures give a shortwinged individual. In Finland temperatures are not usually high enough to dry up reproduction habitats and not even to produce long wings. In Poland the temperature range allows development of both morphs, and in Hungary the summer is rarely cold enough to produce shortwinged individuals in *G. thoracicus*. Of course, evolution toward increased temperature sensitivity may be expected in southern populations of those species living in temporary habitats.

The optimum strategy in a pond at a given moment is monomorphism of either morph. However, the morph of each individual depends on the temperatures prevailing during its larval development and on the photoperiodic and genetic background. When a certain threshold temperature is exceeded, longwinged instead of shortwinged imagos are produced. Hence the morph composition at the site does not perhaps represent the optimal for any given moment, but rather a partial record of the recent illumination and temperature conditions. This strategy is supposed to maximize the average fitness of the metapopulation (see fitness set treatment).

Regional Optima - Permanent Habitats Available

Permanent habitats were defined above on an evolutionary time scale to be such which will be struck by drought less frequently than once in tens or hundreds of years. In the following I shall consider the optimal strategy in spatially heterogeneous environments where the metapopulation consists of a mixture of permanent and temporary habitats (the "harlequin environment" of Horn and MacArthur, 1972).

Many of the environments may be between the extremes of temporary and permanent. However, in the fitness set treatment it suffices to consider only the two extremes which still are able to maintain stable populations, and place the fitness points of the different genotypes apart from each other. The dominant homozygote (and heterozygote) always develops short wings, and its fitness point is always low along the temporary-habitat axis but high on the permanent axis. The recessive homozygote may be long- or shortwinged depending on the environmental switch signals at the time of developmental option - in determining the morph(s) of the recessive homozygote and its fitness point, the results of the previous section may be used. Understandably, the fitness set treatment in spatially heterogeneous environments is somewhat arbitrary for these reasons. The fitness point of the heterozygote lies on the line connecting the two homozygotes if no overdominance can be shown (as in *G. lacustris;* Vepsäläinen, 1974c), or up toward right off the line if overdominance is known (some evidence for *G. lateralis* has been supplied; Vepsäläinen, 1974a).

To find the form of the adaptive function, the grain of *Gerris* habitats must be known. The risk of drying up may, due to weather conditions, be notable during the development of one offspring generation, but insignificant during some other (the variation depends, of course, on the regional climate). That is, the ratio of permanent to temporary habitats available for *Gerris* may vary from one summer to another as there are rainier and drier summers. If the environment is in this way poorly autocorrelated, the spatial heterogeneity introduces itself more or less as alternatives to water-striders, and the grain is coarse. Note also that due to the restricted ability of longwinged migrants to habitat-select, their environment will be coarse-grained, as the metapopulation area includes not only the two extreme habitats of the fitness set but the whole array of intermediate environments (the local risks need not vary as a function of time). I suppose that *Gerris* habitats are for the above reasons so coarse-grained that the adaptive function can be given by a curved line for all genotypes. Adaptive functions which give the natural situation more realistically may be different, as the environment for individuals with the dominant allele (shortwinged ones with a poor dispersal capacity) is often very predictable - only a minute proportion may find themselves in real temporary habitats. I leave most of these speculations out of the fitness sets and adaptive functions, as the details for water-striders are scarce. For obvious reasons univoltine populations are the easiest ones to analyze; the complications due to polyphenism are left out.

In the first analysis in spatially heterogeneous environments (Fig. 5a) the heterozygote has been given average overdominance over both environments. When the frequencies of both environments are moderately high (K_{21}), the optimal strategy is monomorphism of the heterozygote. Due to Mendelian segregation this is not realized, but the fitness set boundary is more toward the origin and the realized optimum strategy, a mixture of the three genotypes, is found at the point of tangency of the adaptive function (K_{22}) with the fitness set. When the frequency of permanent patches decreases, the optimal strategy will be genetic monomorphism for the recessive homozygote (K_1) and the optimal wing-length strategies (determined by environmental switches) are found by the treatment in the previous section. Without over- or underdominance (Fig. 5b) the results are much the same as with overdominance, but the limits for polymorphism are more restricted. With slight underdominance (Fig. 5c) the limits are still more restricted, but genetic polymorphism may be optimal (K_{11}) (see also Levins and MacArthur, 1966). With larger underdominance (Fig. 5c, K_{12}) either homozygote is always optimal depending on the weights of the alternative environments.

The above analysis is pertinent to many *Gerris* species, especially in explaining permanent wing-length dimorphism (genetic polymorphism). We may predict that when water-strider species are arranged by increasing spatial and temporal heterogeneity of their habitats (with regard to the drought risk), the first species should be nearest the strategy of genetically determined shortwingedness, as completely permanent habitats do not include any drought risk (if the two-dimensional fitness set of Fig. 5b is maintained, the adaptive function is a straight horizontal line having its point of tangency at the dominant homozygote). Species which have the main weight at the permanent end of habitats, some of which include a touch of temporarity (Fig. 5d, K_{11}) should express alary polymorphism and the morph frequencies should depend on both the degree of permanency and isolation; permanency and isolation

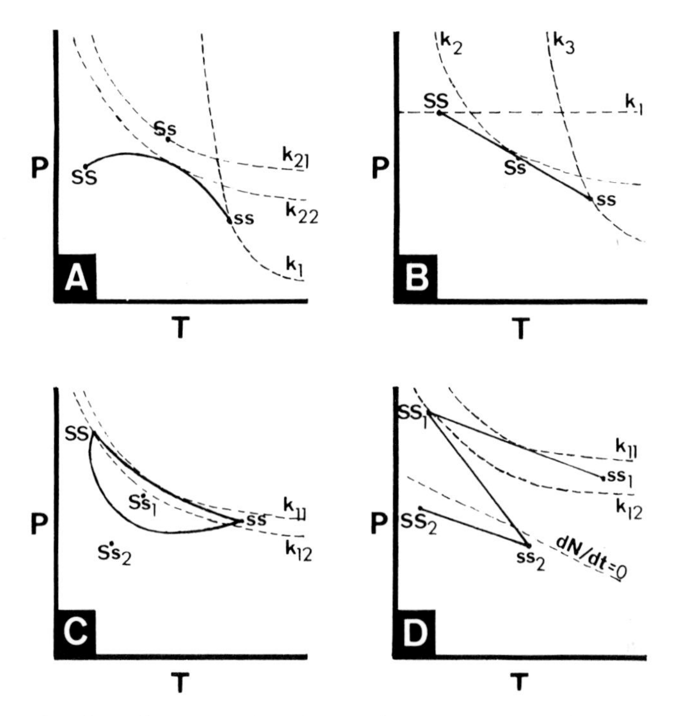

Fig. 5. Optimum wing-length strategies in spatially patchy environments, where the other patch type (*T*) includes heterogeneity in time (see Fig. 4); *P* = permanent patches. Schematic representation - see text for discussion. *SS* = the dominant homozygote, *Ss* = the heterozygote (both genotypes produce shortwinged imagos in nature), and *ss* = the recessive homozygote (the genotype which has the potential for either morph depending on the environmental signals; for this aspect, see Fig 4). (a) Overdominance allows optimal genetic polymorphism in wide environmental conditions. (b) with no overdominance genetic polymorphism may be optimal, when both patch types are moderately frequent (K_2). When only permanent patches are available (or used) monomorphic shortwingedness will be optimal (K_1). High proportion of temporary patches gives *ss*, the recessive homozygote as the optimal genotype, and the optimum wing morph strategy is found by the treatment in Figure 4. (c) Slight underdominance (Ss_1) allows theoretically very limited genetic polymorphism (K_{11}), but greater underdominance (Ss_2) gives only optima of either homozygote. (d) Isolation moves the fitness point of the longwinged morph (the genotype *ss* is affected) toward the origin. The same adaptive function now renders genetic polymorphism in the case of no isolation (K_{11}) and genetic shortwingedness (*SS*) in isolated sites (K_{12}). Isolation together with the poor cold-hardiness of the "shortwinged" genotype (the fitness points of *SS* and *Ss* move toward the origin) decides the northern limit of distribution of *G lacustris* in Finland (with isolation and winter cold the fitness set of the species moves totally to the lower-left hand of the line for $dN/dt = 0$, the limit for maintaining a stable population).

should increase the frequency of genetically shortwinged individuals (in Fig. 5d, isolation moves the fitness point of the recessive homozygote toward the origin, and shortwingedness is now obtained by K_{12}). At the other end of the habitat spectrum, the temporary one (Fig. 5b, K_3) we should find the species showing alary polyphenism (or total longwingedness in the most extreme case) (see previous section). The most euryecious species would be the one which exploits both permanent and temporary habitats and is capable of combining both genetic polymorphism and polyphenism.

The predictions of the wing lengths and habitats of European *Gerris* species agree. Habitat preferences of Finnish and Hungarian species have been described by Vepsäläinen (1973a, 1974a, respectively); for wing lengths, see this paper and Vepsäläinen (1974c, and references therein). I examine the species in two size groups to exclude possible effects of size on the environmental grain. To the large species belong *G. najas*, *G. paludum*, and *G. rufoscutellatus*, arranged by decreasing habitat permanency. In Finland, the stenoecious *G. najas* occupies restricted parts of large or medium-sized streams, which are very permanent but isolated. Populations are totally wingless. In Poland the isolation of suitable habitats is lower and there alary dimorphism exists (Krajewski, 1969). The habitat of *G. paludum* is fairly restricted in Finland, at the northern limit of its distribution, consisting mainly of medium-sized lakes. In Hungary more temporary habitats are included, although lakes are also favored there. The strategy is that of seasonal polyphenism, and the midsummer generation is either shortwinged (in Finland) or dimorphic (in Hungary). *G. rufoscutellatus* prefers more or less temporary habitats and is longwinged in Finland (populations are univoltine); in Hungary the species is rare and only longwinged individuals are known.

The small species are arranged by habitat predictability from *G. lateralis* and *G. lacustris* to *G. argentatus*, *G. odontogaster*, and *G. thoracicus*. The strategies of the three latter species were scrutinized in the previous section, and the predictions were in line with observations on natural populations (the species are seasonally polyphenetic, or in the case of *G. thoracicus* in Hungary, longwinged). *G. lateralis* is either shortwinged or genetically polymorphic. The central area of distribution, defined by the commonness of the species and the low isolation of suitable habitats, is in Central Finland where the frequency of longwingedness (the recessive homozygote) is by far the highest. Isolation increases toward Lapland and southern Finland as does the frequency of the dominant allele for short wings, and many populations are virtually wingless. In Denmark and the Netherlands the species is very rare, and the populations are monomorphic for winglessness (see Vepsäläinen, 1974c).

Finally, *G. lacustris* is along with *G. odontogaster* the commonest water-strider in Europe and prefers the permanent end of the habitat spectrum. It also colonizes, however, small ponds and ditches and is the most euryecious of all European water-striders. In Finland the stability and isolation of *G. lacustris* habitats increase toward the north. There is a steep cline along about 600 km from a shortwinged frequency of about zero in southernmost Finland to almost 100% in the northernmost populations. The known northern limit of the species is about 350 km south from the northernmost collecting localities (Vepsäläinen, 1973a). This has been explained to follow from the increased selection against genetic longwingedness due to increased isolation, and from the increased selection against shortwingedness due to the inferior resistance to cold of the dominant homozygote and/or the heterozygote (shown in laboratory experiments; Vepsäläinen, 1974c) (Fig. 5d; the upper-right-hand boundary of the extended fitness set is at the line SS_2-ss_2). The Hanko peninsula on the south coast of Finland offers an exceptional metapopulation area in the sense that the climate is that of southern coastland but the isolation of suitable habitats is exceptionally high compared with southern Finland in general. The populations harbor a notably high frequency of genetically shortwinged imagos, and in the most isolated univoltine local population the evolution of complete shortwingedness has been observed (Vepsäläinen, 1974c, and unpublished).

South from Finland *G. lacustris* populations are bivoltine, and the high frequency of longwinged imagos in the overwintering generation indicates that the frequency of the allele for short wings is low (as in southern Finland). Part of the longwinged individuals colonize in spring more or less temporary habitats. The offspring, which are virtually all recessive homozygotes, are sensitive to the photoperiodic and temperature switches for wing length just as *G. odontogaster* and the other species described in the previous section. Hence the strategy of *G. lacustris* is the most versatile of all water-strider species. It meets the midsummer drought risks by seasonal polyphenism and exploits the spatially heterogeneous environment widely by fixing part of the offspring to permanent habitats by genetically determined shortwingedness.

Still one more phenomenon of *Gerris* wing-length variation should be explained. In *G. lacustris* the frequency of eclosing shortwinged imagos increases abruptly in late summer in populations where the longwinged overwintering morph is dominating (Andersen, 1973, for Denmark; Vepsäläinen, 1974c, for southern Finland). The determination of short wings is clearly by some kind of environmental switch (see the section on wing-length determination), and consequently the phenomenon belongs under polyphenism. The strategic analysis is that used in the previous section for temporary environments (Fig. 4d). The adaptive explanation is obvious: When the risk of starvation and cold spells before winter grows higher, the adaptive function moves more and more toward the left, and a change from longwinged to shortwinged takes place as the optimal strategy. As the difference in the teneral development of the two wing morphs is often relatively small in comparison to the differences in the commencement of "winter" in different years, the change from longwinged to shortwinged is supposed to be fairly abrupt.

Regional Optima - Resource Limitation

The adaptive significance of the longwinged morph is not restricted to the avoidance of drought-stricken sites. Populations of small ponds are presumably often limited by their food resources, and hence colonization flights of longwinged individuals may be favored (Vepsäläinen, 1974c). The idea of density-dependence was developed further by Järvinen and Vepsäläinen (1976) who devised a simple deterministic model. In this model dispersal implies a risk of failure. Density-dependent regulation of the populations was included by the assumption that the carrying capacity of the environment cannot be exceeded. The equilibrium proportion of the longwinged offspring, $\hat{p}$, was shown to be

$$\hat{p} = \frac{1 - K/PbN}{E}$$

where b is the number of offspring per individual, N the number of reproducing offspring, E the proportion of longwinged individuals which emigrate, K the carrying capacity of the environment, and P the success of emigrants (see Järvinen and Vepsäläinen, 1976, for further explanation). Instability of the environment and isolation were not included in the model, but their effect can partially be studied, because both stability and isolation have an influence on the success of emigrants.

We concluded that density-dependent selection may account for wing dimorphism in *Gerris* populations. This dimorphism is stable, and it is the optimum strategy of the metapopulation. (See also Roff's (1975) more general model.) One interesting prediction from our model is that, due to environmental catastrophes in larger regions, morphism cycles may occur. After the catastrophe vacant sites are abundant and long-wingedness is favored. If the population sites tend to be stable, they are gradually colonized and the selective advantage of dispersal flights and longwingedness decreases. In relatively isolated population sites monomorphism for short wings may be optimal, and here the population may evolve from longwingedness through a dimorphic phase to shortwingedness. Morphism cycles are, of course, also implicit in our previous models (see above). Clearly the concept of morphism cycles bears some resemblance to taxon cycles (e.g., Ricklefs, 1970; Ricklefs and Cox, 1972; Wilson, 1961).

Discussion

The Evolution of Winglessness

The emergence of the group Insecta meant that a new subset of the ecological hyperspace was conquered. As a whole, the new mode of living was a powerful specialization, but the great success was largely built on flight ability. This made possible "seeking pastures new" even as a routine part of the life cycle. Thus the insects made use of habitats which could not be colonized by other organisms. The winged insects date back to the Devonian period which, together with the earlier Silurian, were considerably less stable than the previous periods. The instability continued to the Carboniferous period, known for the extensive adaptive radiation of insects. A whole array of opportunistic roles evolved as effective dispersal made possible the exploitation of capriciously changing environments. It is notable that the land vertebrates also evolved in the Devonian. The main function of wings in insects and limbs in early amphibians was the same, to keep pace with the changing environments. An "urge toward land" need not be postulated. On the contrary, as Romer (1959) puts it for amphibians, the development of limbs seems to have been an adaptation for remaining in the water (for insects, see also Wigglesworth, 1976).

Gradually further specialization took place among insects. These and better competitive abilities were possible only in predictable environments, preferably stable, but at least predictably changing. In a mosaic of habitats specialization includes evolution of means to stay where you are (cf. also Solbreck and Dingle, this volume). In extreme cases inability to fly is favored. In his excellent work on Corixidae, Young (1965a) showed that temperature (probably modified by photoperiod or other environmental factors) has a powerful effect on the development of flight musculature. Young (1965b, p. 167) suggests one way in which environmentally determined flight muscle variation may have evolved. He starts with the "competition" between maturation of flight and reproductive organs (the oogenesis-flight syndrome; Johnson, 1969), and gives a series of corixid species which represent different stages in the evolutionary sequence. Of course the species are not supposed to be on their way toward the "final" stage, but they represent different environmentally conditioned adaptive strategies. Young (1965b) gives as examples *Sigara nigrolineata*, where maturation of flight musculature has been retarded in relation to ovarian growth, and *S. lateralis*, where the flightless morph has retained the capacity for later full development of flight muscles. From here it is only a short step to the fixation of the flightless morph for the total life of the imago. The fixation can be made earlier by repressing the development of wings. The decision can be made relatively late by environmental cues at the time of developmental option (during late third or fourth larval instar in *Gerris*). A further step would be the genetic switch demonstrated in *Gerris lateralis*, which fixes the morph at fertilization and makes the responses of the population more rigid. The sequence of different strategies would thus begin with obligatory development of flight organs (and a compensatory *r*-strategy for losses during migration) in unpredictable environments, while at the opposite extreme would be genetically fixed apterousness (and specialization) in stable and isolated environments.

I do not imply that environmentally determined wing length would be a necessary and even always possible alternative to genetically switched morphs. Andersen (1973) suggested that in semiaquatic Hemiptera permanent (genetic) alary polymorphism is an ancient trait, and that seasonal polyphenism may have evolved quite recently. This might seem an intuitively pleasing proposal, as seasonal polyphenism is especially well developed at temperate latitudes which have experienced glaciation and cannot have had a long uninterrupted history of adaptation. Relevant comparative data on tropical gerrids are scarce. Andersen (1975) supplied sporadic wing-length records on *Limnogonus*, *Neogerris*, and *Tenagogonus* species of the Indo-Australian and Pacific areas. The genera belong to the subfamily Gerrinae and have thus a quite similar morphology to *Gerris* species studied here, and many species colonize intermittent waters. All species show alary dimorphism, which seems to be permanent and genetically determined (Andersen, 1975, implicit, e.g., on p. 11).

The postglacial period has, however, been long enough for genetic differentiation even in northernmost Europe. Instead of giving a historical explanation to the differences between temperate and tropical regions, I formulate the problem in another way: Why is seasonal polyphenism so rare (perhaps nonexistent) in the tropical Gerrinae? The adaptiveness of polyphenism depends on the advantages and costs of the strategy (both absolute and relative to other strategies). Evolution of polyphenism is too expensive in areas which do not supply any environmental signs of predictive value. The environmental cue, to be useful as a switch, should be available at or before the time of developmental option and when both these events precede the environmental selective force (Bradshaw, 1973). The climate of temperate latitudes (with its clear seasonality and pronounced variation of weather) supplies more reliable predictive cues than do the Indo-Australian and Pacific climates. In the tropics seasonal polyphenism might be more expensive than genetic polymorphism, which widens the niche of many Gerrinae species over both permanent and intermittent waters. Temporary habitats are colonized by longwinged imagos after rains (Fernando, 1961), when the probability of producing an offspring generation there is maximal. However, the offspring generation has no suitable cue to decide, at the time of developmental option, between short and long wings, and thus it is monomorphically longwinged ready to leave the site.

Why is the Gene for Short Wings Dominant?

In those few cases where the genetics of wing-length determination in natural populations has been elucidated (e.g., the curculionid *Sitona hispidula*, Jackson, 1928; the carabid *Pterostichus anthracinus*, Lindroth, 1946; *Gerris lateralis* and *G. lacustris*, Vepsäläinen, 1974b), the gene (or supergene) for short wings has been shown to dominate over that for long wings. Recently W.D. Hamilton (unpublished) made a most crucial point and asked why this should be so. I attempt here to answer the question for *Gerris*, but offer it also as a tentative hypothesis for other organisms.

It was shown above that the shortwinged morph is always the local morph, and hence the conservative morph. The function (in the sense of Williams, 1966, p. 9) of the wings is flight, and hence the longwinged morph is the intersite morph, or the colonizing, progressive morph. As migration inevitably implies a risk of failure, emigration from a stable site is by and large selected against (note, however, the effect of resource limitation). Colonization becomes more favorable when the stability of habitats decreases (see previous sections). Hence when emigration of longwinged individuals is favored, the offspring often also face a risk due to instability of the environment. Thus I measure the costs of migration by the proportion of shortwinged individuals segregated to the offspring of the migrants at newly colonized sites. A higher frequency of shortwinged in the offspring means a higher cost, which decreases the advantage of migration.

For comparison of the costs I assume simple Mendelian segregation at one locus with two alleles for wing length, and random mating in the local population. (Environmental switches can here be neglected.) The costs are calculated for an organism with the allele for short wings dominant over the allele for long wings (as in *G. lacustris*), and for a hypothetical case with the allele for long wings dominant. The costs depend essentially on the timing of matings in relation to migration. I give the costs for both pre- and postmigration matings. The question is, what is needed to make the (evolution of) dominance of the gene for short wings universally economical on the above presumptions?

The proportion of shortwinged offspring (Z) at the colonized site is given as a function of the frequency of shortwinged individuals (SW) in the takeoff population of the migrants.

(I) The allele for short wings dominant; premigration matings

$$Z = 1 - \sqrt{1 - SW}$$

(II) The allele for short wings dominant; postmigration matings

$$Z = 0 \cdot SW = 0$$

(III) The allele for short wings recessive; premigration matings

$$Z = \frac{SW}{1 + \sqrt{SW}}$$

(IV) The allele for short wings recessive; postmigration matings

$$Z = \frac{SW}{SW + 2 \cdot \sqrt{SW} + 1}$$

With postmigration matings, dominance is always the less costly determination of wing length, as no shortwinged are segregated. With premigration matings, dominance is the less costly strategy when SW is less than 0.5, and the more costly one when SW is over 0.5. This means that when a shortwinged mutant appears in a population of longwinged individuals, the evolution of dominance is expected (if the mutant allele happens not to be dominant) so long as the frequency of short-winged individuals is below 50% (i.e., the frequency of the recessive gene below 0.71).

The advantage of dominance is enhanced when the flight-oogenesis syndrome is taken into account. The excellent review of Johnson (1969) demonstrates that the syndrome occurs in most insect species studied for migration and maturation of ovaries. Development of flight apparatus retards maturation of ovaries, and migrants are also known to ignore mating and foraging stimuli. It can be generalized that premigration mating and maturation of ovaries is an exceptional phenomenon, and their frequencies are very low in most species. It is easy to see that when at least two thirds of the migrants have the flight-oogenesis syndrome, the advantage of the dominance of the allele for short wings is universal. Accordingly, evolution of dominance is expected to be strong even in species with high frequencies of shortwinged individuals. In *Gerris* dominance is always expected, as all species lay many batches of eggs and several copulations are needed to supply sperm for all batches. Thus in water-striders most of the eggs are always fertilized with sperm obtained at the site of egg-laying. Summarizing, the evolution of dominance of the allele for short wings is expected in *Gerris*, but the general features of most insect species (costs of segregating shortwinged individuals at the colonized site, and flight-oogenesis syndrome) make the explanation plausible also for other species.

How, then, does colonization of very isolated sites by typically short-winged species succeed, as the whole strategy of the species demands transferring shortwinged individuals to the new site? The answer is simple, as only very stable habitats are worthy of colonizing in these species, and in these time is not a critical variable. That is, the turnover rate of populations is very slow, suitable sites will become colonized even with low rate of migration, and the new population has time to evolve. In extreme cases of very high shortwinged frequencies these "costs" may thus prove more or less advantageous, as migrants in all are very scarce. Note also that monomorphic shortwingedness evolves as an adaptation to stable but often isolated habitats. In such populations the costs are formed by the longwinged individuals segregated at the colonized site, and dominance of the allele for short wings evolves.

Life-history Strategies in Relation to Environmental Predictability

I have tried to picture different possibilities for life-history tactics in Figure 6. Each *Gerris* population can be described by combining the proper individual possibilities I to VIII. So, for instance, a Finnish *G. najas* population is described by the line I. On the other extreme, it is conceivable that some Central European *G. lacustris* populations, "the-water-striders-of-all-trades," are covered only by the total set from I to VIII (or at least to VII: cf. Fig. 1). Good evidence for the line VIII has been obtained from Hungarian *G. thoracicus*.

Lengthening photoperiod leads development to lines other than I, III, and IV, and hence to multivoltinism. High temperatures together with lengthening photoperiods increase the proportion of nondiapause (as it accelerates the rate of development) and thus contribute positively to population increase. This is exactly what Johnson (1966) anticipated, and what has been shown by Dingle (1968) in *Oncopeltus fasciatus* bugs. In the above lygaeid, however, the range of strategies is narrower than in *Gerris* as shortwingedness is lacking.

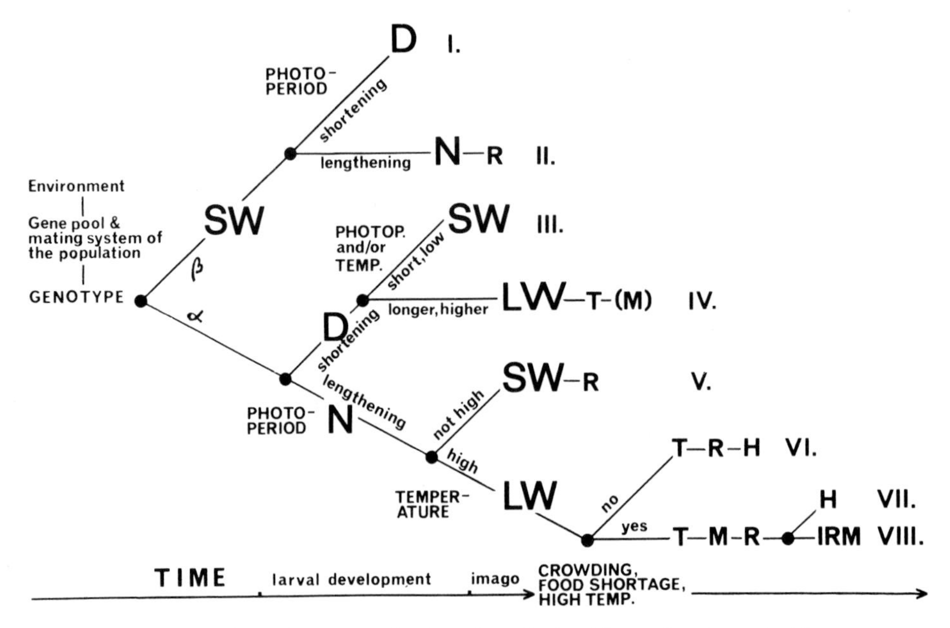

Fig. 6. Individual life-history and wing-length tactics of water-striders as a function of the genotype and environmental factors. The population strategy is a combination of individual tactics. The time scale is for the switch points which are marked with a dot. The switch factor is given with capital letters. The phenotype of the imago is described with symbols (see below) along its line of development. NOTE that the *SW* morph is ready to reproduce *immediately* after the last eclosion (if *N* = nondiapause, which does not overwinter) or immediately after overwintering (if *D* = diapause). *LW*-ness includes always a *delay* (*T* = teneral period), after which migration (*M*) or still later reproduction (*R*) is possible. If migration takes place, reproduction may be *further delayed*. Histolysis of flight muscles (*H*) soon after the commencement of reproduction, and interreproductive flights (*IRM*) are mutually exclusive. α = the recessive homozygote, β = the dominant homozygote or the heterozygote (see Fig. 2). The environmental switches for wing length are omitted from line β.

Multivoltinism is adaptive in the *Gerris* habitats affected by high temperatures; the temporal uncertainty is high as the risk of habitat destruction by drying-up is notable. Abundant offspring and effective dispersal lead to the colonization of a maximal number of sites suitable for reproduction, and thus variation in the extinction rates will not easily destroy the whole metapopulation. As especially in multivoltine populations death from starvation and cannibalism is potentially density-dependent, selection toward migratory habits is enforced. In extreme cases line VIII with interreproductive flights is the strategy of the whole population.

In stable habitats *K*-selection should lead, e.g., to population sizes at the *K*, no necessity for recolonization, slower development, larger body size, longer life span, iteroparity, and specialization (Pianka, 1970). All these trends are discernible (of course compared to other *Gerris* species) in *G. najas*, a univoltine inhabitant of middle-sized lotic waters, monomorphic for winglessness (Vepsäläinen, 1973a, 1973b, and unpublished). From the above, two testable predictions follow. First, the turnover rate (MacArthur, 1972) of *G. najas* populations should be lower than for other *Gerris* species. Second, as chance extinction is lowered more by decreasing mortalities than by increasing the number of offspring (and the risk of overcrowding) (MacArthur and Wilson, 1967; limitation of "ups and downs," Williams, 1966, p. 107), *G. najas* should have a smaller $\hat{r}$ than other species. The first hypothesis can be tested by monitoring natural populations and the second one by laboratory estimation of r for *Gerris* species.

The relations among the life histories (Fig. 6), wing-length strategies (Table 2), environmental conditions, and ecological features of *Gerris* populations are given in Table 3. There is a hierarchy from genetically determined shortwingedness in the stable and isolated sites (category A) through genetically switched dimorphism in a fairly stable but perhaps resource-limited patchily distributed habitats (B), to environmentally switched seasonal dimorphism, capable of tracking relatively unstable conditions by short-term environmental cues (C) or by monomorphic longwingedness even in capricious (Lewontin, 1966) environments (D). The mixed strategy (B + C) meets heterogeneity in space and time.

In Table 3, habitat specialization decreases from category A toward D, and dispersal ability increases. Accordingly, the sequence from A to D should also describe the hierarchy of decreasing competitive ability. The prediction is supported by observations in the field (*G. lateralis* over *G. thoracicus*, *G. odontogaster*, and *G. lacustris*, among others; Vepsäläinen, 1973a) and on developmental times in the laboratory (*G. lateralis*, *G. lacustris*, and *G. odontogaster*, from fastest to slowest; Vepsäläinen, 1973b), and deductions on the distribution and population biology (the fugitive role of *G. thoracicus*; Järvinen, 1976). Also, the habitat range of *G. lateralis* is widest in northern Finland, where other species are lacking or rare (Vepsäläinen, 1973a), and probably this release is due to the absence of many poorer competitors but more prompt colonizers (Vepsäläinen and Järvinen, 1974).

Gerris species - and also the individual tactics in Figure 6 - are easily spread on the r-K spectrum (see discussion in Vepsäläinen, 1974c, Chapter VII), but it is difficult to make progress before the demographic statistics of the species are better known (see Southwood, 1975, 1976; Stearns, 1976). Järvinen (1976) pointed out the importance of measuring extinction probabilities. The set of Finnish *Gerris* species offers an excellent possibility to study the relation between demographic strategies and the predictability of the environment in space and time (see Rabinovich, 1975). To describe qualitatively an extreme *Gerris* strategy in Finland, I characterize that of *G. thoracicus*.

The "Supertramp" of the Miniature Archipelago

We have used in our models several environmental variables, namely heterogeneity in space and time, isolation, and resource limitation. All these are intimately interconnected and exert heavy influence on water-strider populations. I have lately studied *G. thoracicus* in the Tvärminne archipelago in the Finnish Baltic. The species is usually univoltine there, but sometimes a small fraction of the midsummer population develops as nondiapause imagos. The species typically inhabits a set of miniature archipelagos consisting of very small, temporary, poorly isolated, and fairly unproductive rock-pools, which often dry up during the first half of summer - but it reproduces also in brackish-water bays in the Baltic. Under the selection pressure of these factors many special features have evolved.

The overwintered imagos have long life expectancies. Their flight muscles do not histolyze after postwinter dispersal and commencement

Table 3. The relations between the wing-length strategies, life histories, environments, and ecological features of *Gerris* populations. Resource limitation is not included.

Class	Strategy	Environment	Life History (ref. Fig. 6)	Ecological Features	Real Cases
A	Monomorphic *SW*-ness	Stable, isolated	I	Narrow habitat specialist	Finnish *G. najas*
B	Dimorphism, genetic switch	Stable, varying isolation	I+IV (univoltine)	±narrow niche ±good competitors	Many Finnish *G. lateralis*
B+C	Mixed strategy	Heterogeneous in space and time	I-VII(VIII) (bivoltine)	Broad niche, generalist	*G. lacustris* in Central Europe
C	Dimorphism, environmental switch	±unstable, but short-term predictive cues exist; not isolated	IV(+III)(univoltine) III-VII(VIII) (bivoltine)	±fugitive species, powerful dispersal, ±broad niche, *r*-strategy	*G. odontogaster*
D	Monomorphic *LW*-ness	Capricious in time	IV-(V-VII)VIII (multivoltine)	Fugitive species, interreproductive colonization, ±narrow niche, bet-hedging	*G. thoracicus* in Central Europe

of reproduction, as those of most other species do. Interreproductive flights are frequent, and the females disperse their eggs over a long time and many sites. This minimizes the risk of cannibalism and starving of offspring. It also maximizes the number of colonized pools and thus minimizes the effect of year-to-year variation in the drying-up of ponds on the metapopulation. This spreading out of the risk in space and time (den Boer, 1968) leads to maximization of the average fitness in the pond system. The strategy has been called, e.g. by Stearns (1976), bet-hedging, and it includes many features labeled under *K*-strategy in the *r-K* treatment (e.g., small egg batches, iteroparity, long life expectancy of imagos). Bet-hedging seems to be adaptive, if juvenile mortality is highly variable compared to that of adults (see Stearns, 1976). This is the case in *G. thoracicus*, where the persistence of a given rock-pool over the larval period is poorly predictable. Note that the environmental conditions would predict *r*-strategy, using the conventional *r-K* treatment. Bet-hedging, taking features of *K*-strategy in extremely unpredictable and rapidly destroyed habitats, is adaptive in environments where the total population extinction is very much independent of how many eggs are laid and how early in the life of the adult the "big bang" reproductive effort would occur. The point is to average out the catastrophes by keeping the egg-laying individuals alive and active as long as possible (cf. Cohen, 1966). So the *r-K* spectrum obtains an extension over the extremes of *r*-selecting environments: that of bet-hedging (see also Solbreck, this volume).

However, catastrophic summers do occur, and then the offspring number per imago is near zero in the rock-pool system. In such cases the brackish-water refugia of *G. thoracicus* save the population from extinction. Every summer part of the eggs are laid in brackish-water bays of the islands and coasts of the Baltic Sea. In exceptionally drought-stricken summers temperatures are likely to be high, facilitating development in the sea water. By including the two most extreme habitats (judging by permanency and water temperatures) in its array, *G. thoracicus* avoids "putting all its eggs in one basket." The strategy of *G. thoracicus* resembles that of two corixids, *Arctocoria carinata* (Sahlb.) and *Callicorixa producta* (Reut.) studied by Pajunen (1971) in the very same rock-pool system. Their mortality is, however, more efficiently density-dependent and cannibalism is pronounced (Pajunen, 1977). Moreover, they do not live in the brackish water, even if they tolerate it. The differences between the habitat range in the corixids and *G. thoracicus* is presumably due to the somewhat more rapid development of the corixid species (see Pajunen and Sundbäck, 1973), and perhaps also to their more powerful dispersal capacity (see Pajunen and Jansson, 1969). Accordingly, corixids survive by bet-hedging and contracting the development of offspring to complete it before drying of the pond. The gerrids survive in spite of a longer developmental time by extending the habitat niche to brackish water.

Conclusions and Summary

The set of European *Gerris* species (ten of them included here) offers a good opportunity to study the adaptive significance of wing-length variation, among others maintenance of alary dimorphism, by using the comparative method. The local variation of strategies between species is high as is the geographical variation within the species. The wing-length patterns vary from monomorphic winglessness to monomorphic long-wingedness, and from seasonal polyphenism to genetic polymorphism for wing length (Table 2). Accordingly, the mechanisms of wing-length determination include both environmental and genetic switches (Fig. 2). In extreme cases, alleles for short and long wings segregate in the same population, where additionally photoperiod (incremental vs. decremental change of day length) and temperature work as switches for wing length (in nature mainly or totally on the "longwinged," recessive genotype). Photoperiod determines also the reproductive state of the new imago, and all larvae which spend their fourth larval instar (or any previous instar) in shortening illumination rhythm develop diapause (Fig. 2). An example of the wing-length and diapause variation during the year of one *G. lacustris* population is given in Figure 1.

When the adaptive significance of wing-length variation is sought by examination of local populations, monomorphic shortwingedness - due to its greater local reproductive efficiency - is revealed to form the optimum within-site strategy. Longwingedness is the between-sites strategy, as colonization ability is favored in habitats established anew after temporary drying up (fatal to the local population). The relation is obvious between high temperatures which switch the development of larvae to longwinged imagos (in conditions which in lower temperatures would produce shortwinged imagos) and an increase of the risk of drying up of the habitat. The different functions of the two morphs may lead in nature to alary dimorphism by immigration of longwinged individuals and the mixture of two optimum strategies (Fig. 3). Isolation plays an important role, as isolation in connection with even moderate instability of the habitat may prevent colonization - thus all optimal strategies are not possible for natural populations. History is also essential in explaining deviations from optima (see the discussion on morphism cycles; in previous sections).

By fitness set analysis it was shown that alary dimorphism may be optimal in metapopulations living in environments heterogeneous in time (polyphenism; Fig. 4) and space (genetic polymorphism; Fig. 5). By explicitly introducing density-dependence, a most realistic factor, to the models of wing-length strategies, and measuring the fitness of both morphs over all sites utilized, alary dimorphism is shown to be optimal in wide ranges of environmental circumstances (Järvinen and Vepsäläinen, 1976). The present set of models renders many features of water-strider biology intelligible. A comprehensive review of these has been provided by Järvinen and Vepsäläinen (1976).

The diapause determination of *Gerris* by decremental change of day length instead of absolute day lengths is rare among the many arthropod species studied for diapause. This mechanism was explained to be an adaptation to resource-limitation of water-strider populations. As water-striders fail to overwinter after reproduction, it is not advantageous to produce more generations than are allowed by the carrying capacity of the environment, plus the individuals which emigrate (Järvinen and Vepsäläinen, 1976). Differences in voltinism between geographical regions are allowed, among other factors, due to differences in the temperature conditions and the length of the season available for reproduction before summer solstice (the critical point for diapause induction). The long season after summer solstice in southern Europe can be utilized by extending the life and egg-laying period of nondiapause individuals; this also decreases crowding, a potential risk factor in multivoltine populations.

Evolution of flight ability in arthropods is understood as an adaptation to keep pace with changing environments, where the habitat patches favorable for the species disappear and reappear irregularly.

Evolution of flightlessness in insects with flight ability is explained as a reverse process of specialization to stable but isolated habitats.

By measuring the costs of migration and colonization of new, more or less unstable habitats in terms of segregated shortwinged individuals at the colonized site, it is expected that dominance of the gene for short wings evolves. On these presumptions, inclusion of a moderate flight-oogenesis syndrome (with postmigration matings) makes the advantage of dominance universal. On the other hand, in species of stable and isolated habitats migration is selected against (the costs consist now of longwinged individuals), and evolution of dominance of the allele for short wings is expected. Of course, in environments heterogeneous in time, costs of any one system including genetic segregation of wing morphs are too high, and short-term environmental switch signals for wing length are used, if available. Now the costs are presumably built in to the genetic system of the individual, which includes the potential for either one of the morphs, and the developmental path taken depends on the environmental signals at (or before) the time of developmental option. Accordingly, the pattern of environment decides whether it is more adaptive to allocate the energy to one genotype which has a wide response potential, or to many canalized ones. In this respect *G. lacustris* is the most efficient species, as it affords both types of energy distribution.

The benefit of the comparative approach by using a representative set of phylogenetically related *Gerris* species and climatically different regions is evident when the life-history strategies are related to the degree of environmental predictability. The realized strategies of different species of the same genus in different environments can, by using both inter- and intraspecific comparison, be visualized to approximate different optima in different environments. Thus the evolution of strategies in relation to the pattern of environments in space and time may be mirrored. An array of different cases is brought forth, which stimulates further efforts to estimate relevant life-history parameters (claimed latest by Stearns, 1976).

The diversity of individual life-history tactics in *Gerris* is pictured in Figure 6. Population strategies are shown, for example, in Table 3, where they are related to the environments, life histories and ecological features of the species. Monomorphic shortwingedness is realized in the Finnish *G. najas*, a univoltine, narrow habitat specialist of stable, isolated middle-sized rivers. The Finnish *G. lateralis* has a wider but still more or less narrow habitat niche. It is a good competitor of stable habitats. In southern Finland where the isolation of habitats is high, almost totally shortwinged univoltine populations are met, but in central Finland where the habitat niche is broader and isolation smaller, genetic polymorphism is pronounced. *G. odontogaster* is a fairly poor competitor which has a powerful dispersal and a broad habitat niche with a marked weight at the temporary end of the habitat spectrum. Isolation of suitable sites for reproduction is low. Partial bivoltinism is met in southern Finland, and these populations are seasonally polyphenic. The species is a pronounced *r*-strategist compared to the above ones. *G. lacustris* also has a wide habitat niche, but the weight is at the permanent end of the habitat spectrum. The species has efficiently combined a mixed strategy of those met in *G. lateralis* and *G. odontogaster*. Bivoltinism is not as pronounced as in *G. odontogaster*. The longwinged individuals colonize unstable habitats and have the potential for seasonal polyphenism, and genetically shortwinged individuals are selected for habitats. If these happen to be isolated, complete shortwingedness evolves locally in areas where the shortwinged morph may be the less frequent one in all other populations.

All the above species can be spread along the *r-K* spectrum, although more accurate life-history data (on *r* and the time-energy budget of reproductive efforts) would clearly enhance the picture. One species cannot, however, be introduced to the *r-K* theater without equipping it with the two faces of Janus. By this phrase I refer to the many well-known cases, where *r*-selected species resort to migration - an activity which delays reproduction and thus adjusts the behavior toward that of *K*-strategists (e.g., development of longwinged instead of shortwinged, nondiapause imagos of *G. odontogaster*, when the risk of habitat destruction by drying increases; cf. Kennedy, 1975; Southwood, 1975). By *r-K* treatment *G. thoracicus* of the highly unpredictable environments is expected to be effectively *r*-selected. However, its life span is relatively long, reproduction is delayed (due to development of flight musculature and migration), it lays comparatively small batches, and spreads the reproductive effort over a long time. All these features belong to *K*-selected species, but the environment does not predict it. The above observations are explained by the theory of bet-hedging, where the results resembling those of *K*-selection in stable environments are attained in unstable conditions where juvenile mortality varies more than that of the adults (see Stearns, 1976). Real *K*-strategists are good competitors, but *G. thoracicus* is the fugitive species among *Gerris*. Much of its reproductive effort is put into small rock-pools which dry up unpredictably, and complete extinctions of offspring batches are frequent. As the rock-pools comprise a miniature archipelago, interreproductive flights form a special feature of the strategy, and spreading of the risk over space and time (i.e., averaging the risks) is accomplished. As the uninterrupted existence time of suitable rock-pools may in some summers be shorter than the developmental time of offspring of any batch, spreading of the risk over habitats is needed to maintain the populations. This is accomplished by colonizing also stable brackish-water bays. Bet-hedging, leading to averaging out the risks by extended egg-laying and widened habitat niche, is visualized as an extension to the extreme end of the classical *r*-selection circumstances, where the environment of any larval cohort is highly unpredictable; i.e., the changes of habitat are so sudden and unexpected that they cannot be tracked even by any environmental switch signals.

Thus temporary environments consist of two basically different kinds, one where a correlation exists between conditions at the time a choice must be made (e.g., for development of wing length or for the timing of egg-laying) and the subsequent outcome (i.e., number of offspring) and one where there is little or no correlation. The implications for the strategy of the organism are different in each case. The stronger the correlation, the more effectively can reproductive effort be concentrated in a restricted place and time; the weaker the correlation, the more pronounced the bet-hedging strategy over space and time expected (cf. Cohen, 1967 *vs.* Cohen, 1966).

Acknowledgements Most useful comments on the first draft of the manuscript were given by Hugh Dingle, Conrad Istock, Olli Järvinen, Lauri Oksanen, Esa Ranta, and Christer Solbreck. I forward my sincere thanks to them all.

References

Andersen, N. Møller: Seasonal polymorphism and developmental changes in organs of flight and reproduction in bivoltine pondskaters (Hem. Gerridae). Entomol. Scand. *4*, 1-20 (1973).

Andersen, N. Møller: The *Limnogonus* and *Neogerris* of the Old World with character analysis and a reclassification of the Gerrinae (Hemiptera: Gerridae). Entomol. Scand., Suppl. *7*, 1-96 (1975).

Boer, P. J. den: Spreading the risk and stabilization of animal numbers. Acta Biotheor. *18*, 165-194 (1968).

Bradshaw, W. E.: Homeostasis and polymorphism in vernal development of *Chaoborus americanus*. Ecology *54*, 1247-1259 (1973).

Cohen, D.: Optimizing reproduction in a randomly varying environment. J. Theor. Biol. *12*, 119-129 (1966).

Cohen, D.: Optimizing reproduction in a randomly varying environment when a correlation may exist between the conditions at the time a choice has to be made and the subsequent outcome. J. Theor. Biol. *16*, 1-14 (1967).

Danilevskii, A. S.: Photoperiodism and Seasonal Development of Insects (Translation). Edinburgh: Oliver & Boyd 1965.

Dingle, H.: Life history and population consequences of density, photoperiod, and temperature in a migrant insect, the milkweed bug *Oncopeltus*. Amer. Nat. *102*, 149-163 (1968).

Dingle, H.: Migration strategies of insects. Science *175*, 1327-1335 (1972).

Dingle, H.: The experimental analysis of migration and life-history strategies in insects. *In* Experimental Analysis of Insect behavior. Browne, L. B. (ed.). Berlin: Springer 1974, pp. 329-342.

Dingle, H.: This volume.

Fernando, C. H.: Aquatic insects taken at light in Ceylon, with a discussion and bibliography of references to aquatic insects at light. Ceylon J. Sci. (Bio. Sci.) *4*, 45-54 (1961).

Ford, E. B.: Genetic Polymorphism. London: Faber & Faber 1965.

Haldane, J. B. S.: The Causes of Evolution. Ithaca, New York: Cornell Univ. Press 1932 (1966).

Haldane, J. B. S., Jayakar, S. D.: The nature of human genetic loads. J. Genet. *59*, 143-149 (1965).

Honěk, A.: The regulation of wing polymorphism in natural populations of *Pyrrhocoris apterus* (Heteroptera, Pyrrhocoridae). Zool. Jb. Syst. *103*, 547-570 (1976).

Horn, H. S., MacArthur, R. H.: Competition among fugitive species in a harlequin environment. Ecology *53*, 749-752 (1972).

Jackson, D. J.: The inheritance of long and short wings in the weevil, *Sitona hispidula*, with a discussion of wing reduction among beetles. Trans. Roy. Soc. Edinburgh *55*, 665-735 (1928).

Järvinen, O.: Migration, extinction, and alary morphism in water-striders (*Gerris* Fabr.). Ann. Acad. Sci. Fennicae A IV *206*, 1-7 (1976).

Järvinen, O., Vepsäläinen, K.: Wing dimorphism as an adaptive strategy in water-striders (*Gerris*). Hereditas *84*, 61-68 (1976).

Johnson, C. G.: A functional system of adaptive dispersal by flight. Ann. Rev. Entomol. *11*, 233-260 (1966).

Johnson, C. G.: Migration and Dispersal of Insects by Flight. London: Methuen 1969.

Kennedy, J. S.: Insect dispersal. In: Insects, Science, and Society. Pimentel, D. (ed.). New York: Academic Press 1975, pp. 103-119.

Krajewski, S.: Wasserwanzen (Heteroptera) des Flusses Grabia und seines Überschwemmungsgebietes. Bull. Entomol. Pologne *39*, 465-513 (in Polish) (1969).

Landin, J., Vepsäläinen, K.: Spring dispersal flights of pondskaters *Gerris* spp. (Heteroptera). Oikos *29*, 156-160 (1977).

Lees, A. D.: The control of polymorphism in aphids. Adv. Insect Physiol. *3*, 207-227 (1966).

Levins, R.: Mendelian species as adaptive systems. General Systems *6*, 33-39 (1961).

Levins, R.: The theory of fitness in a heterogeneous environment. IV. The adaptive significance of gene flow. Evolution *18*, 635-638 (1964).

Levins, R.: Evolution in Changing Environments. Princeton, New Jersey: Princeton Univ. Press 1968.

Levins, R.: Some demographic and genetic consequences of environmental heterogeneity for biological control. Entomol. Soc. Amer. Bull. (Washington) *15*, 237-240 (1969).
Levins, R., MacArthur, R. H.: The maintenance of genetic polymorphism in a spatially heterogeneous environment: variations on a theme by Howard Levene. Amer. Nat. *100*, 585-589 (1966).
Lewontin, R. C.: Is nature probable or capricious? BioScience *16*, 25-27 (1966).
Lindroth, C. H.: Inheritance of wing dimorphism in *Pterostichus anthracinus* Ill. Hereditas *32*, 37-40 (1946).
MacArthur, R. H.: Geographical Ecology. New York: Harper & Row 1972.
MacArthur, R. H., Wilson, E. O.: The Theory of Island Biogeography. Princeton, New Jersey: Princeton Univ. Press 1967.
Mayr, E.: Animal Species and Evolution. Cambridge, Mass.: Belknap Press 1963.
Mielewczyk, S.: Heteroptera of the waters of the Jelenia Gora basin. Badania Fizjograficzne nad Polska Zachodnia *14*, 35-57 (in Polish) (1964).
Mielewczyk, S.: The influence of warmed up waters on the brachypterism of *Gerris paludum* (Fabr.) (Heteroptera: Gerridae). Archiwum Ochrony Srodowiska *2* 201-208 (in Polish) (1976).
Mitis, H. von: Ökologie und Larvenentwicklung der mitteleuropäischen *Gerris*-Arten (Heteroptera). Zool. Jb. Syst. *69*, 337-373 (1937).
Odum, E. P.: Fundamentals of Ecology. 3rd ed. Philadelphia: Saunders 1971.
Pajunen, V. I.: Adaptation of *Arctocorisa carinata* (Sahlb.) and *Callicorixa producta* (Reut.) populations to a rock pool environment. Proc. Adv. Study Inst. Dynamics Numbers Popul. (Oosterbeek, 1970), 148-158 (1971).
Pajunen, V. I.: Population structure in rock-pool Corixids (Hemiptera, Corixidae) during the reproductive season. Ann. Zool. Fennici *14*, 26-47 (1977).
Pajunen, V. I., Jansson, A.: Dispersal of the rock pool corixids *Arctocorisa carinata* (Sahlb.) and *Callicorixa producta* (Reut.) (Heteroptera, Corixidae). Ann. Zool. Fennici *6*, 391-427 (1969).
Pajunen, V. I., Sundbäck, E.: Effect of temperature on the development of *Arctocorisa carinata* (Sahlb.) and *Callicorixa producta* (Reut.) (Hemiptera, Corixidae). Ann. Zool. Fennici *10*, 372-377 (1973).
Pianka, E.: On r and K selection. Amer. Nat. *104*, 592-597 (1970).
Rabinovich, J. E.: Demographic strategies in animal populations: a regression analysis. In: Tropical Ecological Systems. Trends in Terrestrial and Aquatic Research. Golley, F. B., and Medina, E. (eds.). Berlin: Springer, 1975, pp. 19-40.
Ricklefs, R. E.: Stage of taxon cycle and distribution of birds on Jamaica, Greater Antilles. Evolution *24*, 475-477 (1970).
Ricklefs, R. E., Cox, G. W.: Taxon cycles in the West Indian avifauna. Amer. Nat. *106*, 195-219 (1972).
Roff, D. A.: Population stability and the evolution of dispersal in a heterogeneous environment. Oecologia *19*, 217-237 (1975).
Romer, A. S.: The Vertebrate Story. 4th ed. Chicago: Univ. Chicago Press 1959.
Shapiro, A. M.: Seasonal polyphenism. In Evolutionary Biology 9. Hecht, M. K., Steere, W. C., Wallace, B. (eds.). New York: Plenum Press 1976, pp. 259-333.
Solbreck, C.: This volume.
Southwood, T. R. E.: The dynamics of insect populations. In: Insects, Science, and Society. Pimentel, D. (ed.). New York: Academic Press 1975, pp. 151-199.
Southwood, T. R. E.: Bionomic strategies and population parameters. In: Theoretical Ecology. Principles and Applications. May, R. M. (ed.). Oxford: Blackwell Sci. Publ. 1976, pp. 26-48.
Stearns, S. C.: Life-history tactics: a review of the ideas. Quart. Rev. Biol. *51*, 3-47 (1976).
Steffan, W. A.: Polymorphism in *Plastosciara perniciosa*. Science *182*, 1265-1266 (1973).
Stehlik, J. L.: The fauna of Heteroptera of the mountain High Jesenik. Acta Musei Moraviae *37*, 131-248 (in Czech) (1952).
Tauber, M. J., Tauber, C. A.: This volume.
Vepsäläinen, K.: The role of gradually changing daylength in determination of wing length, alary dimorphism and diapause in a *Gerris odontogaster* (Zett.) population (Gerridae, Heteroptera) in South Finland. Ann. Acad. Sci. Fennicae A IV *183*, 1-25 (1971a).
Vepsäläinen, K.: Determination of wing length and alary dimorphism in *Gerris*

odontogaster (Zett.) (Heteroptera) (Abstract). Hereditas *69*, 308 (1971b).

Vepsäläinen, K.: The distribution and habitats of *Gerris* Fabr. species (Heteroptera, Gerridae) in Finland. Ann. Zool. Fennici *10*, 419-444 (1973a).

Vepsäläinen, K.: Development rates of some Finnish *Gerris* Fabr. species (Het. Gerridae) in laboratory cultures. Entomol. Scand. *4*, 206-216 (1973b).

Vepsäläinen, K.: The wing lengths, reproductive stages and habitats of Hungarian *Gerris* Fabr. species (Heteroptera, Gerridae). Ann. Acad. Sci. Fennicae A IV *202*, 1-18 (1974a).

Vepsäläinen, K.: Determination of wing length and diapause in water-striders (*Gerris* Fabr., Heteroptera). Hereditas *77*, 163-176 (1974b).

Vepsäläinen, K.: The life cycles and wing lengths of Finnish *Gerris* Fabr. species (Heteroptera, Gerridae). Acta Zool. Fennica *141*, 1-73 (1974c).

Vepsäläinen, K.: Lengthening of illumination period is a factor in averting diapause. Nature (Lond.) *247*, 385-386 (1974d).

Vepsäläinen, K.: The determination and adaptive significance of alary dimorphism and diapause in *Gerris* Fabr. species (Heteroptera) (Abstract). Univ. Helsinki Sect. Math. Nat. Sci., 4 pp. (1974e).

Vepsäläinen, K., Järvinen, O.: Habitat utilization of *Gerris argentatus* (Het. Gerridae). Entomol. Scand. *5*, 189-195 (1974).

Vepsäläinen, K., Krajewski, S.: The life cycle and alary dimorphism in *Gerris lacustris* (L.) (Heteroptera, Gerridae) in Poland. Notulae Entomol. *54*, 85-89 (1974).

Vepsäläinen, K., Nieser, N.: The life cycles and alary morphs of some Dutch *Gerris* species (Heteroptera, Gerridae). Tijdschrift voor Entomologie (in press) (1977).

Wigglesworth, V. B.: The evolution of insect flight. In Insect Flight. Rainey, R. C. (ed.). Oxford: Blackwell Sci. Publ. 1976, pp. 255-269.

Williams, G. C.: Adaptation and Natural Selection. Princeton, New Jersey: Princeton Univ. Press 1966.

Wilson, E. O.: The nature of the taxon cycle in the Melanesian ant fauna. Amer. Nat. *95*, 169-193 (1961).

Young, E. C.: Flight muscle polymorphism in British Corixidae: ecological observations. J. Animal Ecol. *34*, 353-390 (1965a).

Young, E. C.: Teneral development in British Corixidae. Proc. Roy. Entomol. Soc. Lond. (A) *40*, 159-168 (1965b).

Migration and Diapause in Tropical, Temperate, and Island Milkweed Bugs

HUGH DINGLE

For when a new insect first arrived on the island, the tendency of natural selection to enlarge or to reduce the wings, would depend on whether a greater number of individuals were saved by successfully battling with the winds, or by giving up the attempt and rarely or never flying. As with mariners shipwrecked near a coast, it would have been better for the good swimmers if they had been able to swim still farther, whereas it would have been better for the bad swimmers if they had not been able to swim at all and had stuck to the wreck.

Charles Darwin, *The Origin of Species*, 1859, p. 104

Introduction

To study evolution is to study variation. What is a truism now was of course not so before 1859. And it was Darwin, also, who combined the study of variation with the comparative method which then as now was the primary method of the evolutionary biologist. No less than with other behaviors and other organisms, the comparative method has revealed the variety and intricacy of escape responses in insects.

Chief among these escape responses are migration and diapause. Indeed the two behaviors are often intimately related behaviorally and physiologically especially in adult insects (Kennedy, 1961). Both involve escape from unfavorable conditions and delay in reproduction (Dingle, 1972), and both may be primed by the same hormones (Rankin, 1974 and this volume). As will be demonstrated below, diapause may actually promote flight, and, of course, flights to and from diapause sites are well-known in a number of insects (Johnson, 1969).

Comparative studies have revealed an enormous amount of variation in both migration (Johnson, 1969, 1976; Kennedy, 1975) and diapause (Danilevskii, 1965; Dingle et al., 1977). Much of the variation in diapause especially is correlated with geographic differences in climate or photoperiod, but much remains to be explained, not least the selective forces which maintain the variation. Understanding of the selective processes becomes crucial in the face of mounting evidence that much of the variation is both environmental and genetic (Dingle et al., 1977; Hoy, Istock, Lumme, Masaki, Tauber and Tauber, Vepsäläinen, and Waldbauer, in this volume). Our understanding of the causes of variation in insect flight and migration is in only the earliest formative stages (Roff, 1975; Järvinen and Vepsäläinen, 1976; Johnson, 1976).

From an evolutionary standpoint, therefore, a comparative study of variation in diapause and migration both within and between species should yield interesting information. The beginnings of such a study are presented here. The insects chosen are tropical and temperate species of milkweed bugs (*Oncopeltus*; Heteroptera; Lygaeidae) occurring in the New World. Studies over the previous 12 years in my laboratory (Dingle, 1965, et seq.) provide a base line from which to start; other reasons for choosing *Oncopeltus* are given below.

Natural History of *Oncopeltus spp.*

The one species of Oncopeltus whose range extends into temperate areas is *O. fasciatus.* It is a common species over much of North America, and its life history and ecology have been studied in both laboratory and field (Dingle, 1968a, 1972, 1974a; Sauer and Feir, 1973; Ralph, 1976; Evans, 1977). In eastern North America this species is a long-distance migrant arriving in the northern parts of its range (e.g., Iowa, Michigan, and New York) in the latter part of June or early July and settling in milkweed patches (mostly *Asclepias syriaca* but also on *A. verticillata, A. incarnata, A. sullivanti,* etc.). Here the bugs reproduce, and their offspring develop on the maturing milkweed plants. From one to three generations are produced depending on the length and warmth of the summer season. The adults of the final generation enter a reproductive diapause as the days shorten around the time of the autumnal equinox and leave the milkweed patches presumably to migrate southward. The bugs cannot survive the northern winters with the result that the first severe frosts kill those unable to mature in time to undertake migration. A similar pattern is followed in California, but here the season is much more extended (March to November) and lack of food seems to be the operative factor promoting migration in the fall (Evans, 1977).

The range of *O. fasciatus* also extends into the tropics and subtropics where it is found throughout the year. The southernmost extensions of the range seem to be Costa Rica (possibly Panama) on the mainland and Guadeloupe in the Caribbean. In spite of extensive searching, my students and I, R.B. Root (personal communication), and F. O'Rourke (personal communication) have failed to find it south of these locations. Earlier reports (cited in Slater, 1964) that it occurred as far south as Brazil are apparently based on misidentifications. We have found populations in South Florida, Jamaica, Puerto Rico, and Guadeloupe (Fig 1), and it is widely distributed in Mexico (Root, personal communication). It is a rare species in Jamaica, but is common to abundant in appropriate habitats in the other locations. In all these areas the primary host seems to be *Asclepias curassavica,* but again other milkweed species, including both vines and forbes, are used as well. Breeding on the Caribbean islands evidently takes place throughout the year, but both our observations and those of Root (personal communication) suggest a possible dry season (winter) diapause in South Florida. The Florida case is particularly interesting and is discussed in greater detail below.

The remaining species of *Oncopeltus* to be discussed here are primarily tropical in distribution. Brief descriptions of ranges and natural histories of each species follow.

O. aulicus - This species seems to be confined to the Greater Antilles and the Bahamas. We have studied it in Jamaica and Puerto Rico where it occurs as widely scattered individuals mostly on *A. currassavica,* but also occasionally on *A. nivea.* It nowhere seems to build up the high local population densities found commonly in other *Oncopeltus* species.

O. cingulifer - This species is perhaps the most common and widespread species in the Caribbean, Central America, and northern South America (Chaplin, 1973; Root and Chaplin, 1976; Blakley, 1977) extending northward as far as southern Texas and Florida. Two morphs occur, the smaller *O. c. antillensis* predominating in the northern half of the range, and the larger *O. c. cingulifer* predominating in the southern half. The latter closely resembles *O. sandarachatus* with which it is often confused even in musuem collections. *O. cingulifer* occurs abundantly on *A. currassavica*

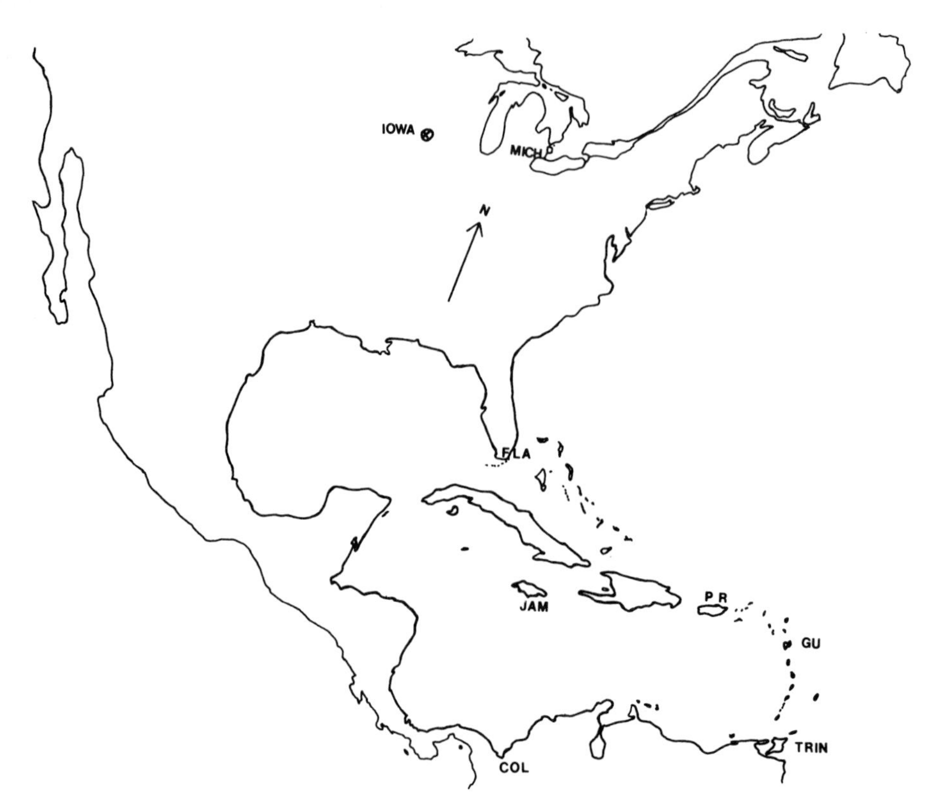

Fig. 1. Sources of populations of *Oncopeltus* spp. used in this study. Col. -- Colombia, Fla -- South Florida, Gu -- Guadeloupe, Jam -- Jamaica, PR -- Puerto Rico, Trin -- Trinidad.

wherever we have found it with the exception of St. Vincent, where we found only a few scattered individuals. We failed to find this species in Puerto Rico and Guadeloupe although it has been reported from Puerto Rico (Slater, 1964). In Jamaica *A. nivea* is also a host plant.

O. longirostris - Little is known of the ecology of this species. It occurs in Central America and northern South America where it seems to be confined largely to coastal lowlands. In addition to other species, it uses as a host the woody exotic *Calotropis procera* (Asclepiadaceae) which has spread through much of the Caribbean region especially in drier coastal lowlands. The long proboscis of *O. longirostis* may permit this species to exploit *Calatropis* which has thick pods and large flowers.

O. sandarachatus - This species seems largely confined to the southern Caribbean (Trinidad, Barbados, northern South America), and Central America. We found it relatively common on *A. curassavica* in Trinidad, although less so than *O. cingulifer*, but rare in Barbados. On the latter island our observations suggest that *A. curassavica* has been largely destroyed by larvae on monarch butterflies, *Danaus plexippus*, thus apparently accounting for the rarity of *O. sandarachatus*.

O. unifasciatellus - This species is widespread over Mexico, Central America, and northern South America (Chaplin, 1973) but seems to be confined largely to high elevations. Indeed it is unable to tolerate high temperatures in the laboratory (Dingle and Caldwell, 1971). Occasional, presumably dispersing, individuals are found in lowlands (Blakley, personal communication). It likewise uses *A. curassavica* as a host and overlaps extensively with *O. cingulifer* (Root and Chaplin, 1976).

All of these species are closely related and belong to the subgenus *Erythrischius* (Slater, 1964) which is confined to the New Yorld. Their similarity is further indicated by the fact that interspecific pairs are quite commonly seen copulating in the field, and hybrids between species can be produced in the laboratory (Chaplin, 1973, and below) and in some cases in the field (Blakley, personal communication). They are certainly ecologically similar to the extent that they overlap considerably on the same host plants although there may be some differences in abilities to use different plant parts (Root and Chaplin, 1976). With the exceptions noted, all are frequently observed as large clusters of nymphs and adults (circa 20-30 individuals on the same plant) on the various milkweed species. Locations where bugs were collected for the studies reported here are indicated in Figure 1. Further details of the ecology of the various species can be found in Root and Chaplin (1976) and Blakley (1977).

Migration and Diapause in Iowa *Oncopeltus fasciatus*

As background for this comparative study, I shall briefly review our work with North American populations of *Oncopeltus fasciatus* which was begun in 1964. Migration has been indexed using tethered flight in which a bug is glued at the pronotum to a small stick and lifted clear of the substrate. Tarsal release or wind currents directed at the bug's head induce flight which is continued by some individuals for minutes or hours. The length of flight serves as an index of migration, and 30 minutes of flight has been operationally considered to be "migratory." Using this technique, we have shown that migratory flight is a function of age (Dingle, 1965, et seq.). Under a photoperiodic regimen of 16 hours light, 8 hours dark (LD 16:8) and at a temperature of 23°C, migratory flight peaks 8-10 days following eclosion to the adult, after the cuticle has been fully deposited and hardened (Dingle et al., 1969). This pattern of maturation with flight peaking at some relatively constant interval following adult eclosion is also characteristic of a number of other insects (Johnson, 1969, 1976; Dingle, 1972, 1974a). What is apparent from all studies, however, is that flight is highly variable; individuals vary from displaying no flight or only a brief burst of flight to flights lasting for several hours (Johnson, 1976).

We still know relatively little about this variation in flight duration, but it is apparent that there are both genetic and environmental influences. Some of these influences are shown for *O. fasciatus* in Table 1. Raising the temperature at which bugs were reared to 27°C resulted in a considerable reduction in the number of bugs making long migratory flights. I have interpreted this to mean that once a bug finds itself in a thermally favorable environment, it will be inclined to stay there (Dingle, 1968b). Starvation increases the number of bugs displaying long flights while rearing under a LD 12:12 photoperiodic cycle results in an increase both in the number of bugs making long flights and the age interval over which long flights take place. This photoperiod effect is reflected in the high proportion of bugs flying for over 30 minutes at LD 12:12 and 24°C (Table 1).

The response to photoperiod is interesting because *O. fasciatus* enters an adult reproductive diapause if nymphs are shifted from LD 16:8 to

Table 1. Effects of Environment and Selection on Flight in *Oncopeltus fasciatus*[a]

Condition	Generation	Sample Size	Number Flying 30+ Minutes	Percent Flying
LD 16:8 23°C	P_1[b]	311	74	23.8
LD 16:8 23°C	G_1	47	30	63.8
LD 16:8 27°C	--	98	10	10.2
LD 12:12 24°C	--	54	38	70.4
LD 16:8 23°C (starved 3 days)	--	148	70	47.3

[a]Data for LD 12:12 24°C from unpublished results of R.L. Caldwell. Remaining data from Dingle (1968b).

[b]P_1 indicates the parental generation from which two males and two females were selected as parents of G_1 (see text).

LD 12:12 (Dingle, 1974b); this diapause terminates reproduction in the autumn in eastern North America and is the prelude to the exodus of the bugs. The "critical photoperiod" at which diapause is induced in Iowa field population is LD 12:12, but it can vary considerably under the influence of selection (Dingle, 1974a; Dingle et al., 1977). High temperatures can override the diapause response. In its photoperiodic responses *O. fasciatus* is a fairly typical "long-day" species (Beck, 1968) exhibiting a facultative reproductive diapause.

What is of relevance here is the relation between diapause and migration which is indicated in Figure 2. The data for this figure were taken from groups of bugs in which the same individuals were flown repeatedly until 30 days posteclosion. In bugs reared on long days at 23°C (N = 311), the usual flight peak occurs at 8 days with little flight occurring after that time. In contrast, bugs reared at LD 12:12 24°C (N = 54) displayed flight activity which began at day 8 and continued for the duration of the tests. Not only did the same individuals fly repeatedly, but presumably because of the time available for flight, many more individuals flew for 30 minutes at least once, the operational difinition for migration. Some 70%, in fact, were "migratory" as compared to 24% in the nondiapause population. As has been repeatedly shown most migrants are characterized by an "oogenesis-flight syndrome" (Johnson, 1969, 1976) or perhaps better "reproduction-flight syndrome" (Dingle, 1974a) in which migration occurs either prior to the onset of reproduction or between bouts of reproduction and is in fact terminated by mating and oviposition. Colonization potential of migrants is thus enhanced (Dingle, 1968a, 1972, 1974a; Solbreck, this volume). Diapause in *O. fasciatus* of course delays reproduction and thus apparently permits migratory activity over a much longer period of the life cycle. Control of the reproductive-flight system in *O. fasciatus* results from differing levels of juvenile hormone (Rankin, 1974; this volume).

A great deal of the variation in both flight and diapause in *O. fasciatus* is also due to genetic influences. Selection for long flight resulted in an increase in the percentage of bugs flying for 30 minutes or more from 23.8 in the parental generation to 63.8 in the offspring (Table 1). Selection in this case was extremely intense since the selected parents consisted of but two long-flying pairs (Dingle, 1968b), but the marked increase in the proportion of offspring flying for long periods suggests

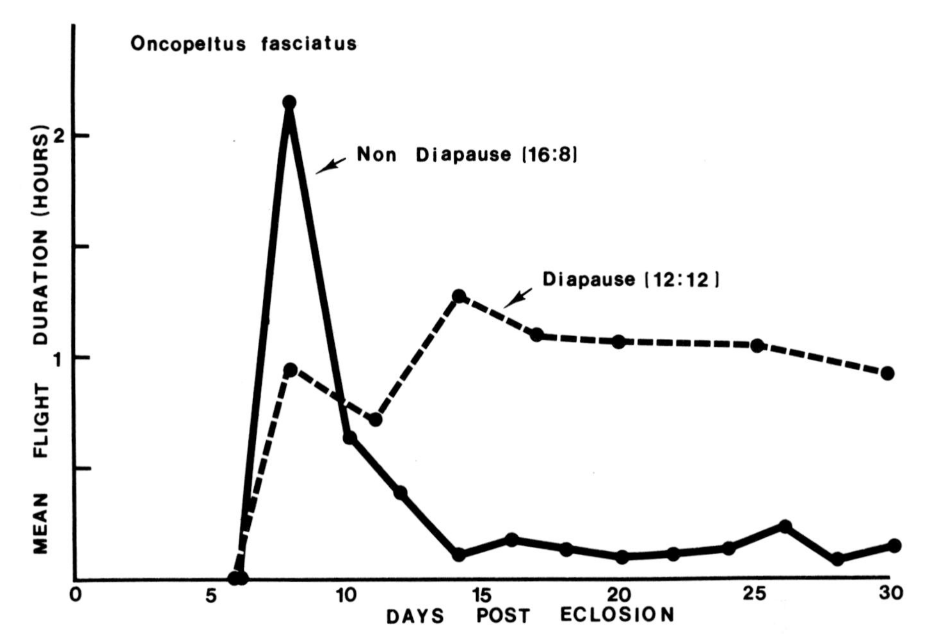

Fig. 2. Contrast in age and flight activity between diapause (LD 12:12) and nondiapause (LD 16:8) females of *O. fasciatus*. Bugs were initially flight-tested on day 8 posteclosion with the same individuals tested at intervals thereafter until 30 days. Nondiapause bugs began reproducing around day 15; diapause not until day 45.

a high proportion of additive genetic variance influencing the trait. Additive genetic variance is the major source of variance contributing to parent-offspring resemblance and is as a consequence, important for evolutionary biology becuase it is the principle variance upon which natural selection acts (Falconer, 1960). In a direct attempt to assess additive genetic variance for flight, Caldwell and Hegmann (1969) estimated the heritability (the ratio of additive genetic variance to total phenotypic variance) for flight in the small milkweed bug *Lygaeus kalmii* to be 0.20 ± 0.06 when offspring scores were regressed on male parent and 0.41 ± 0.05 when regressed on female parent suggesting substantial additive effects. The higher value for the female parent estimate also suggests maternal influences. High maternal effects are characteristic of fitness traits and certainly merit further investigation with respect to how maternal environment bears on migration.

Our studies have also indicated a high proportion of additive genetic variance for the critical photoperiod inducing diapause (Dingle, 1974a; Dingle et al., 1977). Both selection experiments and parent-offspring regression suggest that 70% is a reasonable estimate of the additive variance for the trait, and a nondiapause strain can be produced in the laboratory after selection over a few generations. Bugs at 23°C were initially transferred from LD 16:8 to LD 12:12 where they diapaused as adults by delaying age at first reproduction. Selection for early age at first reproduction in each generation produced bugs which at LD 12:12 displayed reproductive onset at the same age as bugs at LD 16:8. The ability to diapause was not lost since offspring of these individuals transferred to LD 11:13 again diapaused; rather, the critical photoperiod inducing diapause had been shifted to shorter days. We suggested that the high level of additive genetic variance functions as a "genetic rheostat" resulting in eggs of the eastern

North American population of *O. fasciatus* being optimally distributed geographically in the face of varying association between photoperiod and temperature. The genetic variance is probably maintained by seasonally reversing selection pressures and by genetic interchange resulting from migration and exchanges berween adjacent populations on a geographic cline.

A number of other insect species also display what appear to be high levels of genetic variance for diapause (reviewed in Dingle et al., 1977; see also Istock, Hoy, Lumme, Masaki, Tauber and Tauber, Vepsäläinen, and Waldbauer, this volume), and interestingly, heritabilities in the neighborhood of 0.7 have been found for diapause emergence in the fall webworm, *Hyphantria cunea* (Morris and Fulton, 1970) and the corn earworm, *Heliothis zea* (Holtzer et al., 1976). In sum, *O. fasciatus* and a number of other insects share in common high proportions of additive genetic variance for traits associated with diapause.

A Comparative Study of Migration and Diapause

The Choice of *Oncopeltus* spp.

Species of *Oncopeltus* offer several advantages for a comparative study of migration and diapause and interactions between the two syndromes. At least one species, *O. fasciatus*, is a known long-distance migrant and is known to diapause (Dingle, 1965, et seq.). In addition our studies have shown that there is considerable variance for migration and diapause and that there are major components of both genetic and environmental variance involved. These data suggest, therefore, that this species has many characteristics of potential interest for evolutionary studies. In addition, life-history data can be obtained from both field and laboratory allowing appropriate analysis of the behavioral characters in the context of overall life-history strategies (Dingle, 1968a, 1974a).

The occurrence of several species in the genus and the relatively widespread distribution of many over temperate, subtropical, tropical, island, and mainland areas also affords the opportunity for comparing escape strategies both between species and for comparing between populations within species. Studies of the latter sort are likely to be particularly useful in elucidating evolutionary mechanisms as they allow examination of current selective forces and thus the process as well as the products of evolution. *Oncopeltus* possesses many advantages for such studies as there are a sufficient number of species to permit several comparisons yet not so many that it would be impossible to cover the genus reasonably well. There are about 50 species in the genus worldwide of which at least 23 occur in the New World. Of these, 10 are included in the subgenus *Erythrischius* including all the species of this study (Slater, 1964; Slater and Sperry, 1973). All of the species are brightly colored and occur on the upper portions of milkweed plants. They are thus not difficult to census, making field studies relatively easy to accomplish. Finally, the fact that many of the species of *Erythrischius* are known to hybridize both in the field and in the laboratory provides additional interest for evolutionary biology.

Some Predictions Derived from *Oncopeltus* Migration and Diapause

Based on previous studies of *Oncopeltus* and on climatic and ecologic factors prevalent in the tropics and on islands, we can make some straightforward predictions about the responses of the bugs to diapause and migration-inducing stimuli. These and a brief rationale for each can be stated as follows.

1. *Species or populations from largely aseasonal tropical habitats will display no photoperiodic diapause.* Diapause is a mechanism allowing escape from seasonal extremes and in the case of *O. fasciatus* is permissive of migration (Fig. 2). Where seasonal extremes are absent and the host plants flower and fruit throughout the year, as for example, *A. curassavica* does in much of the Caribbean and mainland Neotropics, there should be no selection for diapause and the syndrome should not be present. This seems especially likely in *O. fasciatus* populations in view of the high proportion of additive genetic variance available in that species (Dingle, 1974a; Dingle et al., 1977).

2. *Tropical and island populations or species will contain a lower proportion of individuals displaying long-duration (migratory) flights than temperate populations or species.* The one temperate species of the genus, *O. fasciatus* is able to exploit northern regions successfully because of the ability of this species to escape by migration the oncoming winter which individuals cannot survive. In relatively stable tropical habitats the necessity for such escape is obviated. As in the case of diapause, there is thus likely to be little or no selection for long-distance migration, and its absence should be expressed in the failure of tethered insects to fly for very long. The argument for tropical islands includes the additional element of isolation which has frequently been shown to result in the reduction or elimination of migration (e.g., Darlington, 1943; Vepsäläinen, this volume). The case for islands is analyzed in greater detail in the Discussion below.

3. *Directional selection will reduce the additive genetic variance for diapause and possibly also for flight duration in tropical species from stable or isolated habitats and may result in the evolution of dominance.* We have postulated that one of the contributing factors to the maintenance of high proportions of additive genetic variance for the critical photoperiod promoting diapause in *O. fasciatus* is the regular reversal of selection pressures. This reversal results from an advantage accruing to those bugs which reproduce early in spring and summer, but delay reproduction by diapause and migrate in the autumn, In the absence of the necessity for migration, or where long-distance flight might actually be selectively disadvantageous as on islands, selection should be unidirectional favoring nondiapausing relatively sedentary individuals and additive genetic variance should be progressively reduced. In such cases, these latter traits might be expected to evolve toward dominance as indeed seems to have occurred in other instances of local adaptation with respect to both traits (Jackson, 1928; Helle, 1968; Lumme et al., 1975; Lumme and Vepsäläinen, this volume).

Evidence Bearing on Predictions: Diapause

Evidence concerning diapause was derived from both species and population differences. In addition to *O. fasciatus*, five species of the genus all occurring in tropical or subtropical habits, were tested for their responses to photoperiod. Each species was tested in two photoperiods, LD 14:10 and LD 11:13, and two temperatures, 23°C and 27°C. The photoperiods were chosen as the approximate maximum and minimum (including civil twilight) that could be experienced within the geographic range of any of these species; the temperatures were selected to correspond to those used in studying *O. fasciatus* where it has been shown that 27°C largely overrides the diapause-inducing effects of short days (Dingle, 1974b). As before (Dingle, 1974a,b) the criterion for diapause is a

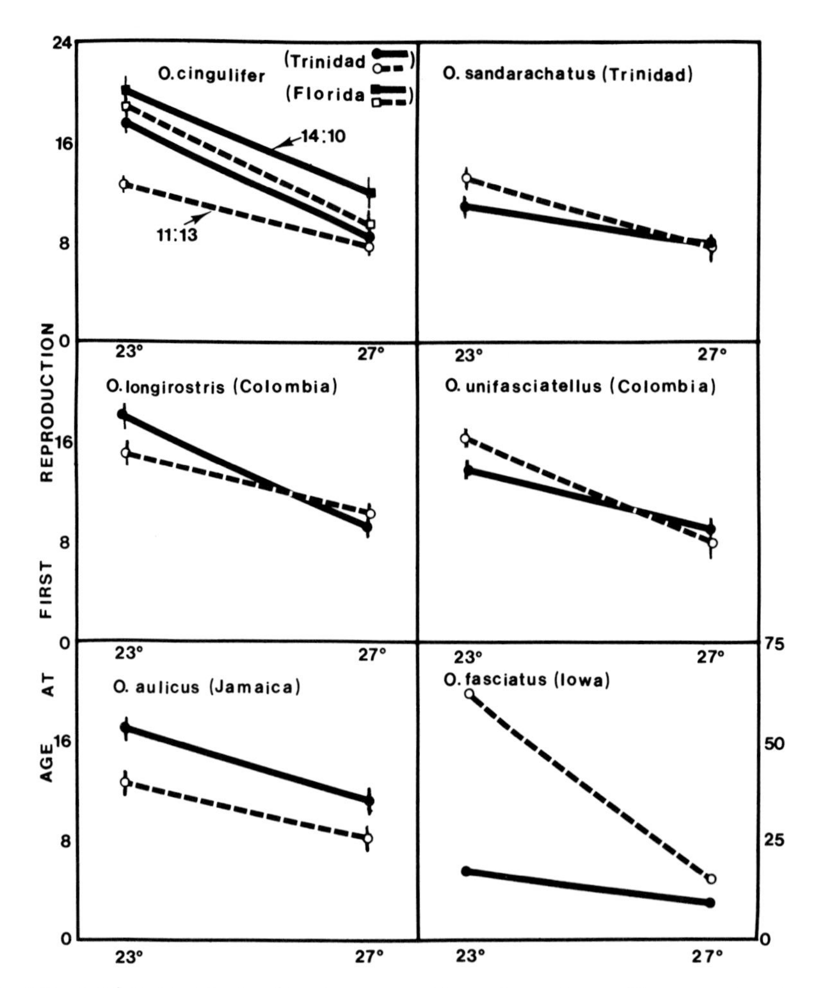

Fig. 3. Interaction plots showing response of various *Oncopeltus* species to four photoperiod-temperature regimens. A difference between points on the same line indicates a response to temperature while a difference between lines indicates a response to photoperiod. Diverging lines suggest an interaction between temperature and photoperiod as in *O. fasciatus* where short days induce delay in age at first reproduction (i.e., diapause) but only at 23°C. Note also on right difference in magnitude of the scale for *O. fasciatus*. Age at first reproduction is in days posteclosion.

substantial delay in the age posteclosion at which females begin to oviposit. Females were monitored by keeping them as pairs with males in individual petri dishes.

The results of the experiments with the different species are shown in the interaction plots displayed in Figure 3. All four conditions are indicated for each species with the mean ages at first reproduction and their standard errors plotted as the points above each of the two temperatures. The two points for each photoperiod are then connected by lines, solid for LD 14:10 and broken for LD 11:13 (note that these are *not* regression lines). A difference in the points on any line thus indicates an effect of temperature, and a difference between the lines an effect of photoperiod. Divergence of the lines indicates an inter-

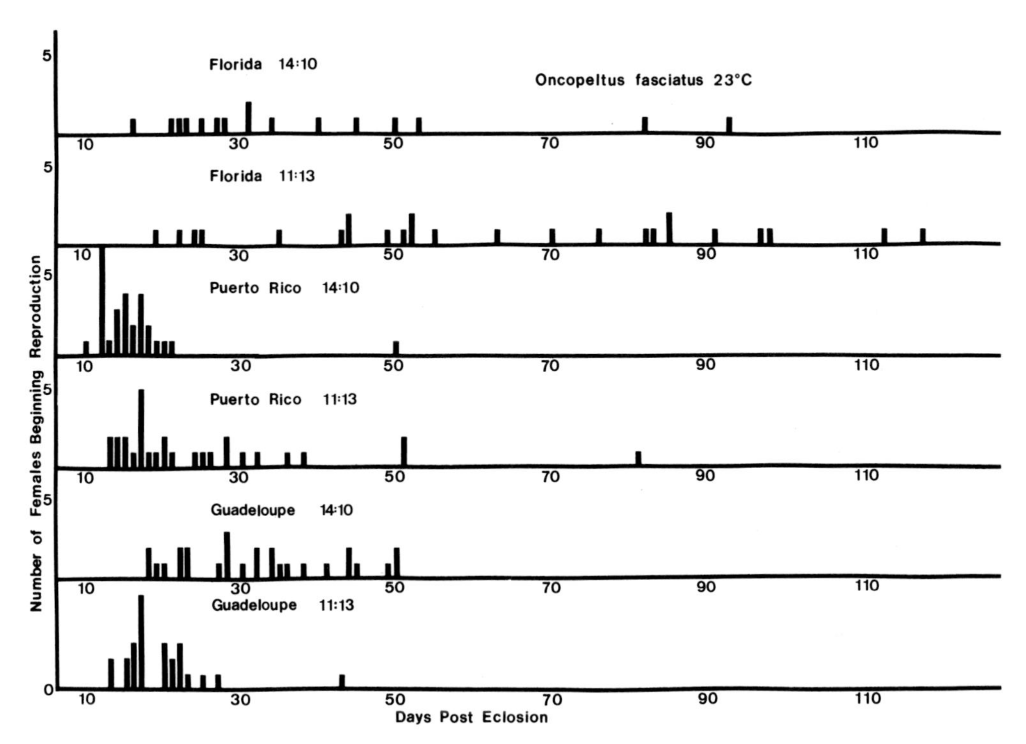

Fig. 4. Frequency histogram showing responses of *O. fasciatus* populations to different light regimens at 23°C. Note responses of Florida sample, especially at LD 11:13.

action effect between temperature and photoperiod (as in *O. fasciatus* where also the difference in scale on the right-hand margin should be noted).

Most obviously the data in Figure 3 indicate a temperature effect on age at first reproduction with higher temperature resulting in lower age in all species at both photoperiods. This is hardly surprising and undoubtedly represents a simple acceleration of metabolic rate at the warmer temperature.

What is more interesting is the influence of photoperiod and of the interaction between temperature and photoperiod. In three species, *O. sandarachatus*, *O. longirostris*, and *O. unifasciatellus* there was no apparent influence of photoperiod alone. The data at 23°C for these species do suggest a slight interaction effect but in view of the low values (circa 2-3 days), I consider it more likely that differences between treatments arose because of variation in temperature between our environmental chambers. At best these chambers are accurate to ±0.5°C and at circa 23°C, variation could easily account for differences of 2-3 days in development time. Note also that the "interaction effect" in *O. unifasciatellus* is opposite to that of the other two species also suggesting that minor temperature variation is a more plausible explanation for differences.

In two species, *O. cingulifer* and *O. aulicus*, there does appear to be an effect of photoperiod. In both species and in both populations of *O. cingulifer* (Florida and Trinidad) *long* photoperiod results in a one- to four-day delay in age at first reproduction. Long days also delay Guadeloupe *O. fasciatus* (Fig. 4). In view of this repeatability the delay is probably real, although small and not qualifying as "diapause." A photoperiod of LD 16:8 results in even more marked delays in *O. cingulifer* (Blakley, 1977). The reason for delay is not apparent although at 14:10 the photoperiod is extreme for these species or populations and extraordinary photoperiods are known to have disruptive effects on development in many insects (e.g., Beck, 1968). No more satisfactory explanation is at present available. The long photoperiod would not correspond with a dry season at any of the locations. In any event, the influence of photoperiod is at most minor, and there is no evidence of a reproductive diapause in response to short days.

In only one species, in fact, do the data indicate a reproductive diapause and that is *O. fasciatus*, the long-distance migrant to North America. For comparison, data for an Iowa population of this species are also displayed as an interaction plot in Figure 3 (note difference in scale on right). This plot clearly shows the interaction between temperature and photoperiod with diapause occurring only at LD 11:13 and 23°C.

In view of the apparent lack of a photoperiodically induced diapause in any of the exclusive tropical and subtropical species, the photoperiodic responses of tropical populations of *O. fasciatus* become of considerable interest. The responses of three populations of this species at 23°C to the two test photoperiods are shown in Figure 4. The responses of the Puerto Rico population are perhaps the most straightforward and about what might be expected. There is no diapause at LD 14:10, as was true also of North American bugs (Fig. 3) and delay in reproduction in only a few individuals at LD 11:13. A residual tendency to diapause might be expected if there were occasional immigrants from diapausing populations or if diapause conferred occasional advantage in surviving an extreme dry season (the dry season corresponds to short days in the Caribbean). In general, however, the Puerto Rican population fulfills the prediction of minimal diapause in a relatively aseasonal tropical environment.

The Guadeloupe population is puzzling because here there is no evidence of diapause at LD 11:13, but a general tendency to delay at 14:10. Again I can offer no satisfactory explanation of this phenomenon except to suppose that it may be the result of exposure to abnormal photoperiods. In a few insects, e.g., *Noctua pronuba*, long days can apparently induce aestivation (Novak and Spitzer, 1975), but in the case of *O. fasciatus* there is no obvious correspondence with any seasonal change occurring in Guadeloupe.

Finally, the Florida population is the most difficult to explain. At both photoperiods a few individuals begin reproducing early, but the majority delay long enough to be considered as diapausing with the delays generally longer on the 11:13 light schedule. Note, however, that there is considerable variation between individuals under both regimens.

Samples from the same three *O. fasciatus* populations were also reared at 27°C at both photoperiods. The results are shown in Figure 5. In this instance the higher temperature virtually eliminates any tendency to delay reproduction in both the Puerto Rico and Guadeloupe populations (compare with Fig. 4). The results for these bugs thus parallel those obtained for Iowa populations of *O. fasciatus* in which little or no diapause was apparent at this temperature. Once again, however, Florida bugs fail to conform. Primarily as a result of a small group of bugs which began reproducing around day 10, and hence displayed no diapause, mean age at first reproduction was somewhat earlier than at 23°C, but significant proportions of the bugs under both photoperiod

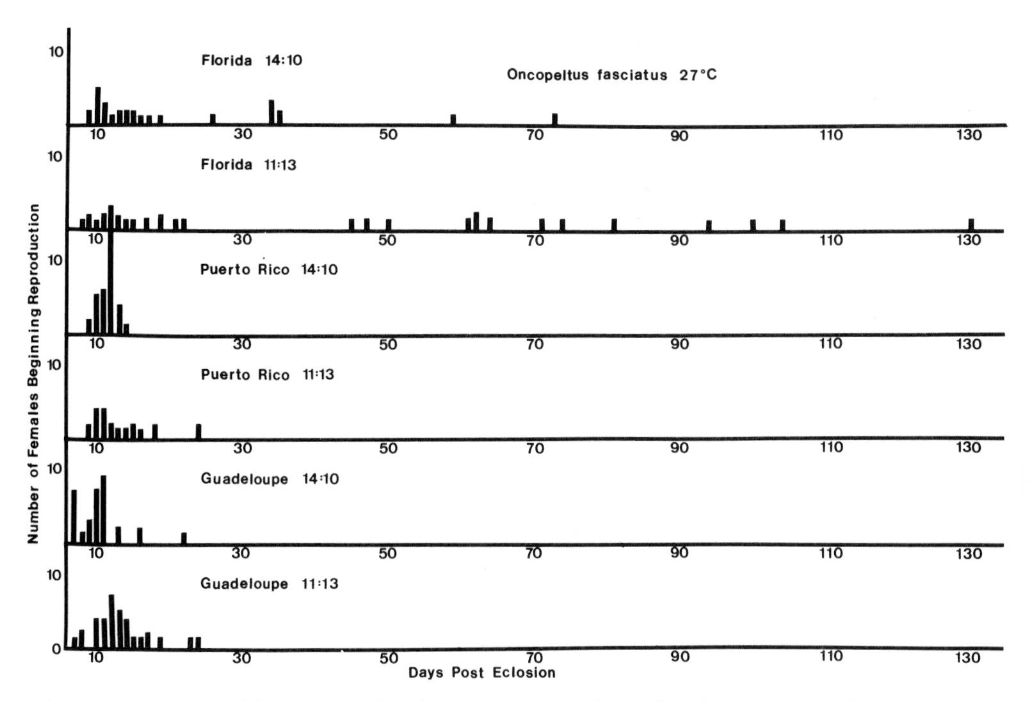

Fig. 5. Frequency histograms showing responses of *O. fasciatus* populations to different light regimens at 27°C. Again note response of Florida sample at LD 11:13.

regimens delayed reproduction. The delay was especially noticeable under short days suggesting a stronger tendency to diapause. Temperature override of diapause is thus apparent, but only in a portion of the sample population. For whatever reason, diapause in the South Florida population is remarkably persistent.

There is no obvious explanation apparent for these disparate results in the Florida populations. I can only conjecture that the response of the bugs is in some way an adaptation to the seasonal cycle in South Florida (Hialeah, the Everglades) where this population originated. Certainly there is a long dry season extending roughly from November to April although there is usually some *Asclepias curassavica* flowering in wetter habitats. It is also possible that milkweed is somewhat erratic during the summer wet season; it is in any event highly patchy in its distribution. It is possible, therefore, that patchiness of the host plants in both time and space results in conflicting selection pressures and a variable population. Studies of *O. fasciatus* (and other species) in Florida environments are clearly in order.

The final diapause experiment that I shall report here concerns an attempt to take advantage of the fact that several of the species of *Oncopeltus* studied will cross to produce interspecific hybrids. In this experiment females from an Iowa population of *O. fasciatus* which showed a high proportion of diapausing individuals on short days were paired with males of *O. sandarachatus*, a species which displays no diapause (Fig. 3). Efforts at reciprocal crosses failed. The rationale behind the experiment was that if offspring resembled either parent species dominance would be indicated while if offspring were intermediate between the parents additive influences would be more likely. (Dominance would also need to be further assessed by backcrosses.)

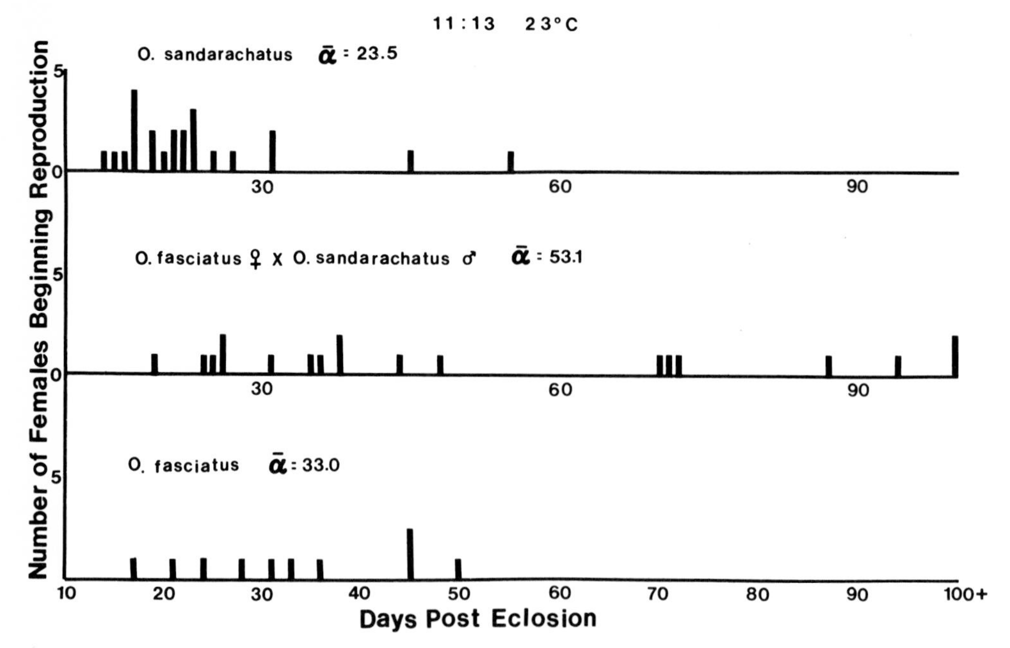

Fig. 6. Frequency histograms of responses of *O. fasciatus* X *O. sandarachatus* hybrids and of samples of the two parent populations tested simultaneously in short days. Reciprocal hybrids were not produced. Note variance in hybrids and delay in age at first reproduction relative to both parent populations.

Dominance effects are known for diapause or nondiapause in other species (Helle, 1968; Vepsäläinen, 1974, this volume; Lumme et al., 1975; Lumme, this volume; Hoy, this volume). The results are shown in Figure 6 which indicates responses of individuals from both parent species plus those of hybrid offspring all of which were reared in the same environmental chamber at the same time. Unfortunately the culture of *O. fasciatus* suffered high unexplainable mortality (the original N was 30 paired females) and the mean age at first reproduction (α) is thus probably underestimated since those individuals with long diapause likely died before reproducing and went unrecorded. The results for the hybrids are as a consequence difficult to interpret, but what is clear is that they do not show a strong resemblance to the nondiapausing *O. sandarachatus* parent. Dominance for nondiapause, seen in other insects (Vepsäläinen, this volume), is thus not indicated. Any further inferences must await more experiments although the present data do not contradict the possibility of additive influences in view of the very high variance of the hybrid offspring or dominance deviation for diapause in view of the late ages of first reproduction of many of the hybrids.

Evidence Bearing on Predictions: Flight

So far our studies have progressed to the point of testing three tropical species for migratory tendency, *O. unifasciatellus*, *O. sandarachatus*, and *O. cingulifer*. The technique for assessing flight was the same as used previously for *O. fasciatus* (Dingle, 1965, et seq.); bugs were attached at the pronotum to a thin stick with a small dab of glue, lifted free of the substrate to stimulate flight (wind was blown at

Table 2. Summary of Flight Characteristics of Three Tropical *Oncopeltus* Species

Species	Mean Flight Duration	Longest Flight	Proportion Flying 30 Min	Proportion Repeating 30 Min	*N*
O. unifasciatellus	9.6 sec	5 min 10 sec	0	0	135
O. sandarachatus	1 min 28 sec	62 min 16 sec	.03	.016	60
O. cingulifer (Lab)	33.4 sec	1 hr 15 min[a]	.022	0	90
O. cingulifer (Trinidad)	14 min 32 sec	1 hr 30 min[a]	.146	--	55

[a]Indicates bug was stopped by experimenter after time interval indicated.

the head if tarsal release was insufficient), and the duration of the subsequent flight recorded. On each test day, bugs were permitted five flights; they were tested every other day from 8 to 20 days posteclosion. With the exception of a few *O. cingulifer* at 27°C, all flight tests utilized bugs reared at 23°C usually on a short-day photoperiod.

The results clearly show that these tropical species display less long duration (migratory) flight activity than do Iowa populations of *O. fasciatus*. Several indicators of flight activity are given in Table 2 which contains results from lab populations flight-tested as described above plus those from a field population of *O. cingulifer* collected in Trinidad. These latter bugs were tested on the day they were collected in the field and were tested only once. Subsequent data on reproduction indicated that they were probably between 8 and 15 days posteclosion. They were, however, likely to be somewhat selected for being dispersers since they were collected by removing all bugs from a milkweed patch and then collecting new immigrants as they arrived.

The first flight indicator in Table 2 is the mean flight duration. This was calculated by summing first the five flights for each day of testing. The mean of these totals, taken over all days of flight testing, then constituted the mean for the individual bug. Finally, a grand mean, which is the value given in Table 2, was taken over all bugs. The longest overall mean flight duration in laboratory-raised insects was thus 1 minute 28 seconds for *O. sandarachatus* while the means for *O. cingulifer* and *O. unifasciatellus* were only 33.4 and 9.6 seconds respectively. These data demonstrate that most flights in these species were of short duration, usually, in fact, only a few seconds.

The remaining data in Table 2 further describe the relatively short-duration flights of these tropical species. The longest flights for any species do not approach the flights of several hours regularly noted in *O. fasciatus*. This statement must be qualified somewhat because the longest *O. cingulifer* flights were stopped by the experimenter, but as the proportions flying over 30 minutes (the operational definition of migration) indicate, even had these flights been much longer, they would have been rare relative to *O. fasciatus*. The failure to repeat 30-minute flights suggests in addition that even those few bugs that made such flights did so generally only once. Finally, the Trinidad *O. cingulifer* tested immediately after capture require brief comment. As indicated above, data on subsequent reproduction suggested that

Table 3. Summary of Flight Characteristics of *O. fasciatus* Populations

Population	Mean Flight Duration	Longest Flight	Proportion Flying 30 Min	Proportion Repeating 30 Min	*N*
Michigan	39 min	9 hrs 25 min	.24	.16	311
Florida	8 min 49 sec	10 hrs[a]	.24	.13	45
Puerto Rico	1 min 23 sec	4 hrs 32 min	.08	.03	60
Guadeloupe	31.7 sec	55 min 14 sec	.045	0	44
Iowa-Diapause	46 min	3 hrs[a]	.70	.54	54

[a]Bugs stopped by experimenter.

these bugs were between 8 and 15 days posteclosion and were likely to be selected dispersers. They were flight-tested only once each. The data for *O. fasciatus* with which they should be compared are therefore those for flight on the initial day of testing. The mean for nondiapause first-flight duration in *O. fasciatus* is over 2 hours (Fig. 2); repeated experiments have also demonstrated that the proportion flying for over 30 minutes approximates 25% in unselected populations (e.g., Michigan population in Table 3) almost twice the proportion of the Trinidad *O. cingulifer*. All the data in Table 2 thus suggest less migratory capability in these tropical *Oncopeltus* species even in those cases where it is likely that flight was selected for by the method of collection.

As with diapause discussed above, the flight responses of different populations of *O. fasciatus* are of considerable interest, the more so in view of the evidence for reduced flight capabilities in other (tropical) *Oncopeltus* species. A summary of various flight characteristics of four *O. fasciatus* populations is given in Table 3. It is immediately apparent that Iowa and Michigan populations display more long-duration flight that their tropical and subtropical counterparts. The mean flight durations are over 30 minutes for both temperate zone populations indicating considerable migration, while means for the southern populations are all considerably less than 30 minutes. Note also that the mean for Iowa-Diapause bugs is underestimated because no bugs were allowed to fly for more than 3 hours; the proportion flying for over 30 minutes and repeating such flights, however, indicates the high migratory potential of these bugs.

Florida bugs do not show as high a mean flight duration as Iowa-Diapause or Michigan bugs, but the proportion flying for over 30 minutes and repeating such flights is roughly equal to these proportions for Michigan. This is, however, somewhat misleading. As Figure 4 indicates, a large majority of Florida bugs reared at LD 11-13 23°C, the conditions under which the flight-tested bugs were reared, were in diapause as indicated by the delays in reproduction. The legitimate comparison, is thus with Iowa-Diapause bugs. As is evident by this comparison, the proportion of Florida bugs exhibiting long-duration flights, .24; is far less than the proportion, .70, of Iowa-Diapause bugs; a similar conclusion is evident by comparing bugs from these two populations repeating long flights.

The data in Table 3 also suggest that there is little migratory flight potential in *O. fasciatus* populations from Puerto Rico and Guadeloupe. This is true with respect to overall mean flight duration, longest flight, and proportions flying and repeating over 30-minute flights. All four measurements also suggest that there is a reduction in flight capability as one proceeds along a gradient of land area and degree of geographical isolation from Florida to Puerto Rico to Guadeloupe. This possibility was tested by comparing means of first, Florida with Puerto Rico bugs and secondly, Puerto Rico with Guadeloupe bugs using the nonparametric Mann-Whitney U test (Siegel, 1956). Differences between the populations were significant at $p < .001$ and $p < .02$, respectively, strongly supporting the hypothesis that a trent toward reduced migratory capability exists along this geographical gradient.

There remains, however, a problem with the Florida bugs. The fact that a majority are in diapause may result in an inflation of the overall mean as a result of the increased opportunities to undertake long-duration prereproductive flights. I therefore compared the longest mean daily flight performances of Florida and Puerto Rico bugs using means from days 8, 10, and 12 posteclosion only, i.e., eliminating all flight data taken after Puerto Rico bugs had begun reproducing. Data for the two populations were compared using the low power-efficiency median test (Siegel, 1956) which still indicated a significant difference at $p < .005$. The conclusion that Puerto Rico bugs show less long-duration flight than those from Florida is thus confirmed.

The generally low level of flight activity in Guadeloupe bugs was further indicated by the number of times bugs failed to fly at all during flight testing. Of the 44 bugs in the Guadeloupe sample, six never flew, although the wings were opened, and nine others flew on only one day of the test period and then in six cases for only a single brief burst of flight. These observations account for one-third of the Guadeloupe sample. That flight thresholds were high was also indicated by the fact that in general it took repeated stimuli (tarsal release, wind on the head) to get Guadeloupe individuals to fly. In contrast to the Guadeloupe sample all Florida bugs flew at least once and with only two exceptions all flew several times during the testing period. In the Puerto Rico sample one bug never flew and four flew only once; all the rest flew on more than one occasion. In addition flight thresholds were low in these latter two populations with tarsal release alone sufficient to induce at least a few seconds of flight.

The Evolution of Migration and Diapause Strategies

Diapause

The prediction that species or populations from largely aseasonal tropical habitats will display no photoperiodically induced diapause is borne out by the results obtained so far for species of *Oncopeltus* which are confined to the tropics and subtropics (Fig. 3). Of the six species studied, only the temperate migrant *O. fasciatus* displays photoperiodically induced diapause, and then not in all populations (Figs. 4 and 5). Diapause in response to changing daylengths is known in species which are primarily tropical, but where such diapause is found, the photoperiod cues various aspects of the seasonal cycle, usually rainfall, as it does in temperate regions. An example of a tropical insect which does respond to day length under these conditions is the African migratory locust, *Locusta migratoria*. The desert locust, *Schistocerca gregaria*, and the red locust, *Nomadacris septemfasciata*, may also do so (Albrecht, 1973). Other tropical insects such as the flesh fly, *Poecilometopa spilogaster* diapause in response to cool temperatures (Denlinger, 1974). Even though short photoperiods coincide with periods of reduced rainfall in the Caribbean and Central America, the climatic data and data on the phenology of milkweeds suggest that the food plants are more or less continuously available or at least, considered over both time and space, are never unavailable for long. Because of the Central American cordillera and the mountainous nature of many Caribbean islands and of northern South America, periods of intense drought are apt to be quite local. The bugs are therefore patchily distributed, but do not require a photoperiodically induced diapause to survive.

Do the bugs have any means of weathering unfavorable conditions? The evidence is that they do, for data from several species of *Oncopeltus* (Rankin, 1972; Blakley, 1977) indicate that the bugs can survive relatively long periods of starvation, at least 30 days, as long as they have access to water. This should give adequate time to disperse to new sites or survive at least short-term climatic extremes or destruction of habitats. In *O. fasciatus* starvation induces lowered metabolic rates which equal those of bugs in photoperiodically induced diapause and are about half that of fed nondiapausing bugs (Kelly and Davenport, 1976). Metabolically, therefore, starvation is the equivalent of diapause. They are also equivalent in that both shut down juvenile hormone production (Johansson, 1958; Rankin, 1974). If lowered metabolism accompanying starvation is the case also in other species, then the evolution of photoperiodically induced diapause in this genus would seem to have required only the appropriate linking of short-day input to the endocrine-metabolic system.

Clearly the link has been accomplished in *O. fasciatus* which successfully invades temperate North America each summer. In this it is similar to other species, such as the silk moth, *Philosamia cynthia*, in Japan (Masaki, 1961), which seem to have been successful in spreading to temperate areas from the south by evolving photoperiodic diapause. The northern limit of the distributions of other species of *Oncopeltus* may then be due, or at least in part due, to the inability to evolve a response to photoperiodic cues. Lack of food would fail as a cue because milkweed seeds are abundant at the end of the summer just at the time when the bugs would need to escape the oncoming winter. If this is a reasonable scenario then the additive genetic variance for age at first reproduction as a function of photoperiod should be low in these species, and it should be difficult if not impossible to produce diapause strains in these species by selection. This genetic variance remains to be investigated. The occasional long flights occurring in these species suggest that migrants could be selected for under the appropriate circumstances.

The Guadeloupe and Puerto Rico populations of *O. fasciatus* also generally fit the prediction of no photoperiodic diapause in the tropics. The slight tendency to delay in short days in the Puerto Rico bugs could be accounted for by occasional immigration. But as indicated above, the delay of the Guadeloupe bugs in long days remains to be explained. If the 14:10 photoperiod is "abnormal" to Guadeloupe bugs thereby resulting in delay, this population differs substantially from Iowa *O. fasciatus* which reproduced in constant light as if it were a "normal" long day (Dingle, 1974b).

Also unexplainable with present information is the occurrence of so much reproductive delay in the Florida *O. fasciatus* populations. The occurrence of such a high proportion of the population delaying reproduction even at 27°C suggests that diapause is more deeply ingrained in the genome than in Iowa bugs where this temperature essentially overrides diapause (Dingle, 1974b). The initial assumption must be that this is an adaptation to some unique aspect of climatic or ecologic conditions in South Florida, but what these are is still to be determined. Until they are, the Florida population remains a fascinating variant within the species.

At this point our prediction concerning the reduction of additive genetic variance for diapause and the possible evolution of dominance remains essentially untested. The initial attempt at a test, involving *O. sandarachatus* X *O. fasciatus* hybrids, fails to indicate dominance for nondiapause since the hybrids did not strongly resemble the nondiapausing *O. sandarachatus* parent. Beyond this, however, the experiment yielded no definitively interpretable results. Questions concerning the comparative genetics of diapause in these bugs therefore remain very much open.

Migration

The three tropical species of *Oncopeltus* so far tested display far less long-duration tethered flight than the North American migrant *Oncopeltus fasciatus*. Habitats are much less seasonally variable for the tropical species and this relative stability should produce less migratory flight. This is the inverse of Southwood's (1962) conclusion that temporary habitats should lead to the development of migration (cf. Carlquist, 1974, p. 488). It should be noted, however, that the absence of extensive long-distance flight does not preclude short-distance dispersal to fill lacunae in the overall population distribution (Taylor and Taylor, 1977). That this is indeed occurring is suggested by the studies of Root and Chaplin (1976) who found considerable movement in a mixed population of *O. cingulifer* and *O. unifasciatellus* in Colombia. The other escape mechanism available to insects, diapause, is also absent in these species (Fig. 3) except to the extent that they can survive periods of starvation with reduced metabolic rates. Since diapause is at least permissive of migration in *O. fasciatus* (Fig. 2), it is not surprising that there is little evidence for long-distance migration in the tropical species which lack diapause. The influence of starvation on flight in these species remains to be tested.

Although the above argument makes evolutionary sense, there are some problems with the flight tests of the tropical species which must be briefly mentioned. First, two of the species populations tested, those of *O. cingulifer* and *O. sandarachatus* were derived from individuals collected in Trinidad. Although Trinidad is extremely close to the continental mainland, some island effects on flight are still possible especially in view of the flight-test results with Florida *O. fasciatus*. Island effects are not a problem with *O. unifasciatellus* since this population was derived from individuals taken in Colombia (although mountain valleys where *O. unifasciatellus* occurs may to a certain extent be "islands" as discussed below), but the *O. unifasciatellus* culture had been in the laboratory for some years and loss of flight potential is a distinct possibility. It it however, worth noting that the Cambridge stock of *O. fasciatus* that I first used for flight testing (Dingle, 1965) showed no flight decrement although it, too, had been in the laboratory for several years. In any event it is probably best to interpret the results of flight tests with the tropical species somewhat cautiously until continental populations derived from freshly collected material can be flight-tested.

More interesting and conclusive are the flight tests of various island and mainland populations of *O. fasciatus*. Here there is progressive reduction of flight as one proceeds from the continental North American mainland through the Florida peninsula to the relatively large and relatively isolated island of Puerto Rico and finally to the still more isolated small island of Guadeloupe. There is much evidence in the literature that isolation leads to flightlessness in a variety of insects (Wollaston, 1854; Jackson, 1928; Darlington, 1943, 1971; Southwood, 1962; Vepsäläinen, this volume); but with a few notable exceptions (e.g., Vepsäläinen, 1974), there are few data on the relation between winglessness and the degree of isolation. The strong correlation between degree of isolation and reduction in flight in *O. fasciatus* is highly suggestive that isolation itself is the major factor involved. A parallel situation is found in another North American lygaeid, *Lygaeus kalmii* (Caldwell, 1969, 1974). In the laboratory 12.8% of *L. kalmii* from cultures initiated with individuals taken in Iowa flew for over one hour; similar tests of *L. kalmii* originating in valleys of the Rocky Mountains of Colorado yielded only 3.4% long flights. Milkweeds, which are also the host plants of this species, are ubiquitous in Iowa, but occur as isolated "islands" in the valleys of the Colorado Rockies. A small isolated population from the hills near Berkeley, California, yielded only bugs which flew for a few seconds.

There have been a number of reasons advanced for the evolution of reduced flight or flightlessness in isolated insect species or populations. Clearly selection must be acting against flight, but the questions are how? and why? Darwin (1859) was the first to tackle the problem while trying to explain the large number of wingless species of beetles found on Madeira by Wollaston (1854). Darwin supposed that winged forms would have been blown off the island thus failing to leave descendants there (see quotation at the beginning of this paper.) In most cases this is now considered unlikely since many wingless insects occur in sheltered habitats, although wind may be a significant factor in the evolution of winglessness or subantartic islands (Gressitt, 1970; Carlquist, 1974). A second hypothesis has been that winglessness evolved in the absence of predators (MacArthur and Wilson, 1967). This seems unlikely for *O. fasciatus* since adults of this species are conspciuous, bad tasting, and seldom preyed upon as adults whether on islands or the mainland. It may also be unlikely in general for insects since a high proportion of island colonists among Coleoptera and Heteroptera, at least, are carnivores (Becker, 1976), and insectivorous birds are certainly likely to be present. Thirdly, Darlington (1943) proposed that wingless individuals may be selected for "inherent simplicity and viability," but simplicity is difficult to define and there are no measurements on wingless forms demonstrating that they are inherently more viable (nor is it clear exactly what should be measured).

The most likely explanation for the reduction of flight in isolated, stable habitats has been summarized by Roff (1975) who examined the question of disperal with a simulation model. In this argument reduced flight evolves under these conditions because, first, dispersal genotypes leave, and there is a net loss because few or no immigrants arrive to replace the emigrants. As the numbers of dispersers become low, they are outbred and outproduced by nondispersers, and the number of dispersing offspring is reduced by mating with nondispersers in those habitats remaining relatively stable and capable of supporting populations of the species. A significant advantage gained by diverting energy from the flight system to reproduction and offspring survival would lead to the evolution of flightlessness. Some flight would be retained, however, if there were an advantage to local dispersal as with *Oncopeltus*. Sufficient isolation to prevent continuous immigration combined with severe environmental instability would lead to extinction (cf. MacArthur and Wilson, 1967). Well-adapted local populations may prevent further invasion by outsiders, a factor which Lack (1976) considers of major importance in maintaining the relative impoverishment of island biotas. Although Lack was careful to restrict himself to birds, the reduced numbers of *Oncopeltus* species on islands (Blakley, 1977 and personal communication) at least suggests the possibility that similar factors may be operating on some insects. If in fact selection has been operating against flight in island populations, then one would predict reduced additive genetic variance for flight (cf. Fisher, 1958; Falconer, 1960). Again this prediction remains to be tested, and, thus, is one of the future projects of my laboratory.

Whatever the mechanisms of selection against flight, it has occurred in a variety of ecological situations. Very few groups of insects have remained immune, and they have responded in a variety of ways. The simplest is displayed by *Oncopeltus* which has evolved reduced flight behavior on islands; other examples of variation in flight behavior are given by Johnson (1976). At the next level is the histolysis of wing muscles once flight is complete (or even loss of wings as in ants, aphids, and termites) as in bugs such as *Dysdercus* (Edwards, 1969; Dingle and Arora, 1973) or forest bark beetles such as *Dendroctonus* and *Ips* which may regenerate these muscles later (Bhakthan et al., 1970, 1971). Other species are polymorphic for short-winged and long-winged forms; morph differences arising from such polymorphism vary from slight in some cicadellids (Rose, 1972) to intermediate as in some carabids (Carter, 1976) to great as in some gerrids (Vepsäläinen, 1974 and this volume). The extreme form of this polymorphism is the trade-off between winged and wingless individuals as is known for example, in gerrids (Vepsäläinen, 1974 and this volume; Järvinen and Vepsäläinen, 1976) carabids (Lindroth, 1945-49), and, of course, aphids, ants, and termites. Finally extreme isolation without selection for even local dispersal may lead to the evolution of flightlessness as in many species on islands including in the subantartic even various flies and midges (Diptera) and wasps (Hymenoptera) (Carlquist, 1974), groups which are notorious for their powers of flight.

Summary and Prospect

Species and populations of tropical and temperate *Oncopeltus* vary considerably in their diapause and migration responses. The tropical species show no photoperiodically induced diapause (Fig. 3), and the three that have been tested, *O. cingulifer*, and *O. sandarachatus*, and *O. unifasciatellus*, display little migratory flight (Table 2). In contrast temperate *O. fasciatus* exhibit a short-day diapause which leads to a delay in reproduction (Fig. 3) and extensive migratory flight which is enhanced by diapause (Fig. 2). *O. fasciatus* populations from the tropical islands of Puerto Rico and Guadeloupe, like the entirely tropical species, fail to show a diapause response under short-day conditions. Florida populations of *O. fasciatus*, rather than displaying an intermediate diapause response, seem to show a stronger diapause than the temperate populations; there is at present no certain explanation for this apparent disparity. With respect to migration, populations show a decreasing tendency for long-duration tethered flights as one proceeds from the North American mainland to Florida to Puerto Rico to Guadeloupe. There is thus a strong correlation between area and degree of isolation and tendency to fly. The evolution of migration and diapause in *Oncopeltus* is evidently profoundly influenced by latitude, degree of isolation, and in the case of Florida, factors not yet understood.

As should be obvious from the *Oncopeltus* data, the varieties of insect flight and diapause provide rich material for the study of evolution.

We have a glimpse of how some of this variation may be evolved, but the complete elucidation of the modes of evolution of both migration and diapause in a given species or population remains mostly in the future. The complexity already evident should provide sufficient challenge to the evolutionary biologist.

Acknowledgments Nigel Blakley and Elizabeth Klausner gave the bugs excellent care at various times and have discussed with me many of the problems associated with *Oncopeltus* life histories. Much of the flight testing was done by Tomi Yeager and Diane Kopec. The manuscript was read in draft by Nigel Blakley, Kice Brown, Joe Hegmann, Conrad Istock, and Elizabeth Klausner. Jeri Dingle prepared the figures. The research in both field and laboratory has been supported by NSF Grants GB-40822 and DEB73-01424-A01.

References

Albrecht, F.O.: The decline of locust plagues: an essay in the ecology of photoperiodic regulation. Acrida *2*, 97-107 (1973).

Beck, S.D.: Insect Photoperiodism. Academic Press: New York 1968.

Becker, P.: Island colonization by carnivorous and herbivorous Coleoptera. J. Anim. Ecol. *45*, 893-906 (1976).

Bhakthan, N.M.G., Borden, J.H., Nair, K.K.: Fine structure of degenerating and regenerating flight muscles in a bark beetle, *Ips confusus*. I. Degeneration. J. Cell Sci. *6*, 807-820 (1970).

Bhakthan, N.M.G., Nair, K.K., Borden, J.H.: Fine structure of degenerating and regenerating flight muscles in a bark beetle, *Ips confusus*. II. Regeneration. Can. J. Zool. *49*, 85-89 (1971).

Blakley, N.R.: Evolutionary responses to environmental heterogeneity in milkweed bugs (*Oncopeltus*, Hemiptera, Lygaeidae). Ph.D. Thesis, University of Iowa (1977).

Caldwell, R.L.: A comparison of the dispersal strategies of two milkweed bugs, *Oncopeltus fasciatus* and *Lygaeus kalmii*. Ph.D. Thesis, University of Iowa (1969).

Caldwell, R.L.: A comparison of the migratory strategies of two milkweed bugs. In Experimental Analysis of Insect Behaviour. Barton Browne, L. (ed.). New York: Springer 1974.

Caldwell, R.L.; Hegmann, J.P.: Heritability of flight duration in the milkweed bug, *Lygaeus kalmii*. Nature, Lond. *223*, 91-92 (1969).

Calquist, S.: Island Biology. New York: Columbia University Press 1974.

Carter, A.: Wing polymorphism in the insect species *Agonum retractum* Leconte (Coleoptera: Carabidae). Can. J. Zool. *54*, 1375-1382 (1976).

Chaplin, S.J.: Reproductive isolation between two sympatric species of *Oncopeltus* (Hemiptera: Lygaeidae) in the tropics. Ann. Ent. Soc. Amer. *66*, 997-1000.

Danilevskii, A.S.: Photoperiodism and Seasonal Development of Insects. Edinburgh: Oliver and Boyd 1965.

Darlington, P.J., Jr.: Carabidae of mountains and islands: data on the evolution of isolated faunas, and on atrophy of wings. Ecol. Monogr. *13*, 37-61 (1943).

Darlington, P.J., Jr.: Carabidae on tropical islands, especially the West Indies. In Adaptive Aspects of Insular Evolution. Stern, W.L. (ed.). Pullman: Washington State University Press 1971.

Darwin, C.: The Origin of Species by Means of Natural Selection. New York: The Modern Library (Random House) 1859.

Denlinger, D.L.: Diapause in tropical flesh flies. Nature, Lond. *252*, 223-224 (1974).

Dingle, H.: The relation between age and flight activity in the milkweed bug, *Oncopeltus*. J. Exp. Biol. *42*, 269-283 (1965).

Dingle, H.: Life history and population consequences of density, photoperiod, and temperature in a migrant insect, the milkweed bug *Oncopeltus*. Am. Nat. *102*, 149-163 (1968a).

Dingle, H.: The influence of environment and heredity on flight activity in the milkweed bug *Oncopeltus*. J. Exp. Biol. *48*, 175-184 (1968b).

Dingle, H.: Migration strategies of insects. Science *175*, 1327-1335 (1972).
Dingle, H.: The experimental analysis of migration and life-history strategies in insects. In Experimental Analysis of Insect Behaviour. Barton Browne, L. (ed.). New York: Springer 1974a.
Dingle, H.: Diapause in a migrant insect, the milkweed bug *Oncopeltus fasciatus* (Dallas) (Hemiptera: Lygaeidae). Oecologia *17*, 1-10 (1974b).
Dingle, H., Arora G.: Experimental studies of migration in bugs of the genus *Dysdercus*. Oecologia *12*, 119-140 (1973).
Dingle, H., Caldwell, R.L.: Temperature and reproductive success in *Oncopeltus fasciatus, O. unifasciatellus, Lygaeus Kalmii,* and *L. turcicus*. Ann. Ent. Soc. Amer. *64*, 1171-1172 (1971).
Dingle, H., Caldwell, R.L., Haskell, J.B.: Temperature and circadian control of cuticle growth in the bug *Oncopeltus fasciatus*. J. Insect Physiol. *15*, 373-378 (1969).
Dingle, H., Brown, C.K., Hegmann, J.P.: The nature of genetic variance influencing photoperiodic diapause in a migrant insect, *Oncopeltus fasciatus*. Am. Nat. (in press, 1977).
Edwards, F.J.: Environmental control of flight muscle histolysis in the bug *Dysdercus intermedius*. J. Insect Physiol. *15*, 2013-2020 (1969).
Evans, K.E.: The annual pattern of migration and reproduction in field populations of the milkweed bug *Oncopeltus fasciatus* in California. Ph.D. Thesis, University of California, Berkeley (1977).
Falconer, D.S.: Introduction to Quantitative Genetics. Edinburgh: Oliver and Boyd 1960.
Fisher, R.A.: The Genetical Theory of Natural Selection. New York: Dover 1958.
Gressit, J.L.: Subantarctic entomology and biogeography. Pac. Insects Monogr. *23*, 295-374 (1970).
Helle, W.: Genetic variability of photoperiodic response in an arrkenotokous mite (*Tetranychus urticae*). Ent. Exp. & Appl. *11*, 101-113 (1968).
Holtzer, T.O., Bradley, J.R., Jr., Rabb, R.L.: Geographic and genetic variation in time required for emergence of diapausing *Heliothis zea*. Ann. Ent. Soc. Amer. *69*, 261-265 (1976).
Jackson, D.J.: The inheritence of long and short wings in the weevil, *Sitona hispidula*, with a discussion of wing reduction among beetles. Trans. Roy. Soc. Edinburgh *55*, 665-735 (1928).
Järvinen, O., Vepsäläinen, K.: Wing dimorphism as an adaptive strategy in water striders (*Gerris*). Hereditas *84*, 61-68 (1976).
Johansson, A.S.: Relation of nutrition to endocrine-reproductive functions in the milkweed bug, *Oncopeltus fasciatus* (Dallas). Nytt Mag. Zool. *7*, 1-132 (1958).
Johnson, C.G.: The Migration and Dispersal of Insects by Flight. London: Methuen 1969.
Johnson, C.G.: Lability of the flight system: a context for functional adaptation. In Insect Flight. Rainey, R.C. (ed.). R.E.S. Symposium 7. Oxford: Blackwell 1976.
Kelly, T. J., Davenport, R.: Juvenile Hormone induced ovarian uptake of a female-specific blood protein in *Oncopeltus fasciatus*. J. Insect Physiol. *22*, 1381-1393 (1976).
Kennedy, J.S.: A turning point in the study of insect migration. Nature, Lond. *189*, 785-791 (1961).
Kennedy, J.S.: Insect dispersal. In Insects, Science, and Society. Pimentel, D. (ed.). New York: Academic Press 1975.
Lack, D.: Island Biology Illustrated by the Land Birds of Jamaica. Berkeley: University of California Press 1976.
Lindroth, C.H.: Die Fennoskandischen Carabidae, eine tiergeographische Studie. Medd. Göteborgs Mus. Zool. Avd. *109*, 1-707 (1945); *110*, 1-277 (1946); *122*, 1-911 (1949).
Lumme, J., Lakovaara, S., Oikarinen, A., Lokki, J.: Genetics of the photoperiodic diapause in *Drosophila littoralis*. Hereditas *79*, 143-148 (1975).
MacArthur, R.H., Wilson, E.O.: The Theory of Island Biogeography. Princeton: Princeton University Press 1967.
Masaki, S.: Geographic variation of diapause in insects. Bull. Fac. Agric. Hirosaki Univ. *7*, 66-98 (1961).
Morris, R.F., Fulton, W.C.: Heritability of diapause intensity in *Hyphantria cunea* and correlated fitness responses. Can. Ent. *102*, 927-938 (1970).

Novak, I., Spitzer, K.: Adult dormancy in some species of the genus *Noctua* (Lepidoptera, Noctuidae) in Central Europe. Acta Ent. Bohemoslov. *72*, 215-221 (1975).

Ralph, C.P.: Natural food requirements, host-finding, and population density of the large milkweed bug, *Oncopeltus fasciatus* (Hemiptera: Lygaeidae) in relation to host plant density and morphology. Ph.D. Thesis, Cornell University (1976).

Rankin, M.A.: The inter-relationship and physiological control of flight and reproduction in *Oncopeltus fasciatus* (Heteroptera: Lygaeidae). Ph.D. Thesis, University of Iowa (1972).

Rankin, M.A.: The hormonal control of flight in the milkweed bug, *Oncopeltus fasciatus*. In Experimental Analysis of Insect Behaviour. Barton Browne, L. (ed.). York: Springer 1974.

Roff, D.A.: Population stability and the evolution of dispersal in a heterogeneous environment. Oecologia *19*, 217-237 (1975).

Root, R.B., Chaplin, S.J.: The life-styles of tropical milkweed bugs, *Oncopeltus* (Hemiptera: Lygaeidae) utilizing the same hosts. Ecology *57*, 132-140 (1976).

Rose, D.J.W.: Dispersal and quality in populations of *Cicadulina* species (Cicadellidae). J. Anim. Ecol. *41*, 589-609 (1972).

Sauer, D., Feir, D.: Studies on natural populations of *Oncopeltus fasciatus* (Dallas), the large milkweed bug. Am. Midl. Nat. *90*, 13-37 (1973).

Siegel, S.: Nonparametric Statistics for the Behavioral Sciences. New York: McGraw-Hill 1956.

Slater, J.A.: A Catalogue of the Lygaeidae of the World. Storrs: University of Connecticut Press 1964.

Slater, J.A., Sperry, B.: The biology and distribution of the South African Lygaeinae, with descriptions of new species (Hemiptera: Lygaeidae). Ann. Transvaal Mus. *28*, 117-201 (1973).

Southwood, T.R.E.: Migration of terrestrial arthropods in relation to habitat. Biol. Rev. *37*, 171-214 (1962).

Taylor, L.R., Taylor, R.A.J.: Aggregation, migration, and population mechanics. Nature, Lond. *265*, 415-421 (1977).

Vepsäläinen, K.: The life cycles and wing lengths of Finnish *Gerris* Fabr. species (Heteroptera, Gerridae). Acta Zool. Fenn. *141*, 1-73 (1974).

Wollaston, T.V.: Insecta Maderensia. London: John Van Voorst 1854.

Escape in Space and Time – Concluding Remarks

T. R. E. Southwood

Migration and diapause are two important strategies for escape in space and time, and various aspects have been well reviewed in the preceding papers. In our discussions we have emphasized habitat unfavorability in terms of food shortage or climatic extremes: the relationship of migration to habitats that are only temporarily favorable in this respect, originally proposed at the 1960 International Congress, now seems well established. Habitats may also become unfavorable because of the development of populations of natural enemies: this is an aspect we have excluded from our consideration. Long-lived reproductives may distribute their progeny within an essentially permanent habitat in different places at different times as a means of escape from natural enemies, e.g., *Heliconius*. Indeed this is one of the reproductive strategies of a *K*-species (Southwood and Comins, 1976).

While recognizing this omission I want to return to consider the framework on which we might define habitat characteristics, which after all provide the evolutionary template on which strategies of escape from "unfavorableness" are forged.

Habitat characteristics may be viewed against the axes of time and space. Heterogeneity in time may be partitioned into durational stability, constancy, and predictability. Durational stability, the number of generations that can breed in a particular location, is defined by H/τ where H = habitat duration and τ = generation time. While constancy is the extent to which the carrying capacity fluctuates. In temperate regions seasonality is the major component in this temporal variablility of carrying capacity. For example, the mirid bug, *Harpocera thoracica* feeds on the opening buds and young catkins of oak; food harvesting can occur on an oak tree for about only three weeks in any one year. The carrying capacity of an oak tree for this species is taken week by week throughout the year and is very variable, mostly being zero but rising to many thousands for a few weeks. However, the habitat has long durational stability and high predictability, and therefore it is not surprising that the species has evolved a strategy of "staying put," in Dr. Solbreck's crisp terms "Here, but later." The rock pools, that form one of the habitats of *Gerris thoracicus*, are - as Dr. Vepsäläinen shows - an excellent example of a habitat with a carrying capacity that is both variable and unpredictable.

Spatial heterogeneity has two components, the patch size of the habitat units and their isolation, the proximity of other habitats which must be measured in terms of the probability of an individual finding a new habitat. Clearly patch size and isolation are important components, but the behavioral attributes of the animal are also of major significance. Birds and at least some butterflies, with their remarkable powers of orientation may be able to navigate to a particular site over great distances (e.g., *Danaus plexippus* to Pacific Grove). As Dr. Rainey's paper so clearly shows, many other insects have evolved an ability to use the kinetic energy of the atmosphere to seek and find new habitats. As Kennedy (1975) has stressed, there is an essential and fundamental behavioral component in these movements; one must liken such insects to a skilled canoeist traveling downstream, rather than to a piece of flotsam. Dr. Dingle's findings, on the limited flight activity of island *Oncopeltus*, and some of Dr. Vepsäläinen's have emphasized the important influence of isolation in the evolution of strategies for escape in space and time. It will surely be possible to calculate, for a deterministic situation, the optimum strategy in terms of movement, diapause (or lengthened life history) or continuous breeding for a species occupying any particular position against this framework of habitat characteristics, durational stability, variation in habitat favorableness and predictability, and in the probability of colonizing a new habitat as determined by their scatter and size (Southwood, 1977).

In reality for any population there will be in addition, the risks from predators and parasites: specialized natural enemies pose a higher risk in the actual habitat, while generalized predators will often be more significant during migration, whether this be by land, air, or water. The minimization of these risks as well as those from climatic and similar factors will be an important determinant for the stage of diapause or migration. Dr. Masaki's observations on crickets overwintering as an egg in northern Japan and as a larva in the south are interesting in this respect.

Of course, the natural world is not deterministic; climatic and other variations lead to a stochastic situation, one in which as Dr. Waldbauer has said, the population should not have "all its eggs in one basket." The considerable polymorphism for migration and diapause shown in many papers must be seen as a response, through natural selection, to this stochastic variation.

As we look to the future, it is perhaps worthwhile to consider some of the new areas that will have to be explored to enhance our understanding of escape in space and time.

1. The measurement of the comparative mortalities suffered during migration and diapause and other alternative strategies. Taylor's studies in Britain show how populations wax and wane in space and time (Taylor and Taylor, 1977); this and other approaches will be needed to endeavor to assess the mortality due to migration in the overall, rather than merely local, population. We will need to move toward something comparable to life tables for species or at least regional populations; no doubt this will have to depend on the development of novel approaches. We must hope that we will be able to quantitatively express the balancing of risks of different strategies.

2. A more quantitative assessment of the inherent unpredictability of the habitat so that we may appreciate the framework against which the polymorphism and variation described in most of the papers in this session has been evolved. Dr. Istock's studies have made an important start in this type of work.

3. A further clarification of the underlying physiological mechanisms and the interrelation of these to the changing continuum of the habitat. Dr. Rankin's paper showed how, as has recently been recognized in population dynamics, these physiological responses are nonlinear, allowing the organisms to switch from one strategy to another as hormone titres pass from one threshold to another.

As we look ahead, we must hope for, and, indeed, expect, striking progress as strategies for escape in space and time are placed quantitatively in their evolutionary, demographic, and physiological frameworks.

References

Kennedy, J.S.: Insect Dispersal. In: Pimentel, D. (ed.). Insect, Science & Society. New York: Academic Press, 1975, pp. 103-119.

Southwood, T.R.E.: Habitat, the templet for ecological strategies. (Presidential Address to the British Ecological Society). J. Anim. Ecol. *46*, 337-65 (1977).

Southwood, T.R.E., Comins, H.N.: A synoptic population model. J. Anim. Ecol. *45*, 949-965 (1976).

Taylor, L.R., Taylor, R.A.J.: Aggregation, migration and population mechanics. Nature, Lond. *265*, 415-421 (1977).

Index

O

P

R

S

THE ROLE OF ARTHROPODS IN FOREST ECOSYSTEMS
Edited by ***W. J. Mattson***

1977. xi, 104p. 28 illus. cloth (Proceedings in Life Sciences)
Proceedings of the 15th International Congress of Entomology, Washington, D.C., 1976.

Here is the first comprehensive volume on the role of both phytophagous and saprophagous arthropods in shaping the structure and regulating the functioning of terrestrial ecosystems. This new book assimilates and synthesizes the many findings on the role of arthropods in each of the three subsystems that comprise the terrestrial ecosystem: the above-ground plant system, the soil-litter system, and the aquatic stream system.

The book addresses the role of arthropods in a variety of areas, such as: plant production, plant succession, inter-species competition, pollination ecology, resource partitioning, and litter decomposition and nutrient cycling. Some dominant themes are the significant impact of arthropods on quantitative and qualitative aspects of system structure, and the sensitivity and responsiveness of arthropods to system structure.

ATLAS OF AN INSECT BRAIN
By ***N. J. Strausfeld***

1976. xiv, 214p. 81 illus. (partly colored) 71 plates. cloth

"Although there are many fine books in entomology, I cannot recall any which are as elegant and beautiful . . . as this truly extraordinary monograph on the anatomy and histology of the brain of the adult housefly, *Musca domestica*. . . . The book has a frontispiece, 20 superb color illustrations, 43 diagrams or black and white drawings, 39 large plates of photomicrographs, and 57 smaller figures of drawings and/or photomicrographs. All of the illustrations are simply stunning. Aside from all of these attractions, the volume is beautifully and lucidly written. Not only has the author created a marvelous scholarly scientific book, but the publishers have lovingly executed it, and together they have made a thing of great beauty."

Bulletin of the Entomological Society of America